WITHDRAWN

OCT 2003

The CAD Guidebook

MECHANICAL ENGINEERING
A Series of Textbooks and Reference Books

Founding Editor

L. L. Faulkner

*Columbus Division, Battelle Memorial Institute
and Department of Mechanical Engineering
The Ohio State University
Columbus, Ohio*

1. *Spring Designer's Handbook*, Harold Carlson
2. *Computer-Aided Graphics and Design*, Daniel L. Ryan
3. *Lubrication Fundamentals*, J. George Wills
4. *Solar Engineering for Domestic Buildings*, William A. Himmelman
5. *Applied Engineering Mechanics: Statics and Dynamics*, G. Boothroyd and C. Poli
6. *Centrifugal Pump Clinic*, Igor J. Karassik
7. *Computer-Aided Kinetics for Machine Design*, Daniel L. Ryan
8. *Plastics Products Design Handbook, Part A: Materials and Components; Part B: Processes and Design for Processes*, edited by Edward Miller
9. *Turbomachinery: Basic Theory and Applications*, Earl Logan, Jr.
10. *Vibrations of Shells and Plates*, Werner Soedel
11. *Flat and Corrugated Diaphragm Design Handbook*, Mario Di Giovanni
12. *Practical Stress Analysis in Engineering Design*, Alexander Blake
13. *An Introduction to the Design and Behavior of Bolted Joints*, John H. Bickford
14. *Optimal Engineering Design: Principles and Applications*, James N. Siddall
15. *Spring Manufacturing Handbook*, Harold Carlson
16. *Industrial Noise Control: Fundamentals and Applications*, edited by Lewis H. Bell
17. *Gears and Their Vibration: A Basic Approach to Understanding Gear Noise*, J. Derek Smith
18. *Chains for Power Transmission and Material Handling: Design and Applications Handbook*, American Chain Association
19. *Corrosion and Corrosion Protection Handbook*, edited by Philip A. Schweitzer
20. *Gear Drive Systems: Design and Application*, Peter Lynwander
21. *Controlling In-Plant Airborne Contaminants: Systems Design and Calculations*, John D. Constance
22. *CAD/CAM Systems Planning and Implementation*, Charles S. Knox
23. *Probabilistic Engineering Design: Principles and Applications*, James N. Siddall
24. *Traction Drives: Selection and Application*, Frederick W. Heilich III and Eugene E. Shube
25. *Finite Element Methods: An Introduction*, Ronald L. Huston and Chris E. Passerello

26. *Mechanical Fastening of Plastics: An Engineering Handbook*, Brayton Lincoln, Kenneth J. Gomes, and James F. Braden
27. *Lubrication in Practice: Second Edition*, edited by W. S. Robertson
28. *Principles of Automated Drafting*, Daniel L. Ryan
29. *Practical Seal Design*, edited by Leonard J. Martini
30. *Engineering Documentation for CAD/CAM Applications*, Charles S. Knox
31. *Design Dimensioning with Computer Graphics Applications*, Jerome C. Lange
32. *Mechanism Analysis: Simplified Graphical and Analytical Techniques*, Lyndon O. Barton
33. *CAD/CAM Systems: Justification, Implementation, Productivity Measurement*, Edward J. Preston, George W. Crawford, and Mark E. Coticchia
34. *Steam Plant Calculations Manual*, V. Ganapathy
35. *Design Assurance for Engineers and Managers*, John A. Burgess
36. *Heat Transfer Fluids and Systems for Process and Energy Applications*, Jasbir Singh
37. *Potential Flows: Computer Graphic Solutions*, Robert H. Kirchhoff
38. *Computer-Aided Graphics and Design: Second Edition*, Daniel L. Ryan
39. *Electronically Controlled Proportional Valves: Selection and Application*, Michael J. Tonyan, edited by Tobi Goldoftas
40. *Pressure Gauge Handbook*, AMETEK, U.S. Gauge Division, edited by Philip W. Harland
41. *Fabric Filtration for Combustion Sources: Fundamentals and Basic Technology*, R. P. Donovan
42. *Design of Mechanical Joints*, Alexander Blake
43. *CAD/CAM Dictionary*, Edward J. Preston, George W. Crawford, and Mark E. Coticchia
44. *Machinery Adhesives for Locking, Retaining, and Sealing*, Girard S. Haviland
45. *Couplings and Joints: Design, Selection, and Application*, Jon R. Mancuso
46. *Shaft Alignment Handbook*, John Piotrowski
47. *BASIC Programs for Steam Plant Engineers: Boilers, Combustion, Fluid Flow, and Heat Transfer*, V. Ganapathy
48. *Solving Mechanical Design Problems with Computer Graphics*, Jerome C. Lange
49. *Plastics Gearing: Selection and Application*, Clifford E. Adams
50. *Clutches and Brakes: Design and Selection*, William C. Orthwein
51. *Transducers in Mechanical and Electronic Design*, Harry L. Trietley
52. *Metallurgical Applications of Shock-Wave and High-Strain-Rate Phenomena*, edited by Lawrence E. Murr, Karl P. Staudhammer, and Marc A. Meyers
53. *Magnesium Products Design*, Robert S. Busk
54. *How to Integrate CAD/CAM Systems: Management and Technology*, William D. Engelke
55. *Cam Design and Manufacture: Second Edition*; with cam design software for the IBM PC and compatibles, disk included, Preben W. Jensen
56. *Solid-State AC Motor Controls: Selection and Application*, Sylvester Campbell
57. *Fundamentals of Robotics*, David D. Ardayfio
58. *Belt Selection and Application for Engineers*, edited by Wallace D. Erickson
59. *Developing Three-Dimensional CAD Software with the IBM PC*, C. Stan Wei
60. *Organizing Data for CIM Applications*, Charles S. Knox, with contributions by Thomas C. Boos, Ross S. Culverhouse, and Paul F. Muchnicki

61. *Computer-Aided Simulation in Railway Dynamics*, by Rao V. Dukkipati and Joseph R. Amyot
62. *Fiber-Reinforced Composites: Materials, Manufacturing, and Design*, P. K. Mallick
63. *Photoelectric Sensors and Controls: Selection and Application*, Scott M. Juds
64. *Finite Element Analysis with Personal Computers*, Edward R. Champion, Jr., and J. Michael Ensminger
65. *Ultrasonics: Fundamentals, Technology, Applications: Second Edition, Revised and Expanded*, Dale Ensminger
66. *Applied Finite Element Modeling: Practical Problem Solving for Engineers*, Jeffrey M. Steele
67. *Measurement and Instrumentation in Engineering: Principles and Basic Laboratory Experiments*, Francis S. Tse and Ivan E. Morse
68. *Centrifugal Pump Clinic: Second Edition, Revised and Expanded*, Igor J. Karassik
69. *Practical Stress Analysis in Engineering Design: Second Edition, Revised and Expanded*, Alexander Blake
70. *An Introduction to the Design and Behavior of Bolted Joints: Second Edition, Revised and Expanded*, John H. Bickford
71. *High Vacuum Technology: A Practical Guide*, Marsbed H. Hablanian
72. *Pressure Sensors: Selection and Application*, Duane Tandeske
73. *Zinc Handbook: Properties, Processing, and Use in Design*, Frank Porter
74. *Thermal Fatigue of Metals*, Andrzej Weronski and Tadeusz Hejwowski
75. *Classical and Modern Mechanisms for Engineers and Inventors*, Preben W. Jensen
76. *Handbook of Electronic Package Design*, edited by Michael Pecht
77. *Shock-Wave and High-Strain-Rate Phenomena in Materials*, edited by Marc A. Meyers, Lawrence E. Murr, and Karl P. Staudhammer
78. *Industrial Refrigeration: Principles, Design and Applications*, P. C. Koelet
79. *Applied Combustion*, Eugene L. Keating
80. *Engine Oils and Automotive Lubrication*, edited by Wilfried J. Bartz
81. *Mechanism Analysis: Simplified and Graphical Techniques, Second Edition, Revised and Expanded*, Lyndon O. Barton
82. *Fundamental Fluid Mechanics for the Practicing Engineer*, James W. Murdock
83. *Fiber-Reinforced Composites: Materials, Manufacturing, and Design, Second Edition, Revised and Expanded*, P. K. Mallick
84. *Numerical Methods for Engineering Applications*, Edward R. Champion, Jr.
85. *Turbomachinery: Basic Theory and Applications, Second Edition, Revised and Expanded*, Earl Logan, Jr.
86. *Vibrations of Shells and Plates: Second Edition, Revised and Expanded*, Werner Soedel
87. *Steam Plant Calculations Manual: Second Edition, Revised and Expanded*, V. Ganapathy
88. *Industrial Noise Control: Fundamentals and Applications, Second Edition, Revised and Expanded*, Lewis H. Bell and Douglas H. Bell
89. *Finite Elements: Their Design and Performance*, Richard H. MacNeal
90. *Mechanical Properties of Polymers and Composites: Second Edition, Revised and Expanded*, Lawrence E. Nielsen and Robert F. Landel
91. *Mechanical Wear Prediction and Prevention*, Raymond G. Bayer

92. *Mechanical Power Transmission Components,* edited by David W. South and Jon R. Mancuso
93. *Handbook of Turbomachinery,* edited by Earl Logan, Jr.
94. *Engineering Documentation Control Practices and Procedures,* Ray E. Monahan
95. *Refractory Linings Thermomechanical Design and Applications,* Charles A. Schacht
96. *Geometric Dimensioning and Tolerancing: Applications and Techniques for Use in Design, Manufacturing, and Inspection,* James D. Meadows
97. *An Introduction to the Design and Behavior of Bolted Joints: Third Edition, Revised and Expanded,* John H. Bickford
98. *Shaft Alignment Handbook: Second Edition, Revised and Expanded,* John Piotrowski
99. *Computer-Aided Design of Polymer-Matrix Composite Structures,* edited by Suong Van Hoa
100. *Friction Science and Technology,* Peter J. Blau
101. *Introduction to Plastics and Composites: Mechanical Properties and Engineering Applications,* Edward Miller
102. *Practical Fracture Mechanics in Design,* Alexander Blake
103. *Pump Characteristics and Applications,* Michael W. Volk
104. *Optical Principles and Technology for Engineers,* James E. Stewart
105. *Optimizing the Shape of Mechanical Elements and Structures,* A. A. Seireg and Jorge Rodriguez
106. *Kinematics and Dynamics of Machinery,* Vladimír Stejskal and Michael Valášek
107. *Shaft Seals for Dynamic Applications,* Les Horve
108. *Reliability-Based Mechanical Design,* edited by Thomas A. Cruse
109. *Mechanical Fastening, Joining, and Assembly,* James A. Speck
110. *Turbomachinery Fluid Dynamics and Heat Transfer,* edited by Chunill Hah
111. *High-Vacuum Technology: A Practical Guide, Second Edition, Revised and Expanded,* Marsbed H. Hablanian
112. *Geometric Dimensioning and Tolerancing: Workbook and Answerbook,* James D. Meadows
113. *Handbook of Materials Selection for Engineering Applications,* edited by G. T. Murray
114. *Handbook of Thermoplastic Piping System Design,* Thomas Sixsmith and Reinhard Hanselka
115. *Practical Guide to Finite Elements: A Solid Mechanics Approach,* Steven M. Lepi
116. *Applied Computational Fluid Dynamics,* edited by Vijay K. Garg
117. *Fluid Sealing Technology,* Heinz K. Muller and Bernard S. Nau
118. *Friction and Lubrication in Mechanical Design,* A. A. Seireg
119. *Influence Functions and Matrices,* Yuri A. Melnikov
120. *Mechanical Analysis of Electronic Packaging Systems,* Stephen A. McKeown
121. *Couplings and Joints: Design, Selection, and Application, Second Edition, Revised and Expanded,* Jon R. Mancuso
122. *Thermodynamics: Processes and Applications,* Earl Logan, Jr.
123. *Gear Noise and Vibration,* J. Derek Smith
124. *Practical Fluid Mechanics for Engineering Applications,* John J. Bloomer
125. *Handbook of Hydraulic Fluid Technology,* edited by George E. Totten
126. *Heat Exchanger Design Handbook,* T. Kuppan

127. *Designing for Product Sound Quality,* Richard H. Lyon
128. *Probability Applications in Mechanical Design,* Franklin E. Fisher and Joy R. Fisher
129. *Nickel Alloys,* edited by Ulrich Heubner
130. *Rotating Machinery Vibration: Problem Analysis and Troubleshooting,* Maurice L. Adams, Jr.
131. *Formulas for Dynamic Analysis,* Ronald L. Huston and C. Q. Liu
132. *Handbook of Machinery Dynamics,* Lynn L. Faulkner and Earl Logan, Jr.
133. *Rapid Prototyping Technology: Selection and Application,* Kenneth G. Cooper
134. *Reciprocating Machinery Dynamics: Design and Analysis,* Abdulla S. Rangwala
135. *Maintenance Excellence: Optimizing Equipment Life-Cycle Decisions,* edited by John D. Campbell and Andrew K. S. Jardine
136. *Practical Guide to Industrial Boiler Systems,* Ralph L. Vandagriff
137. *Lubrication Fundamentals: Second Edition, Revised and Expanded,* D. M. Pirro and A. A. Wessol
138. *Mechanical Life Cycle Handbook: Good Environmental Design and Manufacturing,* edited by Mahendra S. Hundal
139. *Micromachining of Engineering Materials,* edited by Joseph McGeough
140. *Control Strategies for Dynamic Systems: Design and Implementation,* John H. Lumkes, Jr.
141. *Practical Guide to Pressure Vessel Manufacturing,* Sunil Pullarcot
142. *Nondestructive Evaluation: Theory, Techniques, and Applications,* edited by Peter J. Shull
143. *Diesel Engine Engineering: Thermodynamics, Dynamics, Design, and Control,* Andrei Makartchouk
144. *Handbook of Machine Tool Analysis,* Ioan D. Marinescu, Constantin Ispas, and Dan Boboc
145. *Implementing Concurrent Engineering in Small Companies,* Susan Carlson Skalak
146. *Practical Guide to the Packaging of Electronics: Thermal and Mechanical Design and Analysis,* Ali Jamnia
147. *Bearing Design in Machinery: Engineering Tribology and Lubrication,* Avraham Harnoy
148. *Mechanical Reliability Improvement: Probability and Statistics for Experimental Testing,* R. E. Little
149. *Industrial Boilers and Heat Recovery Steam Generators: Design, Applications, and Calculations,* V. Ganapathy
150. *The CAD Guidebook: A Basic Manual for Understanding and Improving Computer-Aided Design,* Stephen J. Schoonmaker
151. *Industrial Noise Control and Acoustics,* Randall F. Barron
152. *Mechanical Properties of Engineering Materials,* Wolé Soboyejo
153. *Reliability Verification, Testing, and Analysis in Engineering Design,* Gary S. Wasserman
154. *Fundamental Mechanics of Fluids: Third Edition,* I. G. Currie

Additional Volumes in Preparation

HVAC Water Chillers and Cooling Towers: Fundamentals, Application, and Operations, Herbert W. Stanford III

Handbook of Turbomachinery: Second Edition, Revised and Expanded, Earl Logan, Jr., and Ramendra Roy

Progressing Cavity Pumps, Downhole Pumps, and Mudmotors, Lev Nelik

Gear Noise and Vibration: Second Edition, Revised and Expanded, J. Derek Smith

Intermediate Heat Transfer, Kau-Fui Vincent Wong

Mechanical Engineering Software

Spring Design with an IBM PC, Al Dietrich

Mechanical Design Failure Analysis: With Failure Analysis System Software for the IBM PC, David G. Ullman

The CAD Guidebook

A Basic Manual for Understanding and Improving Computer-Aided Design

Stephen J. Schoonmaker
Grove Worldwide
Shady Grove, Pennsylvania, U.S.A

MARCEL DEKKER, INC.　　　NEW YORK • BASEL

Library of Congress Cataloging-in-Publication Data
A catalog record for this book is available from the Library of Congress.

ISBN: 0-8247-0871-7

This book is printed on acid-free paper.

Headquarters
Marcel Dekker, Inc.
270 Madison Avenue, New York, NY 10016
tel: 212-696-9000; fax: 212-685-4540

Eastern Hemisphere Distribution
Marcel Dekker AG
Hutgasse 4, Postfach 812, CH-4001 Basel, Switzerland
tel: 41-61-260-6300; fax: 41-61-260-6333

World Wide Web
http://www.dekker.com

The publisher offers discounts on this book when ordered in bulk quantities. For more information, write to Special Sales/Professional Marketing at the headquarters address above.

Copyright © 2003 by Marcel Dekker, Inc. All Rights Reserved.

Neither this book nor any part may be reproduced or transmitted in any form or by any means, electronic or mechanical, including photocopying, microfilming, and recording, or by any information storage and retrieval system, without permission in writing from the publisher.

Current printing (last digit):
10 9 8 7 6 5 4 3 2 1

PRINTED IN THE UNITED STATES OF AMERICA

Preface

It was a black day, indeed, when I walked into my most recent place of employment in south-central Pennsylvania about six years ago. I had just spent 11 years lost in the "other-world" of software development (writing commercial and in-house analytical software for the design of all kinds of gas compression equipment). I had avoided CAD (computer-aided design) like an extroverted skunk going door-to-door trying to sell the benefits of aroma therapy. CAD was just for making those drawing thingies I had heard about. But on this black day, I had agreed to lead the company's full-blown implementation of that "solids" stuff, and the company's designers were as much T-square huggers as I was a tree-hugger (did I mention that I got my engineering degree in Massachusetts?).

Well, the Lord works in mysterious ways, and after about three years, the designers were humming away at 3-D CAD and trying to convince middle managers that it really was the way to go. And those same designers had actually not beaten me to a pulp! Instead, they taught me how things worked in the real world. Together we pumped out a whole bunch of new product, and now there are some construction sites out there that are hopefully benefiting from our little experiment with CAD.

Since I had previously written a book on ISO 9001 (that's another long story), I figured I may as well share what I had learned about CAD with the rest of the world. I guess I have some teacher gene that says I must share what I know for the betterment of society. I do hope this book is helpful to you, and with any luck, you will learn to stop worrying and learn to love your CAD system. And even though little fun is to be had at most workplaces anymore, I hope you will at least find that 3-D CAD can be fun.

Of course, this really important book that I read every day has a story that ends with the moral "of those to whom much is given, much is expected." I, for one, certainly have been given much. First, I have been blessed to have been born and raised in the United States of America. It's not a perfect country, but there is no other nation in the history of the world that comes close to its ability to let people like me breathe free and find their own best way to serve. Secondly, I have been given the best family a man could hope and pray for. Sharon, Melissa, Jennifer, Michael, and Christina were kind enough to let me take the time to do another book. Thanks, guys.

I must express my thanks to some of those folks that helped me learn about this CAD-thingy. Thanks especially to Doug Beckner, Vince Bernabe, and Mike Miller at Grove. I think they actually understood what I was talking about! We implemented the I-DEAS® Master Series CAD system from SDRC (now EDS), so I thank them for their support over the years, too. In particular, thanks to their tech support people under Ron Hickman, Barry Ratihn of their post-sales group, and Rick Miller of their training staff. I also thank Grove's IT department and its latest CAD-burdened victim (I mean employee), Brett Cox. I am not positive, but I think CAD is no good if there are no computers to run it on. Finally, I thank my boss, Ralph Kegerreis, who actually let me get away with all of this. Thanks, Ralph.

Well, there you have it. Have fun.

Steve Schoonmaker

Contents

Preface		*iii*
Chapter 1	Introduction	1
Chapter 2	Computer Hardware Basics	6
Chapter 3	Computer Software Basics	45
Chapter 4	Drawings and 2-D Design	71
Chapter 5	Two-Dimensional CAD	106
Chapter 6	Managing Two-Dimensional CAD	145
Chapter 7	Three-Dimensional CAD	168
Chapter 8	Part Modeling	188
Chapter 9	Surface Modeling	224

Chapter 10	Assembly Modeling	242
Chapter 11	Managing Three-Dimensional CAD	263
Glossary		*285*
Bibliography		*317*
Index		*319*

The CAD Guidebook

1
Introduction

This work is intended to provide a basic understanding or foundation for working with a CAD system. Although there have been many acronyms over the years (CADD, MDA, etc.), in this work CAD is to refer to Computer-Aided Design. Computer Aided-Design, in turn, is assumed to mean various types of tasks related to product design. Some of these tasks would be preliminary design and layouts, design calculations, detailed design, creating 3-D models, creating drawings, releasing drawings, as well as interfacing with analysis, marketing, manufacturing, and end-user personnel. Although this is a broad array of tasks, all of them would be affected by the currently available CAD systems.

1.1 SCOPE AND INTENDED AUDIENCE

Since this work is intended to provide a basic understanding, it is not going to be sufficient for a particular designer or engineer to be proficient with a particular CAD system. Only an appropriate CAD training program is going to be able to provide this proficiency. However, this work would be very effective in conjunction with such a training program. The information in this work can make it easier to understand why the CAD system operates in a certain fashion. This understanding then reinforces the knowledge being learned and shortens the overall learning curve.

This work is also not a complete source of information on specific designer-related activities such as creating complete drawings or Bills of Material (BOMs). However, even those who have never seen or worked with drawings, prints, or BOMs before will be able to follow the information presented. This can be most valuable to those outside of design and engineering that need to support a CAD system (such as Information Technology (IT) professionals). For practicing designers and engineers, they must refer to their own company procedures or other works to complete their tasks for creating and fully documenting their designs.

Similarly, the information on CAD computer systems technology may be oversimplified for IT professionals. They would typically be quite familiar with the basic computer knowledge presented. Even so, some of the computer systems information is specifically relevant for CAD software, and IT professionals who have not dealt with CAD software before would certainly benefit from the information. Although designers and engineers may not be familiar with the computer systems information, they should have no trouble understanding the level of detail presented.

For better or worse, the information presented by this work is heavily geared toward mechanical drawings and mechanical design and engineering. This activity is probably the most prevalent design activity with respect to the manufacturing sector, so it may be the most prevalent use of CAD systems. Civil engineering (or A/E/C) aspects of CAD have a number of similarities with mechanical CAD, so much information presented in this work would probably still be relevant. However, users of CAD for graphics arts or electrical engineering activities may find it is weighted too heavily toward mechanical design.

1.2 APPLICATION

CAD systems have been developing and advancing for decades, and yet they show no sign of becoming completely mature. This is particularly true because of the expanding power and usefulness of 3-D models and their ability to replace and/or enhance drawings. CAD systems are also expanding in their influence over the business of product design and manufacture. Computer systems that are capable of dealing with the demands of CAD are becoming almost ubiquitous; therefore, the opportunities to leverage the CAD system's valuable information is expanding beyond a traditional design and engineering environment. Integration of CAD systems and/or data with the Internet (or more specifically the World-Wide Web) is certainly a sign of this increasing influence, and this trend seems certain to continue.

What this means for this work is that as soon as it is printed, it may be out-of-date. Fortunately, the author feels that the basic theory and application of CAD

Introduction 3

systems will not change that drastically. Even as the systems become more capable, much of the foundation presented in this work will still be relevant. The basic processes and problems that are found in design and engineering will still remain, and the approaches of CAD to solving these problems will probably be the same as well.

1.3 ORGANIZATION

This work is organized to allow most chapters to be read independently. If there are concepts that build on a previous chapter's information, the reader should be able to locate that information by referring to the section headings.

This work is broken into 3 general parts. The first part (Chapters 2 and 3) is geared toward basic information on computer systems that is most relevant to a CAD platform. Chapter 2 covers computer hardware and chapter 3 presents information on computer software.

Chapters 4, 5, and 6 concerns drawings and CAD systems that support drawings. Chapter 4 shows the drawings themselves and how they are used to design products in a 2-D (two-dimensional) medium. Chapter 5 illustrates CAD systems that actually automate 2-D design and the production of drawings. Chapter 6 explains the management of a 2-D CAD environment.

The final part of the work is geared toward 3-D (three-dimensional) CAD. The author's primary experience as a programmer, user, and manager of CAD systems is with 3-D CAD systems, so this part of the work is the most extensive. These systems are the latest and most powerful. Of course, they also demand the most from the users and administrators. Considering the importance of the these systems, this part of the work may be the most useful to the reader. Chapter 7 introduces the topic of 3-D CAD; this chapter should be read if any other chapter on 3-D CAD is of interest to the reader. Chapter 8 refines 3-D CAD with respect to part modeling (creating standard, solid models). Chapter 9 explains 3-D CAD with respect to surface modeling (creating more specialized, free form models). Chapter 10 covers assembly modeling (combining part models into a sort of group 3-D model). Finally, Chapter 11 covers the management of a 3-D CAD environment.

1.4 AUTHOR'S BACKGROUND

The author of this work has had a somewhat varied background leading to a level of expertise with CAD systems. The first experience came in the form of computer programming for CAE (computer-aided engineering) software. The software he helped develop and manage was used for the design and analysis of turbomachinery (pumps, compressors, turbines, and their constituent components). This software was sold under the name COMIG® from the former Northern Research and Engineering Corporation (or NREC). This software was developed at a time that computer graphics (2-D and 3-D) was just becoming standardized, so most of the CAD graphics experience was obtained in a rather "early" environment. Having experience with very large software development projects was also highly beneficial in understanding how software really works and how it is really used (particularly in the engineering analysis community). Hopefully, a sensitivity to the special needs of the engineering analysis field is apparent in this work.

The next experience was as a specialized engineering analyst in the commercial gas compression market. This included creating, supporting, documenting, and verifying of in-house software for reciprocating gas compressors and gas engines. This was at the former Engine Process Compressor Division of Dresser-Rand. This experience was valuable in seeing how engineering work supports actual product development and manufacturing (as opposed to the research environment). Regarding the use of 2-D CAD, Bills of Materials, and Routings, Quality Control was observed. A unique aspect of this experience was being responsible for the development and execution of processes to verify and validate design-centric computer software for the ISO 9001 quality standard (which had become a vital issue to the company). In conjunction with that experience, another published work, ISO 9001 For Engineers and Designers, was published by McGraw-Hill and professional development courses were directed by the author for ASME International. Again it is hoped that a healthy respect for quality systems and practical manufacture is reflected in this work as well.

Subsequently, the author was the project leader for the implementation of a 3-D CAD based design and engineering environment for a mechanical equipment manufacturer. This manufacturer was Grove Worldwide, which designs and manufactures mobile hydraulic cranes and aerial work platforms. This company had been using mainframe 2-D CAD for about 15 years, and they had made the decision to change to a client/server network with fully associative 2-D and 3-D software. The full implementation took approximately 4 years to complete. A design team-based management style was implemented in the same time frame as 3-D CAD, and the net effect of these changes was a very significant reduction in the time to completely design and put into production large cranes (without the use of more personnel).

1.5 CAD IMAGES

The CAD system used in conjunction with the development of this work was the I-DEAS Master Series™ version 7.3 from SDRC® (Structural Dynamics

Introduction

Research Corporation). It was running on the IBM® RS/6000® 43P-140 with AIX® version 3.4. The images were "captured" with the Xwindows screen dump utility and then imported into ImageMagick© version 5.2.4 00/10/01 © 2000 ImageMagick Studio (running on the same IBM platform). In ImageMagick, images were changed to grayscale, cropped, annotated, etc. and then converted to JPEG files.

2

Computer Hardware Basics

2.1 INTRODUCTION

In order to fully master a CAD system, it is important to understand how a computer functions. The intent here is not to become a expert in computer technology, but to build a foundation. With this foundation, a good CAD user or a CAD manager can hopefully evaluate systems and create appropriate design processes. Particularly with the wide proliferation of very capable 3-D CAD systems, it is essential to know how the CAD software is going to interact with the computer hardware; 3-D CAD is one of the most demanding applications available to run on computers. Understanding how the software is trying to utilize the computational resources can make the difference between a productive design process and one that provides little benefit at all.

A basic understanding of computer systems allows designers and engineers to make proper trade-off's in utilizing available computer resources. If a designer knows about graphics accelerators, he or she can tell if a low-end PC can handle a particular task. If the designer knows about how files are written and stored on a computer, he or she can tell what it will take to translate the model to other formats or to other types of computer systems. If the designer knows something about computer networks, he or she will be able to tell how long it will take to

Computer Hardware Basics

transfer a design's files to other computer systems. These situations are common in everyday design and engineering practice. CAD plays a vital role in many aspects of design and engineering. Furthermore, designers and engineers that use CAD are expected to understand how to maximize the benefits of CAD, and understanding the computer systems themselves is an important ingredient in this process.

The remainder of this chapter presents a very basic background in computer hardware. Hardware refers to the physical components of a computer system. However, it must be understood that the computer hardware is of little value without software. Software is the computer programming that runs or "executes" on the hardware. A basic background in computer software is presented in the next chapter.

The hardware is often closely tied to the software that is running on the computer, and in particular, on software called the operating system or OS. Sometimes the combination of the hardware and software is referred to as a platform. For example, two common platforms are the UNIX® platform (the Unix operating system combined with a workstation) and the Windows® platform (the Windows operating system combined with a personal computer (PC)).

2.2 THE SYSTEM

Figure 2.1 shows a very basic model of a typical computer system (such as a PC or a workstation) that would run a CAD program. The PC and workstation platforms have these same basic components. Although in the 1970s, 1980s, and 1990s, it would have been appropriate to consider a mainframe and its terminals, today there are few mainframes running CAD programs.

As can be seen from Figure 2.1, the main components of the platform are as follows:

- Central Processing Unit (CPU)
- memory
- storage
- peripherals (such as monitor, keyboard, and mouse)

All these components are connected within the computer via an electronic highway called a bus so that they can communicate at a very rapid rate (millions or even billions of signals per second). Although the basic components are going to be somewhat familiar to most readers, there are some common misconceptions concerning some of these components. The biggest problem is probably the distinction between memory and storage; this issue is discussed shortly.

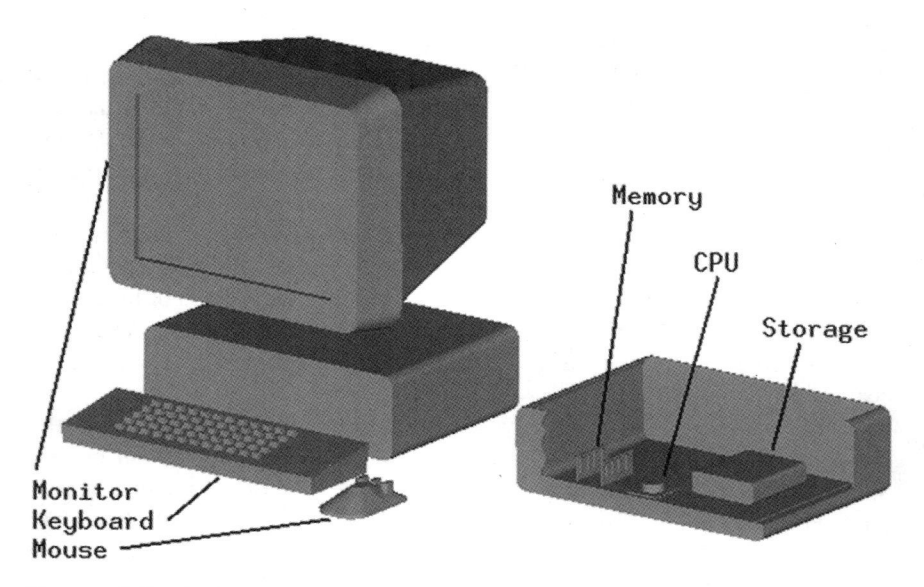

FIGURE 2.1 Basic hardware/components of a CAD platform.

2.3 CPU

The CPU can be easily referred to as the master of the system. It is an integrated circuit (IC or "chip") that really manages the data amongst the components. All the other components, therefore, are designed around the CPU, and the CPU usually has the most obvious effect on the performance of the computer. These devices have been developing and advancing for decades now, and many classes or types of CPUs have come and gone. Although mainframe computers were once built around the CPU (to be shared by hundreds of users via terminals) today each user normally has a CPU dedicated to their use within their personal system.

There are only a handful of vendors that produce CPUs that would typically be used for CAD software. Therefore, there are only a few basic computer systems that CAD users will generally come in contact with. Probably the most popular CPUs (for all applications) are made by Intel®. Intel CPUs have had names such as 8086, 80286, 80386, 80486. After this series of CPUs, a series of Pentium® chips were made (Pentium, Pentium II, Pentium III, etc.), and then the Itanium®. Another popular series of CPUs are made by Motorola®. Their chips have had names such as 68000, 68010, 68020, 68030, 88000, etc. These chips were the basis for the Apple® Macintosh® (Mac) computers. The remaining CPUs found in CAD systems would be from complete computer systems manu-

Computer Hardware Basics 9

facturers such as IBM, Hewlett-Packard® or "HP®," Sun Microsystems™, etc. Their chips would generally be considered "proprietary" since other computer systems makers generally do not use them. This is in contrast to Intel and Motorola which do not provide complete computer systems. Another source of CPUs would be from "clones." These chips are made by companies other than Intel or Motorola, but they are functionally equivalent or compatible with the original chips.

In popular usage, a PC is basically a computer with an Intel or Intel-compatible chip in combination with a Microsoft® operating system. Apple computers (such as the Mac) use Motorola chips, and they would be considered a personal computer, but they would probably not be referred to as a PC. Also, in popular usage anyway, a workstation is generally a very high performance personal computer from a proprietary source, combined with a variant of the Unix operating system. However, with the overall performance of the PC reaching the performance of the workstation, PC companies are now also referring to their computers as workstations, even though they use the Intel-type CPU.

In terms of function, it is only really necessary to understand that the CPU runs or executes very specific, rudimentary logic instructions. These instructions can be referred to as machine code or machine language. Programs such as CAD are not "written" in this language, but computer programs that run on the system must eventually be "boiled down" to this language. CAD software vendors write programs in so-called "high level" languages such as FORTRAN, C, C++, Java™, etc., and programs called compilers convert the program into the machine code instructions.

Probably the most important concept to grasp with respect to the CPU is that it only communicates with the "outside world" (i.e. the CAD software or any data) via memory. The CPU does not really communicate with storage (such as disk drives) directly; instead, the system first brings the data from storage to memory (data going from disk drives to chips), and then the data goes from memory to the CPU. This is important since memory chips are perhaps 1000 times faster in sending or receiving data than disk drives. This, in turn, is important when dealing with computer performance with respect to CAD software.

It turns out that CAD software can be very "intense" for the CPU; CAD demands a great deal of computational power from the system as it makes many mathematical calculations. Therefore managing or optimizing the system's performance is often an exercise in keeping the CPU working on the mathematics and using memory as much as possible (as opposed to just shuttling data back and forth in Storage devices). The CPU communicates with memory via memory "addresses" (a little more detail is presented on this topic in the Memory and Storage sections of this chapter).

The CPU is fed the instructions at a certain fast speed. This happens under the governance of a clock or timing circuitry based on the electronic resonance or vibration of a quartz crystal-based device. This governing speed is known as the clock speed. Clock speeds were once as low as 4.7 Megahertz (or a million "ticks" per second) at the beginning of the CAD for PC era, but now they exceed 1 Gigahertz (or a billion "ticks" per second). At each few "ticks" of the clock, the system can complete a machine code instruction as indicated by the program being executed. This instruction could be moving data around the system (say from memory to a "register" in the CPU); or it could be calculating or manipulating data within the system (say adding 2 numbers together). Although there are other chips or circuits that help the CPU perform these functions, the clock speed and the CPU's basic architecture pretty much determine the computational performance of the computer.

In addition to the clock speed, the performance of the CPU is dependent on the number of "bits" in the system architecture. A bit (short for binary digit) is simply the 1s and 0s, or ONs and OFFs, that computers use and manipulate. Throughout the computer the "bus" transmits these bits between devices. The width or the number of "lanes" in the bus's electronic highway obviously affects the performance of the system. If one system can put 8 bits on the highway at one time, while another can put 32 bits on the highway, then the 32-bit system is going to transmit data much more data in the same amount of time.

Originally, PCs were built upon an 8-bit architecture, with complementary 8-bit CPUs (generally the Intel 8086). When PCs were based on the Intel 80286, they were then using a 16-bit architecture with a 16-bit CPU. Once PCs were based on the Intel 80386, they were using a 32-bit processor (although sometimes with only a 16-bit bus). Once the Intel 80486 arrived, all PCs could be considered completely 32-bit based (although the operating system had 16-bit limitations for many years afterward).

As each of these new CPUs arrived, there was a significant increase in performance since the kind of work PCs generally performed could readily utilize the new architecture's capability. However, once 32-bit architecture was widespread, the amount of data generated by the PC was generally being efficiently handled by the computer; thus, 64-bit and 128-bit architectures only provide relatively small increases in performance.

Another important distinction amongst the competing CPU types is RISC vs. CISC architecture. RISC is an acronym for a "Reduced Instruction Set" architecture. CISC represents a "Complete Instruction Set" approach. The RISC architecture was a hallmark of the workstation-class Unix-based type of computer for many years. The RISC architecture allows the CPU to have long machine code instructions moving more quickly to the CPU, and this allows the CPU to complete a function (such as multiplying 2 numbers) 2 or 3 times faster than the CISC type of CPU. So, the RISC computer can either perform as well as a CISC computer using only a half or a third of the clock speed, or the RISC computer can perform 2 or 3 times faster if both systems are at the same clock speed. But,

Computer Hardware Basics 11

if the clock speeds of the CISC computer are 2 or 3 times faster than the RISC system, then they may have about the same level of performance.

2.4 MEMORY

Unfortunately, many computer users (CAD users included) use the terms memory and disk interchangeably. However, they are actually very different. There is a big difference between running out of disk space and running out of memory. As mentioned earlier, it is important to keep the CAD system using memory as much as possible, since this can have a large impact on CAD system performance. As was seen in Figure 2.1, Memory is made up of integrated circuits or "chips" (somewhat similar to the CPU). However, instead of being the "engine" of the system like the CPU, the memory chips are just a place to temporarily store data that is being worked on.

The memory chips can send, receive, and store at very fast speeds (such as nanoseconds or billionths of a second), while the disk drive can only typically do this on the order of milliseconds (or thousands of a second). Therefore, it is clear that when a CAD program is running, it is important to keep the CPU busy working with the data in the memory chips. However, the data in the memory chips is gone once the power to the system is turned off, so eventually, it is important to get the data stored on the disk drives (which do not lose data without the electrical power). Therefore, memory is considered volatile, since the data in the chips disappears when the power is turned off, and storage (such as disk drives) is considered nonvolatile, since the data on the drives does not disappear when the power is turned off.

Memory can be referred to in a number of ways. Occasionally memory is referred to as core memory, or just "core." In the early 1980s, a VAX 11/780 "minicomputer" which handled 20 users may have had core memory of 16 Megabytes (or about 16 million bytes; each byte made up of 8 ones and zeroes). The VAX's operating system would manage or control the use of that 16 Megabytes for the users on a continual basis. However, the most common term now for memory is RAM. RAM stands for Random Access Memory (meaning it can hold many different kinds of data dynamically). There are other memory chips called ROM chips, but they are generally not relevant to the CAD user.

Although the PCs of the early 1980s had no more than 512 kilobytes (about 512,000 bytes) of Memory, CAD platforms will easily have a gigabyte (about one billion bytes). And, unlike the VAX example, all this memory is basically used by just one user. Of course, the platforms can run more than one program at a time, so there is still some "competition" for the available data space in the memory chips, but this competition for resources should be minimized to keep the CAD program running as smoothly and as fast as possible.

2.4.1 Virtual Memory

It is important to realize that usually there is more "space" in the storage system than in the memory system. For instance, a PC may have an 8-gigabyte (about 8 billion bytes) disk drive, but only 512 megabytes (or about 512 million bytes) of memory or RAM. This makes the "competition" for resources even more important.

When a user decides to run a program (which is probably already stored on a disk drive), the system "loads" the program from the disk to the memory system. Then the CPU begins to execute the instructions that are now found in the memory system. Most operating systems (that are trying to manage or coordinate the data sent to the CPU) have the ability to load and/or unload the programming from the Memory system back to the disk drive. This is called working with virtual memory, and it allows the lower sized Memory system to handle even larger programs, or to handle more simultaneous programs.

For the most part, when a virtual memory computer moves a program's data from one "area" of the memory system to another, it is referred to as paging. However, when a computer moves a program's data out of memory entirely to the slower disk drives, it is referred to as swapping. Of course, the computer will eventually move that data back to the Memory system to finish the tasks associated with the computer program.

The way the computer handles this virtual memory is pretty important with respect to CAD programs. As has been pointed out a number of times, modern CAD programs are large and demand a high level of CPU performance. If the computer running a CAD program does not have enough memory, it prematurely starts to "swap" data and programming to the Storage system. When that happens, a user can notice a very reduced level of performance. This should make sense since it is starting to use the disk drives that may be 1000 times slower than memory.

2.4.2 Memory Addresses

An important concept to grasp is that of memory addresses. The memory chips are often described as having a huge number of mailboxes. Every mailbox has a predetermined identifier or address, and the mail (i.e. data) is supposed to be directed to that mailbox based on the address. So, in addition to having an address, each mailbox can have contents. In other words, at each given address (or mailbox) we expect to find something within it (the data). Addresses might also be referred to as pointers, since they point to where data is located.

These little mailboxes are really transistor-based devices etched into the structure of the chip's chemical layers, but the mailbox analogy is quite helpful. So in a CAD system, a geometric entity (such as a line segment) could be found in memory based on perhaps 4 addresses (symbolizing an X- and a Y-value for each end of the line segment), and the contents of each of these addresses would be the actual values of X and Y at the ends. The addresses could be called "vari-

Computer Hardware Basics

ables" such as X1, Y1, X2, Y2, and the values of these variables could be numbers such as 1.5 mm, 2.5 mm, 12.5 mm, 15.0 mm, etc.

In addition to CAD data (such as the line segment data above), the memory system must hold other kinds of information. First of all, some amount of the operating system program must be kept in memory. This makes sense since the CPU must run the operating system as well as the CAD program, and yet the CPU can only communicate (as usual) via memory. Of course, the computer design and the operating system behavior will affect how much of the operating system must be kept in memory. The more that is kept there, the faster the computer may run, but this means there is less memory available for doing "real" work (like keeping the line segment data). Another kind of information that must be kept in memory is the actual CAD application programming instructions (i.e. the machine code for the CAD program). This is sometimes referred to as the "code segment" as opposed to "data segments." Just as with the operating system overhead, if most of the CAD program is always in memory, then it may run fast. But if too much of the program is in memory, there may be too little left for the CAD data (such as the line segment data) and then it runs slower anyway.

2.4.3 Memory Map

Another important concept to understand with respect to memory is the memory map. Since the CPU only communicates with the overall system via the memory chips, there must be a clear specification in the design of the computer as to how these two electronic components will communicate. This is the memory map. It indicates the predetermined allocation and meaning of memory addresses (Figure 2.2).

Addresses (Hexadecimal):	Addresses (Decimal):	Memory Map Contents
$0000-$01FF	0 to 511	Registers, Counters, Interrupts
$0200-$0FFF	512 to 4095	Drivers, Buffers
$1000-$2FFF	4096 to 12287	Operating System command processor, kernel
$3000-$FFFF	12288 to 65535	User programs (i.e. CAD program and data)

FIGURE 2.2 A very simple memory map for 64 kB of memory (65,536 addresses with 0 counting as an address).

The design of a computer may specify that at address number 640,000, there should be one byte of data (8 ones and/or zeroes) that indicates what color should be shown on the first "dot" or pixel of a computer monitor. Of course, since different computer systems (even from the same manufacturer) may have a variety of options for memory (such as buying a computer with 1 gigabyte of RAM versus one with 2 gigabytes), the memory map must allow for this variation. This is usually accommodated by having all the essentials of running the computer system at the very "low" addresses (perhaps the first few thousand bytes out of the millions or billions of available addresses). At these low addresses spaces you would find data or programming for such things as the keyboard and mouse, since all systems are going to have these devices. However, the memory map would eventually have regions or lists of addresses that are just for applications (such as the CAD program), and the memory map is somewhat "open" at that point. If there is physically enough memory chips on the system, then the addresses in that region would be available. If not, then these addresses would simply not be used (and swapping may start).

The maximum number of addresses possible for the map is set by the number of bits in the architecture. A 32-bit system has a highest possible address number of 4 gigabytes (2^{32} or about 4 billion addresses). So the memory map for the 32-bit system would stop at that point. However, even if the CPU can handle 64 bits, the memory chips or the CAD software may still be meant for 32 bits only. Of course, in time, all the components and software will be able to take advantage of 64 bits, and so the memory map could then extend to 18 billion gigabytes (2^{64}).

It may seem ridiculous to have this much data loaded into Memory for the CPU to manipulate, but CAD software could actually create this amount of demand. It would not be unusual for a mechanical assembly (such as the crane shown in Figure 2.3) to require billions of 3-D surfaces to be totally accurate to the physical product. Since it may take thousands of bytes to describe each surface (color, texture, boundaries, etc.), it would not be difficult to require trillions of bytes to fully realize that 3-D model down to each tiny component (nut, bolt, wire, etc.).

2.4.4 Memory Configurations

In terms of the physical memory chips, they are often found in a wide variety of configurations. Usually, the main circuit board or motherboard contains the CPU as well as the memory chips. These chips are arranged in banks. These banks have a given capacity to hold a certain number of memory chip modules. These modules may be called SIMMs, although there are other types of configurations with different names. The number of these modules in combination with the capacity of a given memory chip will create the total "physical" memory or RAM for the given computer system.

A typical situation would be a SIMM module with 4 chips on it. The chips could have a capacity of 32 Megabytes each, so when they are combined into a

Computer Hardware Basics

FIGURE 2.3 Example of large 3-D model.

single SIMM, there would be 128 Megabytes. Then 4 of these SIMMs could be combined on the motherboard to give a total of 512 megabytes of physical memory. Recall that this then becomes the maximum "address space" that the CPU could specifically utilize to run the system, and then depending on the operating system, this could be expanded to a certain higher number as "virtual" memory. For instance, this 512 megabyte physical memory could be configured to mimic 4 gigabytes of memory as virtual memory.

Table 2.1 summarizes the contrasts between the memory system and the disk system (which is explained in more detail in the next section).

TABLE 2.1 A Comparison Between the Memory System and the Storage System

	Memory System	Storage System
Physical device type	Integrated circuits (chips)	Disk drives Floppy drives
Data behavior	Volatile (data is lost when the system is shutdown)	Nonvolatile (data is not lost when the system is shutdown)
Data speed	Fastest device for the computer (data rates measured in nanoseconds)	Slow devices for the computer (data rates in milliseconds)
Terminology	RAM Main memory System memory Core memory	Disk Drive Hard drive Floppy drive DASD (archaic mainframe reference)

2.5 STORAGE

As already mentioned, the storage system is made up of a device or devices that can permanently store data (it is "nonvolatile"). This is in contrast to the memory system which is erased or empty when the computer is turned off or shut down. In addition, the capacity or "size" of the data that can be stored on a storage device is usually much larger.

In terms of actual devices, the storage system is generally made up of one or more disk drive. These drives are devices that have a spinning a disk within them. Thus, these devices are often also referred to as disks or hard disks. These devices may also be referred to as hard drives, disk drives, or C drives. These terms basically all refer to the same thing. The spinning disk has magnetic material on it that can be altered by recording or playback "heads" (similar to a magnetic tape). These heads are on a electronically controlled arm that can swing over the spinning disk. The whole mechanism is in contained in a vacuum-sealed enclosure that fits into the computer system enclosure. Another type of storage device would be a floppy drive or a diskette drive. These work on the same principle as the hard disk, but they are generally used for transporting smaller amounts of data between computers.

Although the memory system is the only way that the CPU really communicates with the outside world, the storage system is the way that a user really communicates with the computer. Users do not save what the CPU has done, unless it is written to the storage system. This is true for CAD programs, as well. The user does need to be aware of what data is currently stored or saved on the storage system, and the user must be aware of how the CAD program is going to read and/or write new forms of that data. The way that the user generally recognizes the data on the storage devices is through files.

2.5.1 Files

The files that computers work with via the storage system can be thought of as electronic versions of paper files. A paper file is a document, a piece of paper with information written on it, or a set of pages. On the paper file, there will be letters, numbers, and graphics written on it. The computer's files contain the same thing, only in an electronic form. The electronic form really is just a sequence of ones and zeros, and something called the ASCII table dictates which pattern of ones and zeros represent the letters of the alphabet or the numerals 0 to 9, etc. The precise meaning and behavior of computer files on a specific computer system is controlled by the operating system (this is covered in more detail in Chapter 3 on computer software).

The size of files refers to how many bytes (a pattern of 8 ones and zeros) are stored on the storage device for the given file. A typical word processing document could be 50,000 bytes in size. A typical CAD 2-D drawing file could have

Computer Hardware Basics 17

a size of 150,000 bytes. A typical 3-D model of a single part could use 100,000 bytes. A typical model of an assembly of a few hundred parts (such as Figure 2.3) could have a file size of 50,000,000 bytes.

Since the size of these files can be difficult to express in terms of the actual number of bytes, there are metric system "short cuts" for the sizes by the "1000's" (refer to Table 2.2). So, approximately 1,000 bytes is a kilobyte (in the metric system, the prefix for one thousand is kilo). Approximately 1,000,000 bytes is a megabyte, and approximately 1,000,000,000 bytes is a gigabyte. These various short cuts are abbreviated as KB, MB, and GB respectively. In each case, the conversion is approximate. This is because the actual conversion is not based on 1000s, but on 1024. 1024 is 2^{10}, and most of the operations of the computer are based on the logic represented by 2 numbers (ones and zeros), so it is easier to base capacities on the 1024 value. In practice, the capacity can just be assumed to be based on 1000, not 1024, since it only introduces a small amount of error (Table 2.2 shows the actual numbers).

2.5.2 Disk Drive Functioning

It is important to understand how disk drives work and behave since the files that CAD programs generate can be very large, and performance will suffer if the storage system is not utilized properly.

As mentioned earlier, the actual information on the storage system is made up of extremely small regions of magnetism deposited or removed on a spinning disk (referred to as the disk drive). These magnetic regions are located based on a radial and an angular position (see Figure 2.4). In other words, each position can be located by a radius (or track) from the center of the disk and an angular position (or sector) within that circular radius. These locations are grouped and often referred to as cylinders. When data needs to be stored or "written" on the disk drive, the computer finds an "unused" region of the disk and the heads are posi-

TABLE 2.2 Large Byte Count "Shortcut" Terminology

Shortcut	Approximate size (bytes)	Actual size (bytes)
Kilobyte (thousands) or KB	1,000	1,024
Megabyte (millions) or MB	1,000,000	1,048,576
Gigabyte (billions) or GB	1,000,000,000	1,073,741,824
Terabyte (trillions) or TB	1,000,000,000,000	1,099,511,627,776
Petabyte (quadrillions) or PB	1,000,000,000,000,000	1,125,899,906,842,624
Quitealotabytes (quintillions) or QB?	1,000,000,000,000,000,000	1,152,921,504,606,846,976

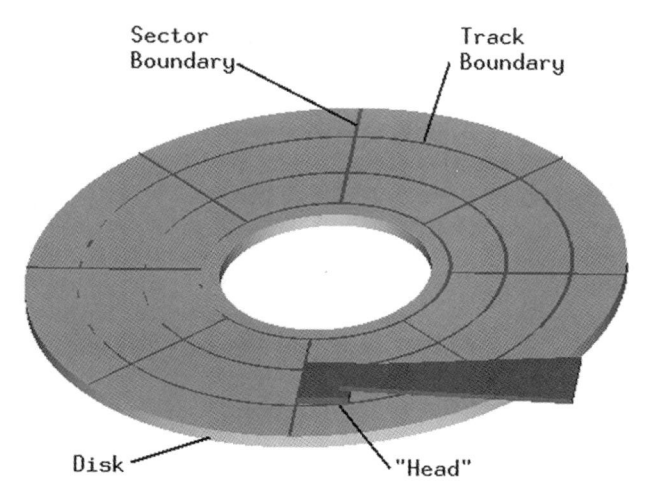

FIGURE 2.4 Disk drive anatomy.

tioned at the region. Then the heads produce a magnetic field at the exact right time to alter the magnetic medium on the disk. This event corresponds to the data in the file being created. When data needs to be retrieved or "read" from the disk drive, the heads are positioned to where the desired data is supposed to be located, and the magnetism on the disk is sensed by the heads. This generates the signals that are then translated back into the data. Believe it or not, this process proceeds at the rate of millions of signals each second. Each second, the heads are sent to the proper location on the disk within very tiny distances, the signals are sensed as the disk spins by at perhaps 10,000 RPM, and then the heads move on again. In that one second, all the text of this book could be read or written, and there are disk drives that can go much faster than that.

A number of issues are raised by the disk drive's behavior. First of all, it is important to understand the idea behind "reading" and "writing." It should be clear that this would be closely linked to data security (i.e. setting up who is allowed to do what with the computer and/or its data). If someone is able to read data from the disk drive, then they would be allowed to see the data stored there. And, if this person is able to write data to the disk drive, then they could create or change data there. A person who is able to do both these operations is said to have "Read/Write" or "R/W" access. On the other hand, if someone is only able to read the data, but not create or alter it on the disk drive, then this person is said to have "Read/Only" or "R/O" access. For CAD programs, this capability corresponds to active or released files or drawings, versus ones that are preliminary or under revision. For instance, if drawings are created and being used in manufac-

Computer Hardware Basics 19

turing operations, then changes to those drawings may be restricted until the manufacturing operation is complete. In this case, the drawings (contained in a CAD program's computer files that are on a disk drive) may be marked or identified as Read Only. The granting of these privileges to access data on disk drives may be controlled by the operating system or the CAD program, or a combination of the two.

Another important issue to consider is how the disk drive finds those tiny regions of magnetic space. Obviously, the computer must have a way of mapping or translating from the name of the data file the user is interested in to the actual physical location on the disk. This is generally accomplished by have a sort of master file on the disk drive. This file is not really accessible to the user, but it is accessible to the operating system which is controlling the computer. For PCs, this master file is often referred to as the FAT (File Allocation Table). On workstations, this sort of information is related to the "inode table" (an inode is a unique file identifier number), or perhaps as the "superblock" data (a "block" is another unit of space on the disk, it is usually 512 or 1024 bytes). This data is referred to as a table since its function is as a lookup table. The system "knows" what data file the user wants to access, so then the table is used to find or "lookup" the proper location on the disk (or following directory structures). The exact workings of this master file and how it leads to data storage and retrieval is beyond the scope of this work, but it is important to know that the master files are there.

2.5.3 Backups

One of the most important issues with respect to the master file is that it can be corrupted. The disk drives can and do fail; they wear or are damaged to the point that they make a "mistake" in reading or writing the users data. If this mistake occurs on the region of the disk where the master file is located, then ALL the data on the disk drive can be lost. The computer can no longer determine the location of all the other files. Obviously, this is a very bad situation that must be avoided. However, since there is no way to absolutely predict when the failure may happen, the only practical recourse is to back up all the disk drive's data. Backup in this case means to copy all the data on the disk drive to some other device or devices so that a backup or reserve copy is available. In many cases, the backup copy is to a more stable or static magnetic material such as magnetic tape in tape drives. These devices are similar to the disk drives in that they use magnetic film to store the data, but there is no spinning disk. Instead, there is magnetic tape. These tapes are then stored in cartridges and filed in places that are protected from physical damage. Since the master file on the disk drives as well as the data on the disk drives must be synchronized or be in agreement, this activity of writing the tapes is often done at times of day when the computer system is not in use.

In addition to the creation and keeping of backup tapes, many CAD systems will rely on redundancy. In this case, the data created and used by CAD users are actually written to 2 physical disk drives. This provides a redundant source of the data. These kinds of disk drives often come in an array or drawer that contains a complete set of disk drives. These sets of disk drives may also be referred to as a RAID array. RAID is an acronym that relates the level of redundancy achieved by the set of disk drives; a number may be used in connection with the RAID array, such as RAID Level 5. Other terms that have to do with redundancy are mirroring and striping.

2.5.4 Formatting

Another issue with respect to the behavior of the disk drive device is formatting. When a disk drive is first made available to a computer system, the master file needs to be created on the device, and the master file needs to know where the available space exists. This is the process known as formatting. There are often different levels of formatting. A low level format would actually have the heads scan and/or check every region of the disk. A normal level format usually just creates a brand new master file. Of course, creating that new master file means the old master file is gone, and thus all the data that was on the disk drive is basically gone (it can't be located any longer).

The use of formatting is generally confined to setting up new computers, but it can be used to improve performance as well. If the master file has been used on a computer system for a long time, the file may become large. Then, even when all the files that are residing on the disk drive are removed, the master file may still be large. In this case, reformatting the disk drive may speed up the rate at which files can be looked up in the master file. Another performance improvement that may be realized from low level formatting has to do with marking physical regions of the disk as unreliable or unusable. In this case, the master file is able to know that certain parts of the disk should no longer be used.

In addition to the performance aspects of the master file, the master file concept can also be used to undelete files. When a file is removed or deleted from a disk drive, the only action that generally happens is that the data in the master file about the deleted file is removed from the master file. However, the physical region (or regions) of the spinning disk that actually stored the data is unchanged. Obviously, if the information in the master file is not completely removed during the delete process, but simply put aside for future use, it would be possible to restore the file that was deleted. Of course, if the physical region of the disk (which would then be considered available for other files) is re-used for new data by the master file, then the "undelete" process is not going to work very well. In general, PCs have had the undelete capability for many years; however, the workstations have never had this capability. One reason not to include this capability is

Computer Hardware Basics

the potential for slowing down the overall speed of data access. If the master file completely "forgets" the deleted file information, then the master file can remain more compact.

Another issue with respect to disk drive behavior is the importance of how the data in a computer file are actually located on the spinning disk. This is especially true for the very large files, and CAD files, particularly 3-D models, tend to be rather large (as mentioned earlier with respect to the memory system it could be 100s of megabytes). With these large files the data is going to far exceed the capacity for storage of one particular region of the disk (that is a particular cylinder or block). In this case, the data is going to be written to a large portion of these regions, and the master file is going to have to keep track of where all those regions are located.

Clearly it would be best if all that data is written to regions of the disk that are "next to each other" or contiguous. In this case, the master file will only have a small amount of information to store; i.e., where the large file starts on disk and how far to go to read all the data. On the other hand, if the large amount of data in a big CAD file is broken into many small pieces, and these small pieces are in regions of the disk that are not contiguous, then the heads above the spinning disk have to move many times (perhaps 100s or even 1000s of times) to reassemble the entire file into the memory system. Clearly this situation is going to be pretty devastating to the performance of the disk drive and thus the entire storage system. This situation is sometimes called "thrashing," although this term may be best considered to apply to the situation of the memory system having to unload data to the storage system, and the storage system has to rapidly attempt to shuttle data to and from memory (or swapping).

One important method to prevent this breakdown in performance of the disk drive is to not let the disk drive capacity become too near to its maximum. As the used regions of the disk grow, the available regions not only become smaller (since less and less free space is available), these regions become spread apart as the larger files create smaller gaps. As a rule of thumb, no disk drive should exceed 80 percent of its maximum capacity. On the other hand, if a disk drive is 90 or 95 percent used, one can be certain that performance is suffering as the heads attempt to find and assemble files.

It was mentioned earlier that reformatting a disk drive will improve performance, and indeed, this will help alleviate the situation of the heads not getting data efficiently due to the poor organization of the data on the disk. However, it was also mentioned that the reformatting will erase all the data in the disk drive. Obviously, this is not always acceptable. In this case, a technique called "defragmentation" can be used. In this case, a special computer program is used to "shuffle" the data around on the disk (without doing the more drastic reformatting). The program reads a file into the memory system, then it deletes enough of the file to create some contiguous space on the disk. Then it rewrites the file from the memory system back to this improved region. This process continues, file by file, until the all the files on the disk drive are in a more efficient arrangement. Of course, one must be certain that all the data on the disk drive has been backed up (as mentioned earlier) before anything like a "defrag" program is used. If the computer system should shut down for some reason during the running of the program, the result could be somewhat catastrophic.

In the chapter on software, the concept of directories and subdirectories for files is presented. This technique of organizing larger numbers of files on a disk drive has some bearing on the performance, as well. It is the master files that need to keep track of the directories, and thus the same arguments of spreading data too thinly on the disk, or making the master file too large and cumbersome can apply to directories. For instance, if there are 10,000 files on a disk drive and they are in a single directory, then the master file for that directory will be relatively large. On the other hand, if these files are spread into 10 different subdirectories, and there is one master file for the top level directory, and then 10 smaller master files for each of the directories, then these master files will be much smaller, and therefore more efficient. Unfortunately, not all operating systems take complete advantage of this sort of efficiency, but it is still to be considered very beneficial for CAD programs, in particular, to keep files in a well-organized system of directories.

2.5.5 Disk Cache

The final concept to be mentioned with respect to the storage system is the disk cache. As already explained, the memory system (which is closely tied to the CPU) operates at a much higher rate of data transfer than the storage system. One way to improve the speed of data transfer for the storage system is to dedicate some memory chips to just communicate with the storage system. This relatively small amount of memory (perhaps a few percent of the overall memory system size) is called a cache or disk cache. It forms a buffer between the main operations of the computer and the disk drives. And, special software may be running within the computer to attempt to keep the data "most likely" to be needed from the disk in that special memory area. This can certainly improve the performance of the storage system, and generally, the more memory that can be made available for the cache, the better the performance. Also, many systems will use this cache technique not only between the disk drive and the memory, but also between the memory and the CPU. These memory-type of caches may also be found on the CPU of the system.

Table 2.3 presents a simple comparison between some common storage system's characteristics. Some of this information will be explained in more detail in Section 3.2: The Operating System.

Computer Hardware Basics

TABLE 2.3 A Simple Comparison of Various Computer System's Storage

System type	Disk drive designation	Example file names
PC	A: (typically floppy disk) C: (main hard drive) F: through Z: (typically network drives)	DRAWING.DAT Model_Of_A_Car.PRT
Unix™ workstation	/dev/hdisk0	drawing.dat model_of_a_car.prt
VAX	USER1: USER2:	DRAWING.DAT;1 MODEL_OF_A_CAR.PRT;32
Mainframe	A (user's read–write area) H (shared "virtual" disks) (also referred to as DASD)	DRAWING DATA A1 MODLCAR PART H/A

2.6 NETWORKING

So far, the basic internal workings of a "standard" computer system has been presented. The essential elements of the system are the CPU, Memory, and Storage systems. However, there needs to be means by which the system is actually put into use and manipulated by programmers and users. This is accomplished by devices known as peripherals. The familiar devices that provide interactivity with the system are keyboards, monitors, pointing devices, speakers, etc. These types of devices will be discussed later in this chapter. One important peripheral for the basic functioning of a computer system is the Network Interface Card (NIC) or network card. This circuit board provides a connection between the computer system and a wider environment of computers and devices. This wider environment is called a computer network or just a network.

Over the years, the relative importance of the computer network has steadily grown. At one time, a CAD-based computer would have been seen as a self-sufficient unit that occasionally communicated with other computers. However, by the 1990s the network was seen as an essential element of any computer system, particularly since the mainframe ceased to be the foundation of a computational framework. Instead, the foundation shifted to many smaller PCs and/or workstations. As that happened, the network was seen as the appropriate means to share resources. At first, this would be just for peripherals such as printers (an office with a dozen PCs could share one expensive printer). Then, the systems started to share data and storage. This meant that just one copy of a PC program could be loaded on a designated central computer system (instead of having to load the PC program to a dozen separate PCs).

Although this new "file server" approach to networking sort of looks like the mainframe system again, there is a significant difference. With the mainframe computer, the data was indeed stored in a central storage system (similar to the "file server" of network), but all CPU power also had to come from that same mainframe computer. With the standard network approach, only the files came from the central storage, while the actual execution of the programs was still done "locally" (i.e. using the CPU in the user's computer). The network approach means that the computational power demands of different users do not affect each other. So, if one user is running a CAD program that fully uses his CPU, it has no effect on other computer (CPUs) on the network that may not be running CAD programs. This would not be the case for the mainframe where all users are "competing" for the centralized CPU capabilities.

2.6.1 Local Area Network (LAN)

The kind of computer network that uses the central computer as the file server and/or the printer server is loosely referred to as a Local Area Network (LAN). The LAN was primarily a local network since the devices that connected to the network had very short distance requirements (such as no 2 computers could be more than 110 yards apart). However, over time, these restrictions became less and less severe, and LANs could grow to cover entire company sites and have 1000s or more computers and devices connected to the network. So, in general, most company-wide networks can just be considered LANs, and most commercial CAD systems in use are dependent on this type of network.

2.6.2 Wide Area Network (WAN)

Another class of network for connecting very wide spread sites (say at various locations in a country) is the Wide Area Network (WAN). At one time, this network would have used a entirely different type of hardware and software, but in many cases, the LAN type of approach can be adapted to the very long distances, but just with a vastly reduced performance. For instance, if phone lines are used to connect the different office sites, the speed at which the network data can be transmitted from one computer to another would be many times slower, but the user may see no difference in how the computer file is stored (it is "transparent" to the user). Since these WANs are generally so slow and CAD programs are generally so demanding on computer system performance, there would be few CAD systems that are really dependent on this type of network (at least for interactive use).

2.6.3 The Internet

Finally, the "network of networks" has arisen called the Internet. It would be easy to think that the advent of this worldwide network was a natural progression from

Computer Hardware Basics

the WANs. But most, if not all, of the successful LAN networks were based on a system called Ethernet™. The Ethernet system was based on a networking protocol called IP or the Internet Protocol or TCP/IP. The way that all the Ethernet-based computers communicated is generally the way that computers on the Internet communicate. Although other types of communication can be achieved with the Internet, virtually all of it is done in the Ethernet style, and therefore, virtually all of the computers on networks communicate via an Ethernet card.

Although the speed of Internet communications can be very limited in comparison to a LAN, and thus the use of the Internet for CAD programs should be ignored just as it would be for a WAN, the advent of the World Wide Web put an interactive and graphical face on the internet. The addition of this graphical interface is made possible by various techniques that optimize the transmission of graphical types of data (despite the slow network speeds). Since CAD programs are very interactive and graphical, these optimization techniques can be applied to the running of a CAD program "across" the Internet (although the user's local CPU again is intended to provide most of the computational resources). Although this may not be the standard approach, it is likely to become more common as the Internet continues to improve.

2.6.4 Network Functioning

As mentioned earlier, the device that connects a computer system to a network is called the network card. This device may actually be a circuit board for the system, or it may be integrated into the main board or "motherboard." In any case, when the computer system needs to communicate with the network (to get a file, a program, or a shared device such as a printer), it will do so with the network card. As with other devices attached to the computer, there are specific locations in the memory map designated for the network data interface. Thus, the computer's software (such as a CAD program) usually does not treat the data from the network any differently than data from a disk drive. However, the way in which the network card actually acquires the data from the network is quite different, and how it is done can greatly impact CAD program performance.

Devices such as disk drives are pretty tightly "coupled" to the computer system in which it resides. The exact size and capability of the disk drive is tracked by the system, and when the software running on the computer creates or accesses the data on the disk drive there is no "competition" for that data. The disk drive is totally dedicated to the operation of that particular computer. This is not the case with the network. The nature of the network is that there can be any number of computers and devices connected to it. A network is intended to be dynamic. Therefore, the computer system that needs to get data to or from the network does not really know what is happening at any given moment, and the resources on the network can not be considered to be dedicated to the computer that wants it. So, the network card needs to be much more involved in the process of retrieving the data. It must communicate with the network with its own programming (usually part of unchangeable software stored on special computer chips called ROM chips), and coordinate the translation of the data to or from the computer system.

The way that the Ethernet-style networking system accommodates the dynamic behavior of the network is through a kind of broadcasting. One can think of the network cards as "listening" to the network cable that it is attached to. It listens for information that belongs to it. Or, the network card may become the "transmitter" sending out data for other computers or devices. When the signals are sent out to the network, the originating computer does not really know if the signal is going to make it or not. If one computer sends out its signal at the same time as another computer, they may collide. Each of these competing computers then waits a short interval (with some intentionally random pause) and then tries again. The message will eventually get to its destination (which is then acknowledged by the computers communicating). It may seem that the collisions will be prohibitive to reasonable performance, but the actual chunks or packets of data that are attempted for each transmission are rather small (with the network card appropriately cutting up the data that the computer wants transmitted), and if the network is working fast enough, the collisions can be of no serious consequence. This "collision-retry" type of network can operate at 10 million (10 Base T), 100 million (100 Base T), 1 billion (or gigabit) bits per second.

Although the network card and network cabling are essentially all that a user needs to know about network functioning, there are many other "behind the scenes" elements to a typical network. Some of these elements would be hubs, routers, bridges, gateways, repeaters, and multiport adapters. Each of these devices are connected to the network and each has a specific task to perform in getting the data from the transmitting computer to the receiving computer or device. Particularly with something like the Internet with its countless computers or nodes, one can imagine there is a great need for shuttling, shuffling, and redirecting of the network data to get it to the right place as efficiently as possible.

2.6.5 Network Performance

There are a number of very important ramifications of networking on programs such as CAD systems. The first issue is network traffic. Considering the way that data is transmitted with the "collision-retry" system, one can see that the performance of the network (in terms of its ability to transmit those large CAD files from a file server to a user's workstation) is going to be greatly affected by how many users are attempting to do the same function. If many users are sending/receiving at the same time, then the collisions are going to be much more frequent,

Computer Hardware Basics

and users will notice that operations take longer. Of course, the operations that take longer will generally only occur when the user is saving a file. Recall that the network approach is to use the user's local CPU as much as possible to do calculations and graphical operations (like displaying a drawing or 3-D model), so network traffic should only be affected when the user is calling up or retrieving or loading a drawing or model, and then when the user is saving a file.

Obviously, if many users are using the same network at the same time (so simultaneous filing is not really avoidable), then the situation can be improved by increasing the speed of the network. Another improvement might be to reorganize the network so that each user's computer has its own connection to a higher speed "backbone" (as opposed to having groups of computers connected to intermediate devices, and then the intermediate devices accessing the file server).

Another consideration for performance of the network is the way the file server is utilized. Consider that even if the network is fast enough to not experience the "collision" situation, but all the drawings and models for CAD users are stored on one disk drive on the file server, then all users must wait for that disk drive "one at a time" regardless of the network's capability. Another important consideration for the network file server is the amount of memory or RAM in the file server. This memory is used to hold or buffer data that is coming from the network for reading and writing (like the cache mentioned earlier for storage systems), so as much RAM as possible should be used. The file server can become a "bottleneck" in that it is the slowest or most constricted part of the network architecture. Clearly it is very important that an appropriate level of expertise be brought to bear on the configuration of the network architecture.

2.6.6 Network Addressing

The way that the network recognizes which information is intended for a specific computer (or "node") is by an address. This is not the same kind of address that is associated with the memory system of a computer. The address system for the network starts with the network card itself. Each network card is given a unique number. This is usually called the Ethernet address. It would be made up of 6 bytes of data (such as 08:00:11:00:AF:08). There is a system to the bytes (a byte being 8 bits) assigned, but that is beyond the scope of this work. The programming that runs on the network card's components uses this number sequence to communicate with the network. This address can be considered part of the "physical layer" of the network architecture.

With the physical address for the network card identified, it can then be translated for use on a specifically designed network (such as a LAN or the Internet). That network card could be placed in different computer systems over time, so there is a higher "level" for assigning the address to associate with a computer system. This higher level address is called the Internet address. This address implements a sort of network layer, and it uses the Internet Protocol (IP). The network layer then communicates with a transport layer, where the Transport Control Protocol (TCP) takes over. The way these layers work is beyond the scope of this work, but terms like TCP/IP are in rather common usage, and it basically means that this Ethernet-style of networking is running the network. In the end, a layer communicates with the actual application (such as the CAD system).

The Internet address has the form of 4 bytes of data. It would be presented in a form such as the following:

128.1.128.157

This address uniquely defines a computer or device on the network in terms that are understood by the software that really controls the network traffic. The devices that "run" the network (such as hubs, routers, servers), or at least the software running on these devices, use the numbers in the address to determine how to sort, redirect, transmit, etc. the "packet" traffic. These devices are not concerned with the data in the packets, just where it needs to go.

There are 3 "classes" to these addresses depending on the kind of network architecture. These classes determine how many networks could be found in the entire site, and then how many specific computers or devices could be found in a given network (in other words the class defines to what degrees a network of networks can be created). The Class A address uses 7 bits to define the network and 24 bits to define specific devices. So, for Class A, there could be 128 networks (2 raised to the 7th power and including zero as an address), and there could be 16,777,216 specific devices on a given network. The remaining bit indicates the class itself. At the other end of the spectrum, a Class C address allows 2,097,152 networks, but only 256 specific devices. Clearly, the addressing scheme used needs to properly fit within the physical network architecture. If the addresses are used to create "sub-nets," then the proper network electronic devices need to be used to keep communications within a specific subnet optimized. For instance, there may be 3 different groups of CAD users at a site, and they do not generally need to communicate data beyond their small group. In this case, the use of special addressing schemes can be an advantage. However, if all the CAD users need to communicate with each other pretty equally, there may be no benefit to this approach.

At some point, the Internet address is translated to a node name or host name. This is a "real" name in the sense that it does not have to be in a numerical format, and users would be more likely to remember it (for instance, serv1, cadmeister, zulu, etc.). The host names are usually kept in a file that resides on each computer system (for instance a "hosts" file).

2.6.7 Peer-to-Peer Architecture

An interesting effect of the Ethernet-style of networking is that it is usually considered a "peer-to-peer" network architecture. This means that all the computers

Computer Hardware Basics 29

on the network are considered "equals" (at least if there were no security schemes applied). Therefore, a user can access other computers on the network to some degree. This is done with programs such as rlogin and telnet. Using rlogin (meaning remote login), one could have a program that does certain calculations run on a different computer. In this case, the user's originating computer will not slow down due to the calculations consuming CPU or storage system resources on that remote system. Obviously, this can cause security problems, but it is a useful capability in a CAD environment where some "batch" processing may be needed for computer programs that run for hours or days. Some other programs that are typically utilized across the Ethernet-type network architectures are shown in Table 2.4.

2.6.8 Network File System

The final subject to be addressed with respect to networking and its importance to the computer hardware configuration for CAD software is the network file system (or NFS® or NTFS). An NFS-type program is basically a "trick" to make network files look like they are local files. For instance, a workstation may have one disk drive attached to it that contains most of the software that runs on that computer (such as the CAD program). When the CAD program runs (using the local CPU resources, of course), it may look for files that are identified by a specific directory and file name. This name may actually be pointing to a file on a different computer on the network (often to a computer called a file server). It is this NFS which makes the network-based files look like they are local files. Set-

TABLE 2.4 Common "Peer-to-Peer" Network Utility Programs

Program or command	Capability
export DISPLAY	Allows graphical user interface to be sent to another workstation as if it were seen on the user's current computer.
ftp	File transfer protocol (ftp) is an Internet based program to transmit files across a network. If a network node is to store files for others to download, then it may be referred to as an ftp site.
ping	An Internet based program to see if a remote computer is responding to the network.
rcp	A program to interactively copy a file across a network following the typical copy command for a Unix workstation.
rlogin	A program to mimic being a local user of a remote computer on the network.
rsh	A program to initiate a system command on a remote computer following the typical process for a Unix workstation.
telnet	A program for gaining access to a remote computer on the network.

ting up the mapping of how local names will actually be redirected to network-based names can be called mounting directories, making NFS mounts or mapping drives (depending on the type of computer system).

The NFS technique can be quite useful in a CAD-based network, but it must be used carefully. Recall that in the earlier discussion of the memory system, there was something called "virtual memory" that allowed a disk drive to make the computer system have more memory capacity than was really available on the memory chips. The part of the disk drive used for this purpose must be as fast as possible, and therefore it should be made available on a local disk drive. Using an NFS type of access for this "swap file" would not work very well. Another poor use of NFS would probably be the operating system files. These are files that the CPU constantly needs, so putting it "across" the network would probably cause problems as well. In most cases, the CAD software would also generally be a poor choice to put on the file server. While the CAD software is running, the CPU may continually need to retrieve information from the disk drive for the software; therefore, a "local" drive would probably be best (unless the central management of the software was an overriding concern).

A good choice for the NFS technique would be data that is only needed occasionally or data which needs to be centrally stored, managed, and protected. The actual drawings or 3-D data files that a CAD program creates are a good candidate for storing on the file server. In this case, the file is really only accessed intermittently (when the user first loads the drawing or model and then whenever

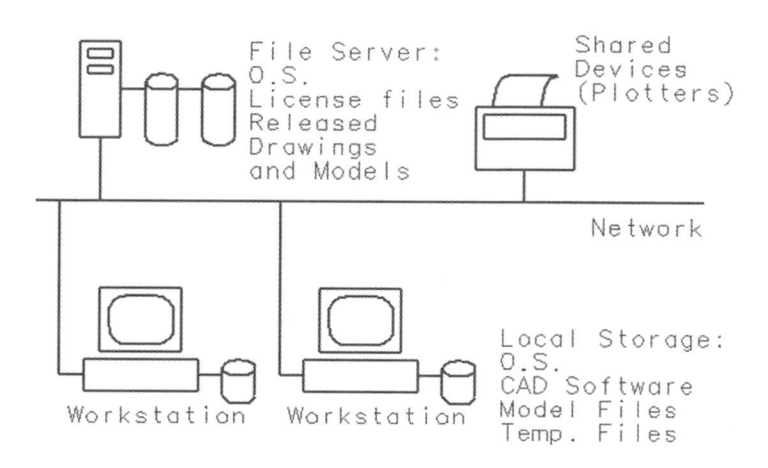

FIGURE 2.5 Typical CAD system network arrangement.

Computer Hardware Basics 31

the user saves it again). If these files are all stored centrally (on the file server), then it is relatively easy to perform system management operations (such as "backups"). Of course, users need to realize that although the files appear to be on a local disk, their use is actually going to generate network "traffic" that affects other users. Figure 2.5 shows a typical arrangement of data and files for a CAD system network.

2.7 PERIPHERALS

This section presents information on the devices that are attached to the computer. These devices are the peripherals: the monitor, keyboard, pointing devices, and hardcopy devices.

2.7.1 Monitors

Probably the most important peripheral for computer systems that are used for CAD software is the monitor and its counterpart, the graphics adapter. The very essence of CAD software is the use of computer graphics. The designer or engineer interacts, calculates, and specifies in graphical terms. Therefore, it follows that the device that creates these images is extremely important.

The monitor is simply the device that shows the images from the software (see Figure 2.1). However, it is not the device that creates the image; this is the graphics adapter discussed in the section below. All that needs to be understood at this point is that the graphics adapter sends the proper electrical signals to the monitor so that it can show the correct image.

The basic characteristics of the monitor are its physical size, resolution, and ability to react to the electrical signals from the graphics adapter. The physical size is generally based on a "diagonal" distance of the viewable area of the screen. This diagonal distance, for instance, would be from the top left corner of the screen to the lower right corner. Obviously, the larger this distance or dimension, the easier it is to view the image on the screen (assuming the graphics adapter keeps the same precision of the image) since the image is larger. For CAD software, the screen should be as large as is affordable (for example, at least 500 mm or 19 inches for most users, or even larger for those with vision disabilities).

The precision of the image is controlled by the monitor's resolution. The resolution is an indication of how finely or coarsely the image can be presented, and it is based on the number of tiny regions on the screen that can be discreetly activated or "painted." These tiny regions are called pixels (or picture elements). If one looks very closely at most monitors the pixels can be seen. Obviously, if a monitor can display more pixels, then it can show a better or more precise picture.

The resolution for a monitor is usually expressed in terms of the number of pixels in the horizontal and vertical directions. For instance, a monitor may be said to have 1024 by 768 resolution. This means that there are 1024 pixels in a single "row" across the screen in the horizontal direction, and there are 768 pixels in a single "column" of the screen in the vertical direction. With a monitor able to display a resolution such as 1280 by 1024, the total number of pixels exceeds one million; this may be referred to as a megapixel system.

Keep in mind, however, that the monitor does not generate these pixels, it merely displays them. It is the graphics adapter that creates them, but there is a limit to the ability of a monitor to display these pixels. Therefore, a graphics adapter that can produce 1600 by 1200 resolution would not be used to full capability with a monitor that is only designed to handle 800 by 600.

The monitor actually generates the display image by a method of horizontal scanning. The electrical signals from the graphics adapter are translated into the control of a beam that excites special chemicals inside the monitor to make the colors appear. This beam proceeds as a horizontal line across the screen, and this line then may scan from the top of the screen to the bottom. The repeating of this beam produces the refresh rate. The rate at which this refreshing occurs (such as a number of times a second) can also have an effect on the performance of the monitor. Another ramification of the scanning behavior of the monitor is that computer graphics data from a computer program is often referenced from the top of the screen, instead from the bottom (which may be more natural in a mathematical sense where the 0,0 point is at the lower left corner). In other words, the location of a pixel at specific X and Y coordinates may have the Y measured from the top of the screen with higher numbers proceeding DOWN from the top; the X coordinate usually then proceeds from left to right, as usual.

2.7.2 Graphics Adapters

As mentioned previously, the graphics adapter is the device which really creates the image to be displayed on the monitor. It is, therefore, probably the most important peripheral for a computer running CAD software. The performance and capability of the graphics adapter could easily have the greatest impact on a designer's productivity. They can be quite complicated devices, but a basic understanding of its functioning should prove quite valuable in optimizing performance.

Graphics adapters are electronic components on a circuit board that can be installed into a computer. Although this circuitry could be self-contained on the "motherboard" of the computer system, usually computer manufacturers permit a variety of graphics adapters for their computer systems (at a wide variety of costs). Therefore the graphics adapter is left as a removable circuit board. Also, the graphics adapters have often been in a cycle of improvement that is indepen-

Computer Hardware Basics 33

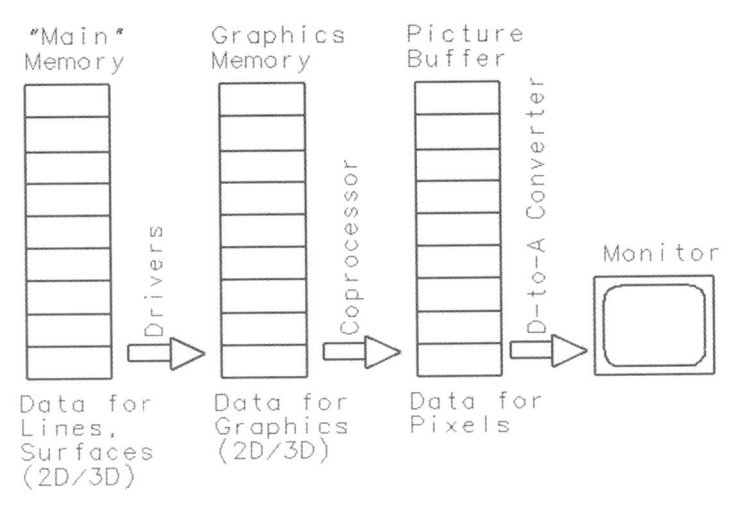

FIGURE 2.6 Simplistic schematic of graphics systems and process.

dent of that for complete computer systems, and leaving them as an expansion or option allows the user to upgrade the graphics adapter without replacing the entire computer. The graphics adapter may also be referred to as the graphics card since these expansion circuit boards are often referred to as expansion cards. Or, the graphics adapter may be referred to as the video card, but this may be misleading since video also refers to analog-type signals like from a videotape or television.

Although the graphics adapter is just an expansion, it could have almost as many components as a complete computer. The graphics adapter may have its own type of CPU that makes calculations and manipulates data in its own memory. This processor may be called a coprocessor to distinguish it from the main processor, or it may be called the GPU (Graphics Processor Unit). The presence of the co-processor has a very large impact on performance. The graphics adapter will also have its own graphics memory systems similar to the complete computer (or a portion of the main memory designated as the graphics memory). In addition, the graphics adapter will have various programs running on it to create, process, and transmit the image to the monitor. The amount of this programming can vary widely depending on the coprocessor and memory.

The basic operation of the graphics adapter is pretty simple. As mentioned above, there is a graphics memory set aside by the design of the computer system (the graphics memory is a set of memory chips of the appropriate size and speed). First, the main computer program (such as the CAD software) determines what image is needed on the screen. Second, data is created for that image in the graphics memory. Third, the graphics adapter takes what has been put into the graphics memory and translates it into an image on the screen as quickly as possible. The graphics adapter is constantly scanning the data in the graphics memory and then creating the signal to scan down the monitor as it refreshing. Since there is a sort of "shuttling" of the data between the computer system and the monitor, and often there is electronic circuitry to temporarily hold or "buffer" the data as it is being shuttled, the terminology of "graphics buffer" or "picture buffer" or even "frame buffer" is often used to refer to the kind of data held by the graphics memory and various memory chips on the graphics adapter itself. The details of these processes is beyond the scope of this work, but the key concept is that there is a special graphics memory area and the graphics adapter is constantly trying to get the data that is symbolized in that memory sent to the monitor. Figure 2.6 shows a simple schematic.

The most misleading statement in the previous paragraph is the second step of the process—creating the data that is put into the graphics memory. In some cases, the CAD software will actually use the main CPU to figure out how to translate the concept of an analog geometric entity (such as a circle) into a set of digital data that can reside in the graphics memory. Obviously, the monitor with a limited number of pixels can not show a perfect circle; it needs to be broken down into a set of pixels that approximates the circle. If the CAD software itself creates this approximation and then sends the data to the graphics memory, then the graphics adapter does not perform this second step, and only has to do the third step (send the image data to the monitor).

However, in other cases, the CAD software will not determine the approximation of the geometry and send the data to the graphics memory. In this case, the CAD software "calls on" the power of the graphics adapter (which would probably have a coprocessor in this case) to make the approximation and create the data in the graphics memory. In this situation, the graphics adapter is doing 2 jobs—creating the data in the graphics memory and sending the image to the monitor. Obviously, this will only work if the programming running on the graphics adapter is compatible with the way the CAD software needs the data to be created. The compatibility issues raised by graphics adapters are generally addressed by the use of industry standards: agreements between computer and electronic device manufacturers to have components behave in certain, predetermined fashions based on some sort of language or protocol. The standards may arise from industry groups such as the IEEE or EIA. Or, they may be considered de facto standards which means that one manufacturer's design is so pervasive, all other manufacturer's change to match it accordingly. A very early example of a de facto standard for graphics was TCS or PLOT10 for Tektronix® 4014 terminals. A more recent standard would be OpenGL®.

At first, the task of creating digital approximation of geometric entities created by CAD software may seem simple. However, this task can be complicated,

Computer Hardware Basics

such as when the CAD program is relying on the graphics adapter to create shaded 3-D models. In this case, the CAD program may be completely relieved from interacting with the user for the manipulation of the 3-D model. Once the CAD program defines the desired 3-D surfaces to the graphics adapter, the user may manipulate and "slice through" the 3-D model by interacting only with the graphics adapter. In this case, the coprocessor may be doing millions of complicated computations per second.

Beyond the presence and capability of the coprocessor, probably the next most important characteristic of a graphics adapter is the amount and configuration of the graphics memory. As mentioned earlier, the graphics memory may be simply a section or region of the computer system's main memory. In this case, there will be a somewhat fixed amount of graphics memory to consider since it is an integral part of the computer design. In other cases, the computer may allow for variation or expansion of the graphics memory. In general, more graphics memory (or Graphics RAM, Video RAM, etc.) indicates a better graphics adapter system.

The graphics memory has enough addresses or locations for storing data within it so that each pixel on the monitor gets a certain predetermined amount of information. If a graphics adapter operates at a resolution of 1280 by 1024, then there are 1,310,720 pixels (1280 times 1024). So, at least these many addresses are needed (or about 1.2 megabytes) in the graphics memory. Of course, each byte can contain any pattern of 8 ones and zeros. These 8 bits can then be used to define specific characteristics of a given pixel. The most obvious characteristic is color.

Looking at an 8-bit per pixel system, then, the emission of colored light is based on 3 primary colors (red, green, and blue), so 2 of the 8 bits can represent the amount of red, 2 of the 8 bits can represent the amount of green, and 2 of the 8 bits can represent the amount of blue (ignoring the other 2 bits). Any of those 2 bit combinations can have 4 different values (00, 01, 10, 11), so there are 4 levels or shades of each individual color possible for this fictitious graphics adapter. Therefore, the total number of possible shades (combinations of the red, green, and blue) is 4 times 4 times 4 or a total of 64.

Note that all the pixels have some sort of definition in their corresponding location in graphics memory, even if it looks like the screen is blank. If the graphics adapter is functioning and blank areas of the screen are black, then it just means that those parts of the Graphics Memory is set to all 0s (no red, no green, no blue/black) by whatever program is running. As mentioned previously, it is the job of the CAD software to get this graphics memory loaded with the right data (on its own or by calling on the coprocessor), and then it is the graphics adapter that looks at the pattern of bits and then sends the appropriate signal to the monitor so that it displays the image desired. This process of turning bits (or digital data) into a "real" electrical signal (or analog) is called a digital-to-analog conversion, and the reverse is called analog-to-digital. This process is performed by "D to A" converters (DACs). This is an important function of the graphics adapter, but it is performed behind the scenes.

Although the resolution in the example of graphics memory is pretty good (about 1 megapixel), the number of colors is not very good. Of course, at one time 64 colors was acceptable for 2-D CAD, but more likely at least 64,000 colors would be needed to show realistic 3-D models for a CAD system (curved surfaces need to show a very gradual change in the color to give the illusion of depth). The way to improve the available colors (or the palette) is to increase the amount of graphics memory so that each pixel has a greater depth of information.

For example, each pixel could go from 8-bit deep data (the example already presented) to 16-bit deep. However, since there are 3 colors, and 24 is divisible by 3, 24 bits is an even better example. Now with the 24-bit color graphics adapter, there can be 3 bytes of data for each pixel, and there can be 8 bits for each color at the pixel. There are 256 different levels of each color possible (there are 256 combinations of 8 ones and zeros), and now the total number of combinations of red, green, and blue becomes over 16 million (256 times 256 times 256). This is basically sufficient to give the needed illusion of shading of 3-D models mentioned, and it is basically sufficient to make a digitized photograph appear real.

So, for the 24-bit graphics adapter with 1280 by 1024 resolution, the total amount of graphics memory used is about 4 megabytes (there are about 1.3 million pixels and there are 3 bytes for each pixel resulting in 3.9 million bytes). 4 megabytes of graphics memory should be considered the "bare" minimum for 3-D CAD systems, though. The quality of the system will be noticeably improved by going beyond the 1280 by 1024 resolution, and there is more than just color data that can be associated with each pixel. If 4 bytes are allocated for each pixel (instead of just 3), then information such as translucency of surfaces or how far away the pixel is from the observer (Z buffer) may be stored with each pixel. This will further enhance the performance and capability of the graphics adapter. Therefore, 16 megabytes or more of graphics memory may be justified.

The last characteristic of graphics adapters that needs to be mentioned is frames (or perhaps called pages). Frames are just a static state of the graphics data being sent to the monitor. Frames are relevant to animations or other operations that need to quickly change the image shown on the monitor. Obviously, if a new frame is sent to the monitor more quickly, the animation is improved. Although animations may generally be associated with multi-media presentations or entertainment, they are of some value to CAD users that are working on mechanical component and/or assembly design. As animations of how to assemble products become more common than assembly drawings, this will become a more important issue.

Computer Hardware Basics 37

Since graphics adapters have processing capabilities of their own (such as the coprocessor), it turns out that the graphics adapter can actually be loading new data to one area of the graphics memory while the monitor is still displaying another set of data. Then, a signal is sent to have the monitor suddenly switch to displaying the area of graphics memory just completed being updated. This operation is based on having 2 frames. One frame is loading while the other is displaying, and then the process is reversed for the 2 frames. As usual, this explanation is an oversimplification, but it can be seen that this sort of behavior is going to require double the amount of graphics memory (for a given resolution and number of colors to display).

2.7.3 Hardcopy

The next type of peripheral to be discussed is hardcopy devices. As the name implies, these are devices that create paper copies from the data in the computer system. This could be printed material (such as a document) or a graphical image. Printing text is not a very significant issue for CAD systems, so this discussion will be generally limited to printing of graphical images.

The most basic distinction between hardcopy of CAD information is 2-D versus 3-D. 2-D hardcopy is going to be centered on the plotting of drawings. Drawings with respect to a CAD system has a very specific meaning; it implies standardized engineering drawings (see Chapter 4). These drawings are made up of 2 dimensional geometric entities such as lines, arcs, characters, etc. The format of these drawing are standardized by organizations such as ASME (American Society of Mechanical Engineers) and SAE (Society of Automotive Engineers). Indeed, throughout this book, the term drawing means 2-D data and the term model means 3-D data was used. Figure 4.1 is an example of 2-D data for a drawing. Hardcopy for 3-D models is quite different (refer to Figure 2.3). The 2-D drawings are really intended to convey specific technical information in a standardized format. The 3-D hardcopy, however, is really intended to convey a sense of realism for a design. Necessarily then, the type of device needed to create these two types of hardcopy are quite different.

The hardcopy device for the 2-D drawings is going to need to draw simple objects (such as lines and circles) in a fast and accurate manner. These hardcopies are likely to be well used, copied repeatedly, and be read by a wide array of individuals from a variety of disciplines. Clarity and the simplicity of the image is paramount. Particularly since it must generally produce excellent copies (or reprographics), this output is almost always just black ink on white paper. Since the device only needs to get information on these black lines, arcs, etc., the data transfer from the computer to the device is not that great (perhaps 500 kilobytes for a typical drawing). This type of device could be connected with simple cables.

The hardcopy device for the 3-D model is going to have to be more sophisticated. This device will almost certainly have to support "full color" (since the illusion of shading mentioned earlier in the section on graphics adapters will need to be preserved). This device will also have to print "all over" the paper instead of just putting down the black ink lines on the white paper (similar to the pixels on the monitor), and even the "background" may need to be printed. Thus, the amount of data to transfer from the computer to the device is generally going to be larger than for 2-D drawings. Assuming the device needs to get as much data as resides in the graphics memory, there could easily be many megabytes of data to transfer to the device. With this demand for data transfer, more robust connections would be used, such as making the device a node on the network and using the relatively fast network data transfer speeds.

The next basic distinction between hardcopy devices would be the type of data actually being generated by the CAD software and being sent to the device. This is a very important issue for making optimal use of CAD system resources. It turns out that there are basically 2 types of data that will "encapsulate" the information for an image that is going to be printed. This collection of data for a particular drawing or model is usually contained in a computer file ("saved" or "written" to the Storage system or disk drive).

These 2 basic types of data are vector and bitmap. A vector is a mathematical concept that indicates position of something with respect to an origin and a direction with respect to that origin. Vectors are discussed in some more detail in later chapters, but for now it is only necessary to realize that a vector can be described by an starting point and an ending point, and that there are mathematical coordinates for these points (i.e. X and Y values). It is essential to realize that for the vector data, it is possible to create the hardcopy of the image regardless of the device chosen. The X and Y values are "analytical" in the sense that any device can handle X and Y data. For example, the vector file can print a line from a point at X = 10 mm and Y = 10 mm to a point where X = 250 mm and Y = 10 mm (which may be the exact locations desired by the user). Of course, if the device is not large enough to handle a 250 mm coordinate, then the data can easily be "scaled" by multiplying all the values by a constant factor, and the drawing or model will look "correct" in the hardcopy format (although as a smaller image).

Another important realization is that the vector data can be considered "perfect"; there is no approximation of the drawing data within the computer file. The coordinates of the points should be exactly what the user of the CAD software intended (such as 250 mm and 10 mm). This means that the use of better hardcopy devices could yield a noticeable improvement in the quality of the image, since each device goes to the source for the drawing data and can then attempt to plot the data. A typical example of a vector data file for a hardcopy is called an HPGL file. This is a type of file created by Hewlett-Packard for support of their plotters.

Computer Hardware Basics 39

The bitmap is quite different. The bitmap is similar to the graphics memory discussed in the section on graphics adapters. In this case, there is no pure mathematical definition of what is to be seen in the hardcopy. Instead, the device that is creating an image generates or dumps the individual bits of data into a file. One can see that this a type of computer file that can be generated directly from the graphics memory. The computer file would simply contain the needed number of bytes of data (i.e. the color) for each pixel. In other words, the bitmap data is simply that—a file that has a map of what each bit should look like. The generation of bitmap files directly from the data in the graphics memory (being displayed already on the monitor) is often called a screen dump. Dumping (in computer system's terms anyway) generally refers to capturing data from memory or RAM (i.e., memory chips) to storage (i.e., a file on a disk drive).

There are a number of ramifications of the bitmap approach, particularly if used in conjunction with a CAD drawing or model. First, since the file can not contain the actual coordinate data that the user indicated, there is a level of approximation "built into" the file. For example, when a circle is approximated by a set of specific X, Y bits (just like pixels), there is going to be some visible jaggedness to the image (as with a graphics adapter and monitor). Of course, the higher the resolution of the data the more smooth the circle will appear, but the hardcopy generally can be no better than what was available when the file was first created (unlike the vector data that can allow different devices to improve the hardcopy results). Resolution for these kind of files and hardcopy devices is generally given as a pitch that is independent of size of the printing area (unlike monitors where resolution is a number of pixels regardless of physical size). In this case, there are a certain number of dots per distance. For inch distances, the resolution becomes so many dpi or dots per inch. 600 dpi (about 24 dots per mm) would be fairly standard, with higher quality hardcopy devices using 1200, or higher.

Another ramification of the bitmap file is size. Notice that the data in a vector type of computer file can be very compact. To define a single straight line, all that is needed is the 2 X,Y values for the starting and ending point coordinates (just 4 numbers or perhaps 32 bytes). For the case of a 240-mm long line (going from X = 10 mm to X = 250 mm as discussed earlier) and a bitmap file being used with a resolution of 12 dots per mm, the amount of data in the file expands to 2880 bytes (using just 8 bits to define a very simple color). Of course, the number of X,Y values in a vector file with many geometric entities can get just as big, but a generalization can be made that file size is a bigger concern for the bitmap files. A typical example of a bitmap data file for hardcopy is a TIFF™ file. To compensate for the size issue, the bitmap file types are often accompanied by compression technology that shrinks the file size.

Considering the 2 types of data files, it has been somewhat customary to consider hardcopy for vector data (2-D drawings) to be called plotting, while making hardcopy from bitmap data (including just printing text) would be called printing. Thus, the devices for the vector data hardcopy would be called plotters, and the devices for other hardcopy would be called printers. For a long time, only the plotters would give the necessary level of detail and accuracy for the geometric entities of the CAD system. Also, generally only the plotters could create the very large output (such as A0- or E-size drawings which are 1200 mm (44 in.) in the horizontal direction). However, the distinction has become blurred since the printers can now create the large output, and the "resolution" of printers of 600 or more dots per inch can be as accurate as the plotters.

Before these very high resolution printers, the plotters were often devices called pen plotters. These devices mechanically moved pens across the paper, or the pen slid in one direction and paper was rotated in another direction. This sort of device created a very accurate drawing based on a fairly direct translation of the data in the vector data file. As mentioned earlier the HPGL file, which was developed for the Hewlett-Packard pen plotters, is still supported in some devices. Indeed, some of the devices that would be considered printers (since they can print bitmaps or text alone also) actually support the HPGL format via a translator built into the printer. Many of these devices actually automatically detect which kind of file has been sent to it (and translates accordingly). Another very popular type of device for plotting (particularly for mainframe systems) was the Calcomp® plotter. This device also had its own file type that may still be supported.

Printers (which generally handle bitmap data, but may have a translator to handle the vector data) are usually based on laser printer technology or print jet technology. As long as the resolution is high enough (perhaps at least 600 dpi), the drawing image should be as clear and consistent as a plotter. The main considerations are likely to be how big the paper output needs to be, how fast the printer needs to go, and how expensive the ink cartridges are.

Of course, printer-type devices are the most appropriate for 3-D models. Thus, a high quality printer can have the advantage of being able to handle both the 2-D and 3-D image data. However, as mentioned previously, a printer for a 3-D image needs to have the ability to handle a large amount of data, and it must be able to produce a large number of colors to provide the shading illusion.

For many years, printers and plotters were attached to a minicomputer or workstation directly. The printers often were connected to a Centronics® parallel port, and plotters were connected to an RS-232 serial port. This approach has generally been replaced by connecting the hardcopy device directly to the computer network. In this case, the files that need to be printed will be spooled or queued on a designated computer that keeps track of the data being sent to the devices on the network (print server activity). The hardcopy device, in turn, must have a fairly large amount of memory to use as a temporary storage area for data

Computer Hardware Basics 41

awaiting printing on the device. This memory is generally referred to as RAM for the hardcopy device. Obviously, a sufficient amount of this RAM must be installed in the device so that there is no bottleneck for the data. Since the 3-D models tend to have the largest file size, even more RAM must be accounted for if the printer is going to be used a great deal for 3-D model printing.

2.7.4 Pointing Devices

The next type of peripheral to discuss is the pointing device. This is a type of device that can direct the user's interaction with the graphical data being presented. This is in contrast with the keyboard which is used to just enter "character" data (numbers, letter, special characters, etc.). The pointing device is an essential part of the CAD system, since a large number of the functions that need to be performed with the CAD software will be via the pointing device.

Some examples of pointing devices would be mice, trackballs, 3-D balls, dial pads, and digitizing tablets. In each case, the user can control the cursor or a pointer that is seen on the monitor. It allows for the selection or "picking" of graphical entities (lines, dimensions, arcs, surfaces). These devices are usually connected to a specially designated connector (or port) for the computer. Unlike most standard computer software, CAD software often requires a 3-button mouse. There are more commands than usual that need to be efficiently communicated between the user and the software.

Interacting with a 3-D model also produces greater demands on the pointing device. Most software uses a dial pad or a combination of the keys on the keyboard with buttons on the mouse to generate the needed control of the graphics interactivity. In order to interact with a 3-D model, the user will need to deal with zoom in and out (the Z-direction degree of freedom), how to shift left and right and/or up and down (panning in the X- and Y-direction), and how to rotate the model in 3 different directions (X, Y, and Z). Furthermore, the user will probably need to "clip" the model. This is a function to cut through or cut into a 3-D model and see inside surfaces after cutting away part of the 3-D model. In all these operations, the user should be able to indicate what degree of freedom needs to be adjusted, and then be able to use the pointing device to dynamically make that adjustment until the precise view is presented on the monitor.

2.7.5 Keyboard

Not much needs to be said concerning the keyboard. It is a fairly standard peripheral that is attached to the computer via a special connector for the keyboard. It allows for input of textual information. The function keys (or other "special" keys such as ctrl, shift, alt) on the keyboard may be used in conjunction with the pointing device to allow for the many operations need to manipulate 3-D models.

One difficulty that may arise with respect to keyboards is how they impact different languages or nationalities. There often are special notations on letters that are unique to specific languages or nations. If a single CAD system is going to be specified for a variety of nations for a company, then the CAD system needs to be able to handle these variations. As with the pointing device however, the computer system and/or the CAD software is likely to have the ability to be customized or adjusted to handle the differences.

2.7.6 Modems

Modems are devices that allow a computer to communicate across media such as phone lines. These devices do not generally have any special impact on CAD software. If a number of CAD software users need to share data, then generally they are connected via a network, not via modem's. If one is transferring CAD data files across a modem, then particular attention must be made to the type of file (ASCII versus binary) and the size of the file. In general, CAD data files are compressed prior to transmission via modem.

2.7.7 Backup Devices

In Section 2.5 on storage systems, it was mentioned that the data in the storage systems should be backed up regularly. This involves making a new copy of the all the data on the daily-used devices (such as disk drives). These new copies are then stored in a safe place (and often another copy kept off site), and these copies can be used to reconstitute a system after a failure in the storage system or undesired deletion of data by users.

Typical devices for "backups" would include tape drives and writeable CD devices. These types of devices are expected to be able to store large amounts of data, be physically stable (and thus able to keep the data stored and correct for long periods of time), and easily managed for finding and retrieving data. The speed with which the data is stored or retrieved is generally less important than the long term reliability of the medium.

These devices may be connected to individual computer systems, or they may be attached in some manner to the overall network. As stated above, the most critical issue is that the data storage is reliable. All the CAD data on the network should be backed up on one of these devices daily, and at regular intervals as much of this data as possible should be duplicated for off-site storage (in case the entire building or site is damaged or destroyed).

Computer Hardware Basics

2.8 ERGONOMICS AND DISABILITIES

The final issue with respect to computer hardware is to mention the requirements for proper ergonomics and adaptive devices. For all users of a CAD system, the computer hardware (particularly the keyboard, pointing device, and monitor) must be arranged and physically accessed in a proper manner to avoid injury. Standards or experts in the field of office ergonomics can be consulted.

For disabled users of a CAD system, there may be adaptive devices available. These devices should be researched and made available as needed or required by law. Although the capabilities of these special devices vary widely, and the demands of the user interface of CAD systems keep expanding, some exciting possibilities include natural language processing and intelligent pointing devices that learn from their user based on the user's physical capabilities.

2.9 CHAPTER EXERCISES

1. With permission, open up and study a computer system (following standard electronic equipment safety procedures such as unplugging all units and grounding oneself). Also refer to manufacturer's documentation as needed. Locate the following:
2. The CPU. Record the manufacturer and the year of manufacture.
3. Storage system (disk drive). Record the manufacturer and determine the access speed or I/O data rate.
4. Memory (main system RAM). Record the number of banks filled, the number of chips per bank, and total amount of physical RAM (most likely all in megabytes). Determine the access speed or I/O data rate.
5. Graphics Adapter (video card). Record the manufacturer and determine the amount of various kinds of RAM located on the circuit board.
6. If a computer network is used at one's campus or company, determine what IP addresses and node names are used for computers in the network.

2.10 CHAPTER REVIEW

1. Exactly how many bytes are in 1 kilobyte?
2. What system does the CPU communicate directly with—memory or storage?
3. Which system is faster in manipulating data—memory or storage?
4. Which system is nonvolatile—memory or storage?
5. Which system would have CAD software installed on it permanently—memory or storage?

6. What is virtual memory?
7. What is the memory map?
8. If a graphics adapter supports a resolution of 1024 by 768 pixels, 8 bits per color, and 2 frame buffers all in the graphics RAM area of the memory map, how much Graphics RAM is required (to the nearest megabyte)?
9. If a data file sent to a print device contains the X- and Y-values for the endpoints of lines, does this file contain vector data or bitmap data?

3

Computer Software Basics

3.1 INTRODUCTION

The next step in building a foundation for understanding how CAD systems are best employed is to present some basic information on computer software. As with the Chapter 2 on computer hardware, the intent is not to present a thorough discussion of computer software, which is an immense topic. Rather, there are some important issues that are operating behind the scenes on the computer that have a significant impact on CAD software and its users.

Recall from the previous chapter that the combination of computer hardware and the basic software that controls that hardware is often referred to as a platform. This chapter is going to present information on this controlling software called the operating system (OS). However, keep in mind that some computer hardware may only run one operating system, while other computer hardware may run more than one. For example, the computer hardware built around the Intel-based CPUs may run both the latest Windows software (as an operating system) or a brand of Unix software. However, other manufacturers (so called proprietary systems) may have their hardware run only their own individual operating systems. Fortunately, most of the features of these different operating systems are similar, so the discussion presented in this chapter will be generally applicable. Of course, this situation can change drastically and quickly since the computer systems market tends to be quite dynamic.

Figure 3.1 shows a simplified schematic of the "layers" of computer software that are generally found on a computer that could run a CAD program. They can be considered layers since the upper layer programs can not function without the lower layer program. Recall from Chapter 1 that the CPU is the device that executes or "runs" all the instructions of the computer, and recall that the CPU has a rather limited number of "machine code" instructions that it can actually perform. It is the task of the software layers that run on a computer to translate a "real" job to be performed (such as CAD) into these machine code instructions.

The lowest layer of software can be considered the low-level instructions that actually make something happen on the computer. Besides the "language" for the machine code instructions, this layer of software contains programs such as the BIOS (a basic input/output software that allows for variety of device configuration), and drivers. Drivers are a pretty generic term for pieces of "code" or software that the computer must have resident in the memory system in order to translate functional operations to the actual digital signals that make a device function. There isn't a great deal that needs to be understood about this lowest layer of software, except to know it is there.

The next higher layer of software is the operating system. This is probably the most important layer for the long-term, reliable functioning of the computer. An important attribute of the operating system is that it is the layer between the machine layer and the layers that a user really interacts with. It is the layer that

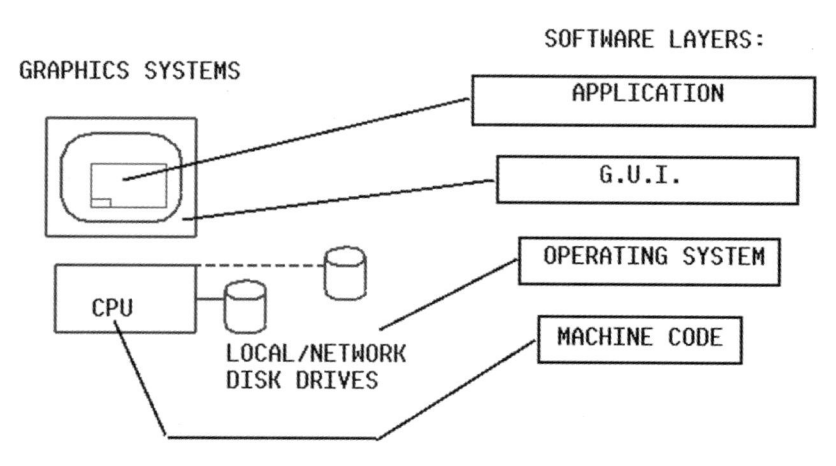

FIGURE 3.1 "Layers" of computer software.

Computer Software Basics

"understands" the functioning of the hardware as well as the application such as CAD. Therefore, it really controls and coordinates the functioning of virtually all the software on the computer. Keep in mind that the operating system itself is usually not a single computer program. It is a collection of a wide variety of programs coordinated by a master command interpreter or kernel program. The command interpreter program is often kept in the memory system (for ready use), while the other pieces of the operating system (such as a program that prints files) may go in and out of memory on an "as needed" basis. To the user, though, this transition is transparent; it seems that the operating system is doing all these functions seamlessly (or at least it should).

The next higher layer of software is called the Graphical User Interface or GUI (pronounced "gooey"). As the name implies, it uses graphics (as opposed to text commands) to interface with operating system. In some cases, the GUI could be independent of the operating system (the GUI called Xwindows runs on a variety of computer systems); while others are more limited (Microsoft Windows is virtually exclusive to the PC-class computers which use an Intel-compatible CPU). Indeed, Microsoft Windows (the most standard Windows program) is not really distinguishable from the operating system. However, at one time Microsoft Windows was a program that ran "on top of" an operating system called PC-DOS® (which did not use graphics at all). At this point, it should be assumed that all CAD programs will be running on top of a GUI of some kind.

The top-most layer of software is the actual application that the user wants to use. In this case, the most relevant example is a CAD program. Other examples would be "office productivity" programs (such as word processing, spreadsheets, etc.) and analytical programs such as Finite Element Analysis (FEA). It is important to realize that the CAD program may not function without the GUI, and the GUI is not going to function without the operating system, and the operating system is not going to function without the BIOS and driver programs. Clearly the CAD program is quite dependent on a proper overall configuration of the platform.

Keep in mind, however, that this concept of software layers is a simplification; it is just helpful to think in these terms when a specific troubleshooting or performance issue arises. For example, the user interface of CAD software may be broken down into a number of different windows, and in this case, the GUI of the system is going to be used to create these graphical windows. However, another part of the user interface of the CAD software may skip the GUI layer, and even the operating system layer, and send instructions directly to the graphics adapter device via the driver in the bottom layer.

The next section of this chapter begins a detailed discussion of the operating system layer of software.

3.2 THE OPERATING SYSTEM

As mentioned above, the OS of a computer system is clearly the most important layer of software of which a user needs to have some knowledge (besides the CAD application itself). Anything a CAD user learns concerning the operating system is time well spent.

There are many important functions of the operating system, and as many as possible are presented here, but this is really a basic overview. The following are some generic functions generally expected to be performed by the operating system.

Memory management: Brings applications (such as the CAD software) from the storage system (i.e. disk drive) to the memory system (i.e. the RAM chips).

File management: Creates, modifies, copies, deletes the files in the storage system.

User management/security: Controls users and what users have access to which functions of the computer system.

Command interface: A set of commands and/or language that can be used to control and automate functions.

Network management: Controls access to network resources (at least at a high level).

Device management: Controls what peripherals are active and their configuration.

Queue management: Controlling functions (such as printing) that need to manage data streams to off-line operations.

The next sections provide some information on the each of these topics.

3.2.1 Memory Management

The primary means by which the operating system controls the computer is via memory management. When a user wants to run a program (such as the CAD software), the user somehow indicates this to the operating system (such as clicking on an "icon" or typing a command). The operating system then "looks" at the appropriate location on the storage system (whether a local disk drive or on a network), finds the file for starting the CAD software, and then copies the executable data found in that disk file into the memory system. The operating system then signals to the CPU where in Memory to find the data and programming, and the CPU then performs the functions indicated by the new code in memory. Thus, although the CPU does the processing of data, the operating system tells the CPU what function is desired by the user.

Of course, there are many other functions that the computer performs just to keep itself operating. Although a user may be running CAD software, there is plenty going on behind the scenes, and the operating system is generally coordinating this unseen activity. Usually this sort of management is configured when the computer is first set up, and then the operating system simply sustains this management activity without intervention. However, each of these unseen activities requires resources (such as memory) when they run or execute.

Often the way the operating system manages memory (between the competing programs) tends to be a distinguishing characteristic of the operating system. In the past, operating systems such as PC-DOS only allowed one program to run at a time. Now most operating systems, such as Unix and Windows, do allow more than one program to be running at the same time. This behavior can be referred to as multiprocessing or multitasking (although this can be an oversimplification). Older derivatives of Windows allowed all programs to enter and leave the Memory system in a sort of cooperative manner, but Unix and its derivatives, as well as newer versions of Windows have one small master program that coordinates and controls the other programs vying for resources. This small master program is called an operating system kernel. The kernel based systems tend to be more stable since the master program is more likely to be able to get rid of rogue or badly-written programs, while systems without a kernel can crash more often.

3.2.2 User Management

Beyond the behavior associated with multitasking, there is somewhat of a distinction between operating systems depending on whether they are multi-user or not. In this case, the operating system is not only able to coordinate the functions of different programs, it is able to manage different users. These are computers that have "logins" or "logons." This kind of operating system is also going to have different classes of users. Some users are going to be able to modify the system itself, while other users are going to only be able to run the system, while yet other users may have even fewer privileges. In many cases, the ability to control users is blurred between a network and an individual computer. Unix and mainframe operating systems are able to manage individual, independent users accessing a single computer. This is how they were originally developed, since computer terminals were used by a group of people. Windows based computers tend to only exhibit multi-user behavior when the PC is connected to a network, and privileges are granted in consideration of the network resources. Controlling this function can be referred to as user management.

Depending on the operating system, there can be a fairly sophisticated set of privileges that can be granted and/or revoked to certain users. The most important function to control is the privilege of modifying the computer system or operating system itself. This is known as the administrative or "admin" privilege. For most operating systems, the name of user that has this privilege is derived from "admin" (SYSADM, NETADM, etc.). On Unix systems, the user with this privilege is called "root" (since Unix systems have an interesting "tree" organization scheme based on files or pseudofiles and access to the "root" of the tree would mean any part of the tree could be changed). Thus, on Unix systems, having "root access" would mean that the computer system or the operating system could be fundamentally altered.

Besides administrative access control, user management often involves segregating users so that there are only certain parts of the computer system that they can affect. In this case, different users may be able to access files in one "area" of the storage system that has been designated for their use, but they may not access files in other "areas" (which could be part of someone else's reserved space). This sort of control would be typical for a CAD user on a typical computer system. They have the privilege of working on the computer system, and they have some space in which to do their work, but there are areas of the computer that are off limits in some way. In some cases, this user may have whole systems hidden from his view, while in other cases, the user can "see" other resources, but can not actually use them. This space on the computer system where a general user can create and destroy data is generally called "home" or the "home directory" or a "C drive."

Another aspect of user management that can be configured in many operating systems is called groups. In this case, there are types of users that are grouped together. As a group, then, they would have access to certain privileges such as being able to read and write each other's files. This sort of arrangement could be used to control different design teams working on a CAD system. Only members of a given team would be able to alter the team's files. However, "higher end" CAD systems would probably offer this capability as part of the CAD system itself (instead of relying on the operating system).

3.2.3 Command Interface

Operating systems generally have a command "interpreter." This is a type of computer language that works "directly" with the operating system. A part or "module" of the operating system is "invoked" to interpret the commands of the user. These commands can be interactive, meaning the user is typing the command and then the computer acts on the request. The commands for the interpreter could also come from a file. This is an important function for automating tasks that the user does not want to keep typing manually.

In many cases, the use of this automation process is "hidden" from the standard CAD user. Instead, a CAD administrator or someone from the computer systems department would probably use the command interface to define functions, and then the standard CAD user would be given a means of requesting the function (such as through an icon or pull down menu). Of course, even standard

Computer Software Basics

CAD users can derive some benefit from being at least somewhat familiar with the command interpreter capability. This would facilitate understanding what the operating system is capable of doing, and it could facilitate assisting the administrators of the system.

There are a number of names and capabilities associated with the command interpreter and its language. For workstations, the most common reference is to the shell program or shell scripts or just scripts. The shell is seen as a program that is sort of a shell around the real program that is controlling the CPU, memory, and storage, and it provides a more direct interface between the user and the operating system. These shell programs come in a number of "flavors" based on a succession of versions of the Unix operating system. Some versions of Unix generally were coming from AT&T's Bell Labs and its descendants as System IV, System V ("System 5"), etc. Other versions were coming from the University of California at Berkeley as BSD (Berkeley Standard Distribution). One flavor generally had a shell program called "sh"; the other had a shell program called "csh." Later, versions of this operating system somewhat merged the shells into a "ksh." The programs that can be written to work with these shell programs can be quite powerful, and the "scripting" of unix is a renowned capability for the operating system.

There are also a number of terms applied to the dominant operating systems for PCs. Most early PCs worked with the program called PC-DOS (for Disk Operating System), and the PC-DOS program had a command interpreter. Mainframes and early microcomputers also had programs called DOS and/or OS, but this book will be assuming that DOS means the Microsoft PC operating system program. The PC-DOS command interpreter could be programmed through a type of file called a batch file; it ended with the .BAT file extension (batch programming was another mainframe term for many years, but again that will be ignored in this work). Thus, it is not unusual to have the use of the command interpreter capability be referred to as the "DOS prompt" (even though Windows has long since replaced DOS), or for files that drive many command interpreters to be referred to as "batch files." Since Microsoft has controlled most, if not all, of the development of the PC command interpreter capability, there is far less confusion about their content (as would be the case with the sh/csh/ksh situation with Unix). However, since the PCs were not generally driven as much by scripting (for interactive commands as well as shell scripts), the PC would generally be considered less capable with respect to the command interpreter in comparison to Unix.

3.2.4 Network Management

As mentioned in the chapter on computer hardware, a significant portion of a computer's function is in conjunction with a computer network. The network is the way that computers communicate with each other at high speed. This is only possible if the operating system is "aware" of its connection to the network, and if it is, in turn, "known" to the other devices on the network.

At the operating system level (which is a fairly high level with respect to networking), most computers communicate with each other and with users via node names. Since most network architectures are based on allowing many computers to communicate to many others (as if they are all equals), one can envision a number of islands with wires connecting the islands in a vast web of connections. In this sort of a logical arrangement the islands are called nodes. Thus computers on a network are often given node names. These names identify the computer to other users and other computers, although the node name is usually mapped to be equal to a unique numbering system (refer to Chapter 2 on computer hardware for the addressing schemes).

Beyond just knowing the "node names" of the other systems on the network, the operating system needs to define what privileges are granted to users that access the system from across the network. Usually there are reduced privileges for these "remote" users, as opposed to the local user or users. With most operating systems, there is usually a means of "crosslinking" or "linking" or "mounting" data on disk drives from one computer on the network to another computer. This is extremely valuable in setting up computer software like CAD systems for a group of users. In this case, the drawing and 3-D model files can appear to the user to be stored on their personal computer, but in reality it is being kept on a secure computer file server with proper backup systems. This can be done by making an entire disk drive appear to be a local device, when it is really on a file server computer (the typical PC network approach). This can also be done by making a directory or folder appear to be a local directory, when it is really a directory on a file server (the typical Unix network approach). In either case, the operating system must be set up properly to permit this sort of data sharing.

Another issue that may arise with respect to network management is license management. Some commercial software programs, and "large" CAD software in particular, are controlled by a set of licenses stored on the network. The number of licenses is predetermined by the company's purchase of the software, the more the customer pays, the more licenses that are available. Then, as the CAD system is used by more users, the number of available licenses decreases. Since these users are usually accessing the program from the network, it is a network function to keep track of how many licenses have been paid for, and "check in" and "check out" the licenses to users across the network. This process is usually controlled by a special kind of program called a "daemon" that is somewhat detached from the operating system once it starts. It "polls" or regularly communicates with the various computers on the network and manages the process of letting individual computers run certain functions. The daemon or "license daemon" usually keeps track of the licenses that the CAD system allows by a "password" file. This file has long strings of numbers and letters that are created from encryption of a particular customer's system (usually using the Ethernet address of the network interface circuit board on the license "serving" computer).

3.2.5 Device Management

The chapter on computer hardware discussed the peripheral devices that are typically found on computers that run CAD software. The presence and behavior of these devices must be managed by the operating system. Each time a new device is added to a computer system, the operating system needs to know it is there so that its function can be integrated with the other devices on the system.

There are a number of levels of operation that are important with respect to device management. The lower levels do most of the work of having data properly flow to or from the device, but it is the operating system that actually allows users or application programs (such as CAD software) to make use of the hardware. Just because a new disk drive is physically attached to a computer does not necessarily mean that a file could be created on it.

Hopefully the device management operations of the computer can be preset for a specific CAD software system or platform. After this initial configuration, there may be no need to alter the configuration. But, support for the computer hardware in the operating system is a somewhat never-ending story, particularly for the PC class systems. In this case, there are so many manufacturers of so many different devices that can be attached to the PC (which can be advantageous in a general sense), that it is difficult to guarantee that the system will function satisfactorily over a long period of upgrades. Generally with the unix workstation computer systems, there is a much more limited choice of devices, but then there tends to be fewer problems with sustaining stable operation since the limited list of devices is often carefully checked out or verified for good behavior by the computer workstation manufacturers. In fact, the choice of peripherals may be limited to just components from that particular manufacturer.

3.2.6 Queue Management

On a computer system, a queue is a capability that involves performing specific tasks in a self-sustaining background mode. This means that the computer can work on a task behind the scenes while the user goes on with other activities. The most common use of the queue system is printing or plotting. In this case, the user requests that something (such as a drawing) be printed on a hardcopy device. The operating system receives this request from the program (such as the CAD software) and then it sends the data to the device. The queue system is also capable of keeping track of multiple requests. The requests are serviced one at a time (as if standing in line or as the British say "standing in a queue").

Managing the queue systems is usually a major function of the operating system. There are many choices in how these queues are configured. For instance, some printing or plotting may be done on devices that are locally attached to a computer system. In other cases, the devices may be on a print server, or the devices may be attached directly to the network. In all these cases, the operating system for the local system as well as the file server must be set to handle the printing and queuing operations. For most CAD users, the typical arrangement is for there to be various smaller printers that are available for check prints or perhaps for screen dumps; while the network would also have available production quality devices for large plots of drawings. The operating system would be used to set up different queues on the various computers in this approach. Typically, the administrator for the operating system would give these different queues different names, and a certain amount of disk drive space would be allocated for the queues to hold data. Then these queues are then available for programs on the computer system to utilize.

Another issue to consider for queues is the type of data format sent to the devices. Some printers would accept simple ASCII data files, while others would only accept specific plotter languages.

Although the use of prints or blueprints is generally being replaced with electronic formats and viewers, the same queue management issues are going to arise. In this case, there has to be proper digital archives for keeping and tracking the electronic format files, and the access of these archives is going to be controlled via the operating system.

3.2.7 File Management

The final issue to discuss with respect to operating system software is file management. This is clearly the most important type of management that users need to understand to maximize effectiveness with the computer system.

As mentioned in Chapter 2 on computer hardware, computer files are the basic vehicle for holding data on a computer for "permanent" storage. The files are kept in what can be called the storage system, but for most computer systems the storage system is going to be a disk drive device. So, the operating system is generally going to keep track of all the data on a disk drive or a set of disk drives.

One of the most obvious characteristics of file management is going to be how the disk drives are designated. On some operating systems (such as Windows), the drives are given letter designations (e.g. A or C). In this case, the A drive is going to be a diskette device that can be used to transfer files between computers. The A drive may also be referred to as a "floppy drive," but the diskette case has not been floppy for quite some time. To refer to a specific file on this type of operating system, then, it can begin with the letter designation for the disk drive as shown by the following:

Computer Software Basics

A:\USER\GUY\FLOPPY.DAT

Note that the full designation for the device is followed by a colon ":". This format was also typical on the VAX/VMS operating system. On this system, the following might be typical:

USER1:[USER.GUY]FLOPPY.DAT;1

Note that in both these examples, the disk drive designation (A: or DISK1:) is followed by two directory designations. Directories or folders are a very essential element of the operating system. They allow a great number of files on a particular disk drive or file system to be well organized. In the first example (PC-DOS), the "backslash" or "\" is used to delineate the directories. In the second example (VAX/VMS), the square brackets and a dot are used to show the directory names.

3.2.7.1 Directories or Folders

Directories can be seen as a tree structure (refer to Figure 3.2). At the top there is a first level or "tier" of directories that form the most basic division of the filing structure on the disk drive or file system. Some of the directories of this first level would typically be used to segregate files that are part of the operating system software from files that are subsequently loaded to the computer as applications. The applications may then be further segregated as those loaded for all users versus those loaded for specific users. Then, there may be directories at the first level

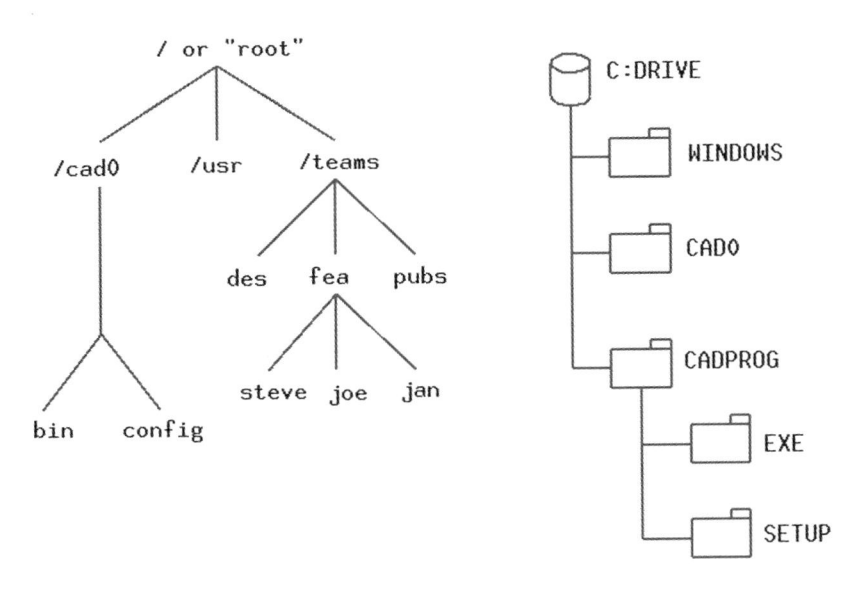

FIGURE 3.2 Typical directory and Subdirectory structures.

that are for various users to use as they see fit. This would be more typical for multi-user systems that are being shared. In most cases, though, individual users have their own computer, so most likely there would only need to be one level of segregation between a single user and the operating system (i.e. the Windows directory containing all the operating system files).

At the next level of segregation, a specific application (such as the CAD software) may utilize its own scheme for organizing files. Typically, there would be a level of directories for such things as the executable or binaries that actually run on the computer, the sample cases, the help files, and customization or configuration files. Exactly how the disk drive or drives are broken into a system of directories is a pretty involved process; a full description can not be given here.

There is no real difference between these directories and folders in the graphical user interface (GUI) based systems (such as File Manager, Windows Explorer, File Tools, etc.). Most users are going to be able to work with the folders exclusively, but users should be aware that the more fundamental topic is considered to be directories, and many IT professionals will continue to refer to them as directories and subdirectories.

Note that on the unix-type operating systems, directories become an even more important issue. On these systems, there is no designation for a disk drive device (like A: or C: or USER1:). Instead, the entire operating system and all the applications are melted into a single directory structure. At the top of this directory is the root file system. All files (regardless of what disk drive they are on) are then given a consistent name within the overall structure. If there are different disk drives, then they are given a specific location in the directory structure called a "mount point." For instance, the following may be a directory name in the operating system's structure (note that unix-type systems use a forward slash or "/" for the directory delineation):

/usr/local

A unix operating system may actually be configured so that any reference or use of any file that starts with this set of directories actually means use a specific disk drive. When it has been configured this way, one would say that the disk drive has been "mounted" to this specific directory. It is also possible to make a mounted file system to a disk drive that is not on the local computer. In this case, the directory may be mounted to a file server on the network. As mentioned in an earlier chapter, this arrangement on a unix-type computer is referred to as an NFS mount, where NFS refers to Network File System which is a sort of disk drive emulation that makes Ethernet type of networking behave like a disk drive.

Using the unix-type operating system, the full designation of a file could then look like the following:

/usr/local/user/guy/hardata.dat

Computer Software Basics

When it comes time to write a file to a diskette, then one needs to know the proper directory designation for the desired device. For a floppy drive, a typical directory might be the following:

/dev/fd0

Now when the user or the CAD software needs to write a file to the floppy drive, it simply needs to point to the specific directory name (such as referring to the file as /dev/fd0/write_this_file.txt).

3.2.8 File Format

Although it is not necessarily a function of file management, there is an important issue with the way different files are structured on various operating systems. These issues often relate to the "inter-operability" of CAD software. Since there are many different CAD programs used on different operating systems at the innumerable companies that do design and manufacturing, it is inevitable that files will need to be exchanged between these systems. A great deal of frustration with this process can be avoided by learning a little bit about file structure.

First of all, it must be understood that files are just collections of bit patterns. These bit patterns are organized in groups of 8 bits (i.e. 1s and 0s or ONs and OFFs or voltage HIGHs and LOWs). These 8 bits are known as a byte, and it turns out that one can easily create 128 different patterns of the 1s and 0s using 7 out of the 8 bits (some systems reserve that last 1 or 0 for special purposes). These 128 different combinations have been standardized to cover just about all one needs to put information into a file (things like letters, numbers, punctuation marks, etc.). These standardized combinations are known as the ASCII codes or the ASCII table. Referring to Table 3.1, one can see that the upper case letter A is code 65 (in "decimal" form), and the letter B is code 66, and so on.

Clearly having a standard set of patterns like this is essential to exchanging files between different systems. The only computer system that does not use the ASCII codes are the older IBM mainframe computers; they used something called EBCDIC. However, this system would not be at all typical in CAD applications currently. If this type of data file is encountered, then an EBCDIC to ASCII conversion would have to be performed at some point.

3.2.8.1 ASCII Files

Even assuming that all data files on the CAD systems are based on ASCII codes, there is still more that needs to be understood concerning the file data format. Often files are referred to as "ASCII files." Actually, this does not refer to the file being just being based on the ASCII codes at all. Instead, it means that the file has a certain use of the codes.

Referring to Table 3.1, it can be seen that there is an ASCII code 13 that does not indicate a letter or number. It indicates a carriage return or a "CR." For

TABLE 3.1 Selected ASCII Codes

ASCII code (decimal)	ASCII code (hex + 7-bit binary)	Description
10	$0A=0001010	Line feed (LF)
12	$0C=0001100	Form feed (FF)
13	$0D=0001101	Carriage return (CR)
27	$1B=0011011	Escape (Esc)
32	$20=0100000	Blank (white space)
65	$41=1000010	Uppercase A
66	$42=1000010	Uppercase B
67	$43=1000011	Uppercase C

some readers (particularly older ones), this will make some sense. The carriage refers to a piece of a typewriter or teletype machine. These were machines that were used to create documents and transmit information prior to digital computers. The carriage was a part of the machine that moved across the paper as letters were struck onto the paper. To return the carriage meant that it moved back horizontally to begin a new line. Normally, this would be followed by a line feed ("LF"). This line feed meant that the paper advanced vertically under the carriage so that the new line of text would be below the previous line. So, when these teletype (or "tty") machines were used as computer peripherals, it would be normal to write the ASCII code 13 and then ASCII code 10 (for carriage return and then line feed) to start a new line.

Like it or not, this ancient history is written on virtually every computer in existence. This is because in a file referred to as an ASCII file (which is a common type of file indeed), one or both of those CR/LF codes are present at the end of each "line" of a file. It is these codes which indicate to the operating system that a line has ended, and the next one is starting. When users edit an ASCII file, they see each of the lines in a structure like a page in a book (refer to Figure 3.3). However, the computer operating system sees the file as a long string of these ASCII codes; some of the codes are the data that a user expects to see, and then

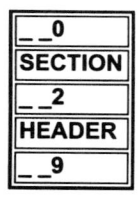

```
|_ _0    |
|SECTION |
|_ _2    |
|HEADER  |
|_ _9    |
```

FIGURE 3.3 A typical ASCII file as seen by the user.

Computer Software Basics

some of the codes indicate the end of lines (refer to Figure 3.4) or other control codes or unprintable characters. Often, the computer system or operating system will refer to these lines in the file as records. So these new line codes can also be referred to as an "end of record." Finally, these type of files may sometimes be referred to as text files, flat ASCII files or even just flat files, but "ASCII file" is probably the most universal terminology.

An important concept to understand in an ASCII file is that it is sufficiently well structured that virtually any operating system can understand it directly, since all the data is in these well constructed records. There is no real translation needed between different operating systems for ASCII files. So, an ASCII file is generally transferable between a unix system and a Microsoft Windows system. The only possible difficulty is that one operating system uses only one of the CR/LF codes and the other uses both, but most data transfer operations between these systems will not have a problem with this.

One way to try to tell if a file is an ASCII file is to attempt to edit it. Most operating systems have programs available that allow a user to view the data in a file directly. The operating system's file editor program is just going to read the data in the file and display it on the screen using end of record codes to know when to break up the file into lines. Two examples of file editor programs would be "vi" or "dtpad" for unix and "Notepad" for Microsoft Windows. One problem that may arise when editing these files is record length. If a file is meant to be edited, then each line will not exceed say 80 or perhaps 132 characters (columns of letters, numbers, etc.). In this case, all the data should be easily viewed.

CAD programs often have an option to create neutral files that can be used to translate drawings from one CAD system to another. Since they are used to transfer between systems, they are an obvious good candidate for use of the ASCII file structure. So, CAD file types such as DXF™, IGES, STEP, etc. are ASCII files.

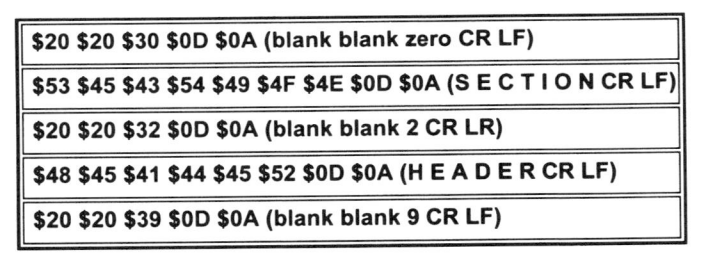

$20 $20 $30 $0D $0A (blank blank zero CR LF)
$53 $45 $43 $54 $49 $4F $4E $0D $0A (S E C T I O N CR LF)
$20 $20 $32 $0D $0A (blank blank 2 CR LR)
$48 $45 $41 $44 $45 $52 $0D $0A (H E A D E R CR LF)
$20 $20 $39 $0D $0A (blank blank 9 CR LF)

FIGURE 3.4 Bytes for an ASCII file as seen by the operating system (in hexadecimal).

3.2.8.2 Binary Files

Some files managed by the operating system are not ASCII files, even though they use the standard ASCII codes in them. These files are generally referred to as binary files. The important characteristic is that they do not have the special end of record ASCII codes. Instead, the bytes just are continuous data without breaks. It some cases, these streams of bytes will be structured into a standard length that is easily moved around by the operating system (say 4096 bytes or 4 Kilobytes). In other cases, the operating system is just expected to read the data out of it arbitrarily. Indeed, a benefit of the binary files is that operating system can more easily "jump around" in the file and take data in and out of it as needed. Another benefit of the binary file structure is size. Since there are no "end of record" bytes in the file, more "real" information can be stored in the file.

But the penalty for binary file structure is that it is not as universally transferable between different operating systems. The binary format is more dependent on the particular operating system. In fact, different computer systems may be referred to as binary compatible or not. This indicates whether the computer systems could exchange files without concern of files being binary format or not. The incompatibility generally arises from the fact that bytes are always made up of 8 bits, but in looking at the pattern of 1s and 0s, one has to decide whether the most significant digits are on the left or the right. Although it is universally accepted that digits to the left in a "normal" number are more significant, there are some operating system advantages to considering them to be on the right. This byte order issue is expressed by saying an operating system is "big-endian" or "little-endian." Particularly between different unix-based systems, there is no standard for this issue. However, in considering various generations of PCs, this is not a problem. All PC-DOS and Windows operating system computers can share binary files without a problem.

Most of the data files used by a CAD system are actually in binary format (so that they are accessed more quickly, use less disk space, etc.). This would include file types such as DWG, PRT, and ASM. Also, image files such as TIFF, GIF[SM], and JPEG are binary files.

3.2.8.3 Compressed Files

The final issue for file structure is file compression. Compressed files are files that have been reduced in size, without losing any of the data that is contained in the file. This type of file is an advantage when transferring data between systems or companies, and transferring CAD data files in this way is a very common requirement.

Files are compressed when a special computer program (which may be a standard feature of the operating system) analyzes the data in the file and figures out areas of the file that use space inefficiently. For instance, if there is a section of a file that contains the ASCII code for the number 0 repeated 50 times, then the file compression may replace this 50 bytes of codes with just a few bytes of codes (a few bytes indicating that it was a number 0 being repeated and a few bytes to indicate that it was repeated 50 times). Once a file is compressed, it usually has a different file extension. On a Microsoft Windows system, the compressed file would likely have a file extension of ZIP; on a unix-based system it would likely have a file extension of Z. For ASCII files (such as DXF files for CAD programs), the compression process can typically reduce a file size by 80 percent or more (so a 1 megabyte file could become a 125 kilobyte compressed file); this can dramatically reduce the amount of time it takes to transfer drawings between computer systems or companies. For binary files (such as a DWG file), the reduction is much less dramatic. These kinds of files may only be reduced by 25–50 percent.

Note that once a file has been compressed, the result is a binary file. So, after compressing a DXF file (which is an ASCII file that can be easily copied), the compressed file is no longer an ASCII file (which can not be copied as universally). Once a file has been compressed, it is typically transmitted between systems via e-mail or ftp. Obviously, once the file has been transmitted, the file must be decompressed. This is done by reversing the process used to compress the file. If it was compressed on a Windows system, then the receiving computer system should also be running Windows. Some of the file compression programs have versions on both Windows and unix-based systems, so in some cases, there could be different operating systems on the different computer systems.

3.2.9 Unix

One of the most popular operating systems for CAD software is generically referred to as unix. It was originally developed in the 1970s for minicomputers. It was written in such a way that it was easy to convert it to run on different computer hardware. Therefore, it became a standard offering on systems from a variety of computer manufacturers such as the former Digital Equipment Corporation or DEC®, Hewlett-Packard (HP), International Business Machines (IBM), Silicon Graphics® (SGI®), and Sun Microsystems. Since a variety of companies have owned the rights to the name unix over the years, each manufacturer created its own name for it. On DEC systems it was called Ultrix, on Hewlett-Packard it is called HP-UX®, on IBM it is called AIX, on SGI it is called IRIX®, and on Sun Microsystems it is called Solaris®. Another unix-like operating system is called Linux®. This version is known as "open source" and can basically be considered a public domain version; it is not owned by any specific computer manufacturer, but the manufacturers may offer computers with Linux instead of their own unix.

There are unique aspects of unix on each of these systems, but in general a user that has run CAD software on one of these platforms would not have any difficulty working on another.

As mentioned previously, one unique feature of unix is that everything is a file. The entire system is organized into a single "grand" directory structure. All of the devices, access to the network, all of the software, etc. are all treated as files. The top of the directory tree structure is the root directory, and the special user on the system that has total access to all files is therefore called root or the root user. If one has root access, then one has complete access to the computer. Refer to Figure 3.2.

A file that represents a device that could be found on the computer is in a subdirectory of the root directory called "dev." So a disk drive (or a partition of it) could be called /dev/hdisk0. Any operation to read or write data to this special file will be sending or receiving data from the disk drive (keeping in mind there really is no data file called /dev/hdisk0). In another example, accessing a drawing file on a CD-ROM may be accomplished by copying files from a directory called /dev/cd0.

Another unique feature of unix is the ability to link files from one directory to another. For instance, each user accessing a unix file server (where all CAD drawings are kept) may be given a standard home directory such as /users/team1/fred. When this user fred starts the CAD software, the software may look for a standard file in his home directory that sets defaults for the CAD software (such as the text height of notes in drawings). However, most companies prefer to have all users use the same defaults. In this case, a unix administrator can have each user have what looks like a real copy of this defaults file in their home directory, but instead this "dummy" file is really linked to a master copy of the file on the file server. Other operating systems tend to only permit this sort of mapping or linking on a directory-by-directory or drive-by-drive basis, not file-by-file.

Another unique feature of unix is the ability to have a variety of security options to protect files. Files can be protected (say against accidental deletion) based on individual files, individual users, groups of files, groups of users, or on all files and all users. These combinations can be applied anywhere in the overall directory structure, so there is great flexibility in setting the security characteristics. Also, it is possible to have users temporarily be promoted to higher levels of privilege based on parameters within a specific application (such as a CAD data manager program). This promotion process is based on a unix feature called "setuid." If the CAD software includes a sophisticated data manager, then directories can be created that contain all of a company's important or "released" drawings and models, and access to delete these files may be controlled only by administrators within the CAD software (even though users can create and/or modify files in this directory with their temporary read/write privilege). If users

Computer Software Basics

circumvent the CAD system and use operating system commands instead, again they will be prevented from deleting the released files (even though with the CAD system they can create or modify files in this directory).

3.2.10 Microsoft Windows

Another operating system that is commonly used for CAD systems is Microsoft Windows. This program dates from the 1980s; it was written for the Intel-based personal computers. Originally it was just a Graphical User Interface (refer to the section below on GUI) that ran on top of the Microsoft MS-DOS® operating system. With the version called Windows 95, Microsoft Windows became the real operating system that controlled the PC (instead of MS-DOS).

One drawback with older versions of Microsoft Windows was that it had to run programs that were rather old and were written for 16-bit systems (see the chapter on computer hardware for information on 16/32-bit architecture). Unix had been only working with 32-bit systems by this time, so unix generally worked better with the larger CAD software (since the larger CAD software ran on the 32-bit systems).

A unique characteristic of Microsoft Windows is that it is entirely a graphical system. Basically all of the administrative tasks can be accomplished without the use of text-based commands. On unix-based systems, many functions are still performed with text-based commands (although many administrators prefer this approach anyway). As discussed in the section below on GUI, unix-based systems therefore require a separate program to be running on the workstation to handle the graphical user interface (but again, this can be an advantage in some circumstances).

Another unique characteristic of Microsoft Windows is that it is not tied to any specific computer manufacturer (as most of the versions of unix are). This means that a much wider range of choices for computer hardware is available. In some cases, however, this is a disadvantage since programs are not necessarily guaranteed to work with all of the thousands of choices of hardware. Particularly with larger CAD software systems, there will be only a limited list of devices (graphics adapters, monitors, input devices, etc.) that will be supported by the CAD software vendor. If a company does not use the listed devices, then the CAD software vendor may not be required to fix problems that may occur with the software.

3.3 THE GUI

As shown in Figure 3.1, the next layer of software on a computer system "above" the operating system is the Graphical User Interface (GUI). This is shown as a separate layer of software, but in some cases it is not necessarily a separate program. On unix-based systems it is a separate program, but not necessarily on Windows (although it can still resort to the DOS prompt running either in a window or in full screen).

The GUI is somewhat taken for granted, but many computers can still interact with a user without graphics. This mode can be called the DOS mode or the terminal mode or the command line mode. In this situation, the user simply types in character-based commands (instead of using the mouse to point and/or click) and the computer responds directly to the commands. It is a very simple and generally more powerful means of controlling the computer, but it is not necessarily easy for users to handle. Therefore, generally only administrators resort to the command line mode. For most users, the operating system will be inseparable from the GUI.

For most unix-based systems, the GUI is called Xwindows. This program was created as part of a Department of Defense–sponsored project which not only provided a GUI, but was also closely tied to the use of a network. This allows some rather interesting possibilities. For instance, Xwindows permits operations such as "exporting displays" where one computer provides the computational power, but another computer actually shows the graphical results. It is possible, therefore, to remotely login to a computer in a different country, and make the GUI operate in that location, or back in the original location. Also, since Xwindows is running on top of the operating system, it can be managed as a program without affecting the operating system itself. This means that it can be highly and easily customized for users without affecting the basic operation of the computer system.

No doubt because Xwindows was government sponsored, it is a rather open program. An important version of Xwindows was 11. So Xwindows is often referred to as just X11. The source code for X11 is generally available to anyone. This means that someone could create their own version of Xwindows. Indeed, some computer manufacturers created their own versions of Xwindows that are proprietary or only available from them running on their own systems. However, anyone that has used Xwindows on a system should be able to understand any of the other versions.

In addition to the prevalence of Xwindows on unix-based systems, many of the unix computer manufacturers have standardized the Xwindows interface even more closely. This system is called the Common Desktop Environment (CDE). The CDE-standard systems all have the same icons and customizations so there is very little difference between the different manufacturer's systems.

For PCs (at least any system running the Microsoft Windows operating systems), the GUI is Microsoft Windows itself. Thus, the operating system itself is providing the GUI. This has the advantage of very clear standardization. All PCs are going to look basically the same to users. However, this has the disadvantage of problems with the GUI being able to affect the computer as a whole. As long

Computer Software Basics

as the operating system is stable, this is no problem. But if a program running on the desktop starts producing errors and uses devices or memory it is not supposed to (due to failures or poor programming techniques), then it will not be possible to stop just the offending program by just entering commands under the GUI program.

For CAD users, the GUI is an essential component of managing data and programs so that drawings, models, etc. can be created efficiently. Usually, the CAD program is running in a window of the GUI program. This means that the CAD program can be manipulated on the computer desktop easily. It can be started, stopped, minimized, maximized, etc. Understanding that the CAD program is running as a part of the GUI allows CAD users to be more productive. For instance, the CAD program can be run side-by-side with other programs important to design, analysis, and design management (such as office productivity software or Bill of Material programs).

3.4 THE APPLICATION PROGRAM

Again referring to Figure 3.1, the top layer of software is now reached. This is the real program that users are generally interested in using (such as the CAD program). This type of program is often called an application or an application program.

Note that standard operating systems allow multitasking to the degree that more than one instance or copy of the application program can be running at the same time (within the limitation of the CAD system's licensing). For CAD users, this means that different windows could be working on or viewing different drawings or models. This is a significant advantage; it is the instancing that permits this sort of operation. However, it is important to keep track of exactly what each window is doing with the CAD data. Problems can arise if more than one CAD session is trying to work on the same drawing or model at the same time.

Application programs, in particular, are programs that are written in high level languages. Indeed, if a user is going to write their own programs, they will likely use these same sorts of languages. For many years, the most important high level language for programs such as CAD was FORTRAN. FORTRAN stands for Formula Translation, and it was designed to solve scientific and/or engineering problems. Obviously, CAD programs are closely related to these activities. However, as programs became more interactive, and the graphical interface became more and more powerful, other languages became popular (since the user interface is not really an application of engineering formulae). Probably the most important language in this highly interactive stage is C and its descendant, C++ (called "C plus plus"). C++ is generally the preferred language for GUI based programming. Of course, some parts of CAD programs may still use FORTRAN or C for highly computational activities (such as doing analytical calculations) and leaving the C++ for the interface activities. Some other popular languages that have probably been used for CAD programs are Pascal and ADA. Also, the rise of platform-independence via Web browsers have been instrumental in the rise of another language called Java.

These high level languages can be a significant issue for CAD users. In many cases, users or their companies are interested in customizing or expanding the capabilities of the CAD application. This process of adding new operations or customizations is often performed by something called an API (Application Programming Interface). This is a common feature of CAD software. Examples would be creating a command for the CAD system that would create a company standard drawing format or note that shows the name of the company, the person that created the drawing, the part number, etc.

To perform this customization, one refers to the documentation that is provided by the CAD software vendor. This will indicate how source code files are generated or edited. These files will have commands that follow a standard language and technique created by the CAD software vendor. These commands are standardized for all companies that buy the software, but each company can create applications that suit their own special needs. These programs can be pretty simple, or they can be quite complex. The simplest case is usually the macro. A macro is usually created by having the user click or type a command to record keystrokes or a sequence of commands. When the user is finished recording, the macro can be played back. This playback can be used over and over again, and thus the functionality of the CAD software can be expanded. And, the macro files can often be edited to further expand the functionality of the macro. The form and syntax of the macro file often follows the patterns of high level languages. The macro files may have elements of FORTRAN, C, Lisp, or BASIC programming.

Some CAD programs also provide access to the same routines and procedures that the CAD program itself uses. This means that the user would use a high level language (such as C++) to create their own source code, and this is then "linked" into the CAD software routines. In some cases, the API for a CAD program would consist of the standard procedures that can be accessed in the CAD software itself. These type of sophisticated programs would often also be "linked" into the GUI software since these programs are likely to provide their own interactive graphics as well. Creating these kinds of applications linked to the CAD software routines are most likely not going to be created by "normal" users. A fair amount of computer programming experience is assumed with the API.

The final topic to present with respect to the application program layer is licensing. Licensing refers to the way that copies are restricted by the software vendor. As mentioned earlier, users can often create multiple copies or sessions of the CAD program on the GUI. However, most software vendors restrict how many copies can be started by all users at a company. The programming that keeps track of how many copies are being used at any one time is generally re-

Computer Software Basics

ferred to as a license manager. This is another application running on the operating system that is watching all the computers on the network to see if the application is being used and how much.

Some CAD software contains many modules or pieces. Sometimes these pieces are licensed or bought separately. The license manager will track each of these pieces independently. Some license managers are very sophisticated applications in themselves. Since they run "detached" from a single computer they are referred to as "daemons" (of course, they are also referred to as "demons"). Sometimes the license manager is referred to as the "license daemon." Normally, users have no interactions with the license manager. The password or config file that has encrypted information on the number and type of modules that can be run by users is usually taken care of by system administrators.

The chapters on 2-D and 3-D CAD management have more information on the overall administration issues related to the CAD application.

3.5 THE WEB

Although it is not shown in Figure 3.1, the World Wide Web, the program written by Tim Berners-Lee in 1989, offers another sort of layer of computer software. The Web connects a global network of computers that use Internet protocols. With the addition of a "browser" program the user "sees" the data on the various computers in the network. The user can also "move" between these computers by selecting "hyperlinks." These are tags or mark ups that have a visible graphic for the user to click on, and then the tag indicates what address (server, directory, file name, etc.) that the user will be redirected to.

One could consider the browser program to take the place of the "application" layer, and then the global network providing yet another possible layer. When users access the Web via network connections, they are often going through a gateway provided by an independent firm called an ISP (Internet Service Provider).

Although CAD systems frequently use the Web browser and the data on the Web for transferring information between companies and CAD systems, it is possible to put the CAD software itself onto the computer on the network. In this case, the CAD software is not really loaded on the local user's system. Instead, the user sees the CAD software via the Web browser, and the network downloads the pieces of the CAD software on a demand basis. When software is "served" to a user in this fashion, it is generally referred to as an ASP (Application Service Provider).

The ASP approach is problematic for CAD software, however. One difficulty is the size and scope of the software. CAD programs tend to be very large and complex. They have a great deal of functionality and capability that is expected to be useful for a broad range of users. This sophistication tends to guarantee that the software will need a large amount of computational and network power, and the Web does not always provide enough power. Of course, CAD users are generally only using one of these functions at a time, so with enough "granularity" to the software, it could perhaps be served in small enough pieces that the user is not affected by the poor performance of the network. Another difficulty is the large data files. Even meager 3-D models can produce rather large data files. The size of these files can be another impediment to getting good response time from the ASP approach.

An advantage of the ASP approach is outsourcing of the maintenance and administration of the CAD software. In this case, a company can be relieved of the very significant demands of managing and administering the complicated CAD system. On the other hand, relying on an outside source for something as vital and proprietary to a company as the information in the CAD system may be a real problem. So the ASP approach may be the best solution for some companies, but not others (probably based on the overall size of the company).

3.6 DATA ACQUISITION SOFTWARE

Another type of software that may be relevant to CAD systems and users is data acquisition. This is a class of software that is used in laboratories and shop floors. They are meant to measure physical parameters and record the information on a computer system. The physical parameters may just be geometric such as lengths, diameters, and distances. The physical parameters may also be such things as pressures, temperatures, accelerations, deflections, etc.

The two likely scenarios for interfacing a CAD system with data acquisition are analytical verification and manufacturing verification. For analytical verification, the product geometry contained in the CAD system may be used to generate simulations of the product behavior (attempting to predict whether the product will fail in service, how fast it can go, etc.). This computer-based analysis often requires the information from the CAD system. Subsequently, when a physical prototype is built, the system then can be instrumented to acquire physical data from the prototype. Finally, this data (created with the data acquisition system) may need to be re-imported back into the CAD system. This new data or geometry can then be compared with the original design geometry or the analytical models for design verification (as discussed by quality system standards such as ISO 9001).

The other potential scenario for data acquisition systems being an issue for a CAD system is manufacturing verification. In this case, parts or assemblies that have been manufactured are checked against the actual design (as documented in a drawing or as modeled in a 3-D model). Once again, there is a need to take physical measurements (this time geometry-based measurements) and then "im-

Computer Software Basics

port" the data to the CAD system. These physical measurements are usually made by a device called a Coordinate Measuring Machine (CMM). The CMM has a heavy table or other grounding mechanism to make sure that the item to be measured is not disturbed. Then there is usually a probe of some kind that is used to touch the part at very specific locations. For instance, a CAD model may indicate that at a point on a surface where X is 5 mm and Y is 20 mm, the Z value of the point is supposed to be 10 mm. The probe is moved with respect to the ground to the very precise X and Y value (such as 5 and 20). Then, the probe is moved along the Z direction until it touches the part. This value of Z can then be compared to the 10 mm theoretical value from the CAD model. Obviously, if the values agree closely (say within a fraction of 1 mm), then the model can be considered verified.

Data acquisition software is rather unique. It runs on a computer system at a low level. It needs to communicate directly with devices on a signal basis (as opposed to using files or other traditional digital data), so at some point the signals (referred to as electrical analog information) need to be converted to digital data (such as storing the physical characteristic as a real number in a real data file). This conversion is performed by circuits referred to as Analog-to-Digital Converters (or just "A-to-D"). The actual electrical signals connected to the prototype or the devices that detect the location of the probe on a CMM are referred to as transducers. As the analog signals from the transducers are sent to the computer controlling the data acquisition, the computer often samples a number of the wired connections from the transducers. These separate connections are often referred to as channels. This may allow many transducers to be used on a single test with a single computer. However, the amount of time that the computer software analyzes the data from a particular transducer is reduced depending on the number of transducers being sampled.

Data acquisition software is often closely tied to particular computer hardware (CPU as well as transducers and data acquisition or DAC circuit boards). The entire process of acquiring and formatting the data from the devices involves knowing exactly how the devices function. Also, specific tests that a company performs are often highly customized, so the combination of transducers and software may only be applicable to a particular company.

When the data acquisition software and/or hardware is customized for a particular company, special programming techniques may be used. In particular, a Programmable Logic Controller (PLC) may be used to drive a specific sequence of steps or operations in the test (including sending data directly to a CAD system). These PLCs often employ a type of programming called Relay Ladder Logic (RLL). This programming would typically be done by specialists in the data acquisition or test engineering fields (as opposed to a CAD user or administrator).

3.7 CHAPTER EXERCISES

1. Determine the machine code instructions for the CPU chip that was found in the computer in the Chapter 2 Exercises. Record how many instructions are used by the chip.
2. Determine how many different operating systems are in use at your campus or company. Record the version numbers, maintenance levels, service packs, etc. for these operating systems.
3. Create and edit an ASCII file with the computer system's file editor program.
4. Using computer system documentation as needed, create a shell script, command shortcut, or batch file that starts the CAD software loaded on the system.
5. Record what disk drive or storage resources available for the computer system is actually accessed via the network.

3.8 CHAPTER REVIEW

1. List each of the software layers that are found on a typical computer system that runs CAD. Describe each of the layers.
2. Why can't a binary file be reliably transferred between different operating systems?
3. In an ASCII file, how does the operating system "know" when each line in the file ends?
4. What operating system is likely involved if a file is identified as the following: /u/sharon/config.sjs?
5. What operating system is likely involved if a file is identified as the following: C:\CADMAN\CONFIG.SJS?

4

Drawings and 2-D Design

4.1 INTRODUCTION

This chapter presents information on the standards of "drafting" or mechanical drawings. This is certainly an essential background for understanding 2-D CAD and CAD systems in general. The production of these drawings is generally still the end "result" of the CAD system for many companies.

Generally, if one is interested in fully creating a design (and not just creating drawings), then 3-D CAD is certainly the more appropriate choice. Although 3-D modeling is discussed in later chapters, keep in mind that drawings and 2-D design still has a role to play in 3-D CAD systems. First of all, 3-D models may be used as the "source" of geometric information in drawings (created automatically by the software). Secondly, the first step in the creation of 3-D models is often to create 2-D "sketch" geometry on a plane. So, the information in this chapter can provide a foundation for 3-D CAD use, as well.

4.2 BACKGROUND

For perhaps a century or more, the creation and use of paper drawings was the best and perhaps only means of truly documenting a design. These drawings took various forms ranging from small handmade sketches to large sheets of paper many meters long using drafting machines. Eventually, the overall format as well as the smaller details were standardized across entire nations, and to some degree internationally.

Although the author is convinced that paper drawings and 2-D electronic data eventually will be completely eclipsed by the sophistication of 3-D models (particularly when portable 3-D model viewing devices are cheaply available), it remains an necessity that those dealing with design work be familiar with drawings. Indeed, once 3-D models are the only medium, many of the concepts in drawings will continue to be found in annotated 3-D models.

There are a number of types of drawings. A basic list is shown in Table 4.1. These drawings can generally be categorized based on the engineering discipline they are related to. For instance, electrical schematics would be typical for electrical engineering, A/E/C and survey drawings for civil engineering, plant drawings for industrial engineering, and, of course, mechanical drawings for mechanical engineering. There are specialized CAD software packages that are geared toward specific disciplines, but many CAD packages can be used for more than one of these disciplines. It is worth mentioning that for electrical engineering, in particular, that there is a class of software that is referred to as E-CAD. These packages are usually meant for electronic, printed circuit board (PCB), or integrated circuit (IC) development activities. They also may have the ability to simulate the electronic behavior, and create schematic drawings based on the component layout and design. However, E-CAD is beyond the scope of this work, although there are occasional references to it.

Mechanical drawings are one of the most important types of documentation used by the manufacturing sector. These drawings have been standardized and refined to serve the needs of design, manufacturing, purchasing, and legal depart-

TABLE 4.1 Some Basic Types and Uses of Drawings

Type	Use
Mechanical drawings	Document the design and to some degree the manufacturing processes of individual mechanical parts (often called "details") as well as the assembly or arrangements of parts.
Schematics	Used by the electrical, hydraulic, and pneumatic disciplines. They document the wiring or piping of a system as well as the constituent componentry of the circuit.
A/E/C	Used in various types of construction projects (architectural, civil engineering, and job sites).
Plant engineering	Used for building and/or facilities management.
Surveys	Document various types of property.

Drawings and 2-D Design

ments of countless firms involved with manufacturing products worldwide. Furthermore, they are "living" documents to a great extent since they are not just created for a particular project (such as with many architectural or civil engineering projects). The mechanical product may be manufactured for many decades and mass produced into millions of copies that must be revised and "change controlled" throughout its life cycle.

Due to the complexity of the mechanical drawings, their tendency to dominate the attention of manufacturers, and since the author is a mechanical engineer, mechanical drawings will form the foundation of the information presented by this work.

4.3 TYPICAL MECHANICAL DRAWING

Figure 4.1 shows a typical mechanical drawing. The mechanical drawing is used for a number of purposes, and there may be some variation among different companies, but they all have some standardized elements. It is essential that anyone working with drawings (3-D CAD users included) be aware of all these elements. The basic elements of the drawing are explained in the following sections.

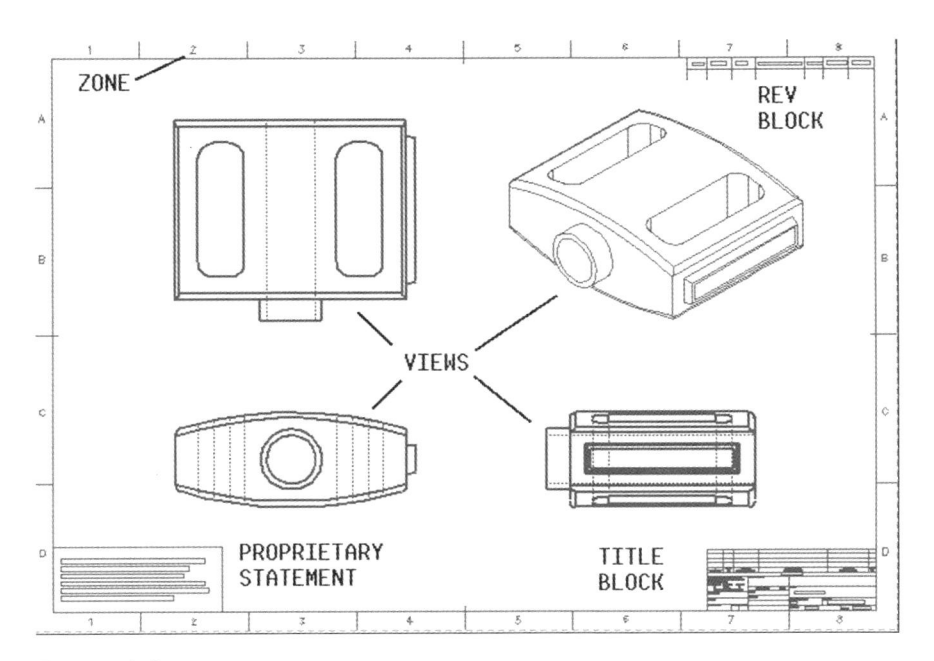

FIGURE 4.1 Typical mechanical drawing.

4.3.1 Format

The format refers to the "border" or "frame" that bounds the drawing information. It includes a number of important pieces including the title block, the proprietary information statement, zone indicators, and the revision block. There are industry standards covering formats such as ASME Y14.1, Y14.1M, and ISO 5457.

4.3.2 Title Block

The title block is considered part of the format and supplies the basic identification information for the drawing. It includes the name of the item shown in the drawing, the drawing number, the revision level of the drawing, the name of the company that owns the drawing, the size of the drawing, the sheet or page number, the person that created the drawing, etc. It is usually located at the lower right part of the format. Since the title block is part of the format, it is covered in the standards listed for the format.

4.3.3 Proprietary Information Statement

The proprietary information statement is a note that indicates the security of the drawing. This note will often state that the drawing is the property of the company, that the drawing contains information that is to be considered a trade secret, and that the receiver agrees to abide by this trade secret status once they accept a copy of the drawing. The trade secret status provides very restrictive security for a document. For instance, if someone steals the information in the drawing and another company profits from the information, that other company can be liable for paying back 3 times what they gained from the information (referred to as triple indemnity). Therefore, anyone dealing with drawings in any way (including CAD users) must be aware of this provision, and be certain to never violate it. The proprietary information statement is often located as part of the format, although it is a company standard (not an industry standard).

4.3.4 Zone Indicators

The zone indicators are letters or numbers shown in the margin or border of the format with lines or tick marks delimiting them. The combination of the vertical and horizontal zone indicators tells the reader to look at a particular region or zone of the drawing (such as C-5 or D-3). This is most often used in conjunction with the revision block where the revisions are said to be made in a particular zone. The larger the drawing, the more important this issue becomes. Since the zone indicators are part of the format, refer to the standards listed for the format.

Drawings and 2-D Design

4.3.5 Revision Block

The revision block is a section of the drawing that indicates the revision or history of the design. As mentioned earlier, mechanical drawings are expected to document a particular state of the design. This state may indicate the geometric properties of the item, part attributes such as the material of which it is to be made, the acceptable tolerances, and vendors approved by engineering to be able to manufacture the item. This is all very important information, and knowing when it was changed and by whom is vital to establishing control over the total design. The drawing for one small part can have life or death consequences for an entire machine such as an airplane or an automobile. Anyone working on drawings (particularly CAD users) must keep this in mind. Therefore, revisions must be carefully documented. Since the revision block is part of the format, refer to the standards listed for the format.

4.3.6 Bill of Material or Parts List

The Bill of Material (BOM) or Parts List section of the drawing is a note that may or may not be part of the format. It is associated with a type of drawing that shows an assembly or group of individual parts (often called details). The bill of material will identify each of the individual parts and their quantity in the assembly. There is no widely adopted industry standard for paper-based Bills of Material; it is generally a company standard. However, many companies now rely on computer software to handle all Bills of Material.

4.3.7 Views

The part or assembly or machine that is actually documented by the drawing is usually shown in the drawing (although not always). The part of the drawing where the reader actually sees the design or the geometry is contained in groupings called views. They are called views because the design is shown from a variety of view angles. For instance, the reader is shown what the design looks like from the front, and the right side, and the top. This would correspond to the Front View, the Right View, and the Top View respectively. Each of these views is expected to be placed on the drawing in a particular location so that the reader knows which view angle is being expressed. Unfortunately, this view placement is not standardized globally. In the United States the views are placed in one arrangement (called "first angle"); while in Europe the views are placed differently (called "third angle"). Some relevant standards for views and their projections (which are explained in Section 4.9) are ASME Y14.3, DIN 6, and ISO 128.

4.3.8 Notes

The notes on a drawing are simply instances of text or wording. They are used for a wide variety of tasks. A note may indicate information that has no relationship to the geometry of the design (i.e. the name of a company approved to build the design). On the other hand, notes may have a relationship to specific features of the design. In this case, a note may have an arrow with it that points to a hole, and the note may indicate that the hole is supposed to be drilled and tapped (a hole drilled and then screw threads cut into the sides of the hole). Of course, there are countless other tasks for these notes. Also, an entire drawing may be just notes. This is often referred to as a *word drawing*. Standards that can be referred to for notes include ASME Y14.34M and Y14.35M.

4.3.9 Item Callouts

The *callouts* on a drawing are a way of identifying different components or parts on a drawing that actually shows more than one part. These separate parts are often called *details*, and they are usually numbered. These numbers, then, are usually recorded in the Bill of Material as *items*. The callouts then become a means of connecting the information in the Bill of Material with the physical design. The callout numbers can be shown inside geometry such as a circle (called a "balloon" for obvious reasons), or they are a larger size number with an underline. In any case, the number should have an arrow pointing to the part of the drawing that clearly shows that particular part.

4.4 STANDARD DRAWING SIZES

Drawings are created in standard sizes. Table 4.2 shows a table of these sizes. Note the difference between identification systems of inch and mm (metric) drawings. Inch drawings use different capital letters; the mm drawings use a number after the letter A.

Before CAD systems were developed that could electronically store and retrieve drawings, drawings were kept in flat files (extra wide file cabinets) that were arranged by drawing size. For instance, there would be C-sized cabinets for C-size drawings, D-sized drawers for D-size, etc. The drawings within the drawer would then be identified by the drawing number shown in the title block. Therefore, to locate a particular drawing amongst hundreds or thousands of drawings, one would look at the letter and number combination (at least for inch drawings). So a drawing referred to as D46372819 would be found by looking at the D-sized cabinets and then using the number to find the appropriate drawer in the cabinet. Within that drawer, one would sort through the drawings to find that one drawing. This approach (with the leading letter designation) may still be found in the part

Drawings and 2-D Design

TABLE 4.2 Standard Drawing Sizes (refer to ASME Y14.1, ASME Y14.1M, ISO 5457)

Inch or U.S. drawings	Vertical	Horizontal
E	34.0 in.	44.0 in.
D	22.0 in.	34.0 in.
C	17.0 in.	22.0 in.
B	11.0 in.	17.0 in.
A	8.5 in.	11.0 in.
mm or metric drawings		
A0	841 mm	1189 mm
A1	594 mm	841 mm
A2	420 mm	594 mm
A3	297 mm	420 mm
A4	210 mm	297 mm

numbering system after a CAD system is installed in a company (although the drawing size is no longer needed to find the drawing anymore).

Note that sizes A, C, E and B, D are meant to be multiples. That is, 2 C-size drawings next to each other equal an E size drawing, 2 A-size drawings equal a C size, etc. There are larger sizes of drawings beyond the E size, but they are not very typical in mechanical design. There is also roll size drawings. In this case the E size is kept in the vertical direction, but the drawing rolls out longer than the E size. The horizontal dimension of the roll or R-size drawing could be 60 in., 72 in., 84 in., etc. (each size another 12 in. longer). There are similar roll sizes for the mm type of drawings.

4.5 TITLE BLOCK DETAILS

As mentioned earlier, the title block is a region of the drawing that contains vital identification information such as drawing number, revision level, etc. This type of information is central to proper document control and management. Document control, in turn, is vital to meeting the requirements of quality systems (such as ISO 9001) and is essential to having a viable system for managing the activities of any design department. Although the title block is a feature of a drawing, this sort of document management information forms the basis for CAD based engineering data management systems (even 3-D CAD). Just because the data management system is automated does not mean that items such as revision level are no longer important. There will still be revision control, and attributes such as revision levels will still be used to maintain revision control.

A close up example of a standard Title Block is shown in Figure 4.2.

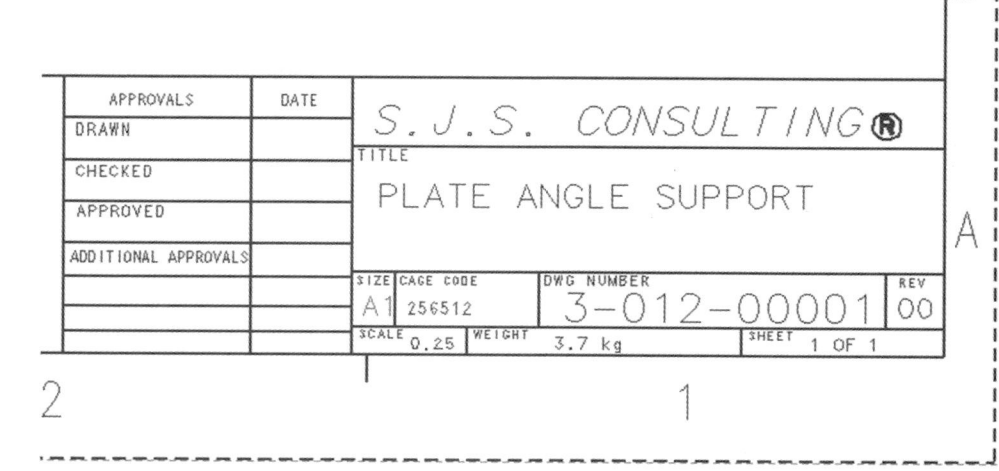

FIGURE 4.2 Standard Title Block (reference ASME Y14.1).

4.5.1 Title

As shown in Figure 4.2, the title is the basic naming for the drawing. It should be as succinct and descriptive as possible. Often these names are standardized to be a noun followed by a progression of classifying adjectives. The following would be an example:

PLATE ANGLE SUPPORT

This title indicates that the drawing is for a plate that has a bend, and that it is meant to be a support of some kind.

Note that the title and all other notes on a drawing are generally in upper case letters. One probable exception is for metric or SI units. Units such as kilogram and millimeter are abbreviated as kg or mm.

Also note that the drawing title is often reflected in automated engineering data management systems. A Bill of Material program will probably expect some sort of name for items (in addition to the drawing or part number), and the drawing title is often used for this purpose. However, this means that the drawing title

Drawings and 2-D Design

may have to conform to typical software data restrictions. This may include a limit on the total number of letters or characters in the name. It may also include restrictions on letters or characters that can be used. For example, double quotes ("), dollar signs ($), and slashes (/ or \) are typical delimiters between character data and may cause difficulties for the computer software that is integrated with the CAD system. Therefore, these characters may not be permitted in the title.

Finally, many CAD systems allow data such as the drawing title to be automatically entered and/or exported from the CAD software. In this case, the title is not entered directly by the user into the Title Block. Instead, the user is prompted for this information with a dialog box, and the user enters the information into fields. Then, the program inserts the field information into the Title Block automatically, and it often makes sure that the character limitations are met by disallowing restricted characters. Also, when data management programs need to obtain the Title Block information, the programs can then access the title directly (without having the user view the drawing).

4.5.2 Company Name

The company name is usually found above the drawing title (Figure 4.2). The address of the company is also often shown in this space. If there is a trademark logo for the company, it should also be shown. Anything that firmly establishes the ownership of the trade secret or intellectual property shown in the drawing should be used. Drawings are still an essential means of establishing ownership over the design.

4.5.3 Drawing Number

The next item in the Title Block would be the drawing number. This is probably the most important identifier from a document or design management viewpoint. Most of the automated data management systems are going to use the drawing number as the most important or primary "key" for finding and accessing information on this drawing or the item it documents (part, assembly, schematic, etc.).

4.5.4 Revision

After the drawing number, probably the next most important information is the revision or revision level. As seen in Figure 4.2, the revision level is usually to the right of the drawing number. Often when a drawing is requested the revision level is included (e.g. a specific image of a drawing may be completely specified by requesting a number such as 1-111-2-00003 and "Rev" 3).

The revision is often a number or letter that is sequentially increased as the design documented by the drawing is changed. For instance, the first version of the drawing released for use could have a revision level of 00, the next version would have a level of 01, then 02, etc. However, the change process for a drawing is not arbitrary; instead, it follows some very important revision or "interchangeability" rules. Some changes to a part are substantial enough to warrant a brand new drawing with a brand new number, instead of just revising the existing drawing. Whether the new number is needed is often dictated by whether the new revision of the part can be used interchangeably with all previous existing revisions (that may be in stock).

Another common system for identifying revision level is with letters. In this case, the initial release of the drawing may show no letter at all. Then the first revision would be A, the next revision B, etc. Since letters like I and the number 1 look similar, often there are letters that are skipped to prevent confusion.

In conjunction with the revision level, details of the changes made to the drawing are found in the Revision Block section of the drawing (see Section 4.6).

4.5.5 Predominant Scale

As discussed later in this chapter, drawings represent the physical or geometric properties of an object using views. These views are drawn to a specified scale. If an object is 100 mm long and the line drawn in a view to represent that length is drawn 10 mm long, then the scale is 0.1 or 1/10 or 1:10. Usually most or all of the views on a sheet of the drawing use a given value of scale. The value of this predominant scale is shown in the Title Block under the heading of SCALE (Figure 4.2).

4.5.6 Weight

It is not unusual for the weight of an object to be a vital piece of information for the object represented in the drawing (obviously a mechanical drawing in this case). The standard Title Block includes a space for WEIGHT between the SCALE and SHEET (Figure 4.2).

4.5.7 Sheet

The standard Title Block includes a space for the sheet or page number (Figure 4.2). If one sheet of a drawing is not enough space to properly document the design, then additional sheets are created (using the same drawing number and revision level). This is referred to as a multiple sheet or multisheet drawing. In the space for SHEET, the sheet number is shown as well as the total number of sheets. For instance, 1 OF 4 indicates that the sheet is the first of four total sheets filed under the particular drawing number.

4.5.8 Size

The standard Title Block includes a space for the drawing size. The drawing size is the letter or number/letter combination for standard drawing sizes discussed earlier in this chapter. See SIZE in Figure 4.2.

4.5.9 CAGE Code

The United States Government has established a numbering system to identify the type of item or business associated with a manufacturer. At one time this was known as the FSCM or Federal Supply Code for Manufacturers. More recent copies of the drawing format standards indicate this number now is the CAGE Code.

The CAGE Code is a five character (numbers or letters) code. It is shown in the part of the Title Block under the heading of CAGE. Since this code is often related to the type of company doing the design or manufacturing (but not the actual item shown in the drawing), it usually can be set as part of the standard CAD Title Block just once. Designers or CAD users then do not need to be concerned with it.

4.5.10 Drawer, Checker, Approver

Part of the standard Title Block indicates information about who created the drawing and how it was checked or approved. This information is found in the blocks that say DRAWN BY, CHECKED BY, and APPROVED BY.

The DRAWN BY space indicates the person that originally created the drawing. Along with the name, there is a DATE field to indicate when the drawing was officially created. This is not necessarily as important when CAD systems control drawings, however. The CAD database and/or operating system usually also keep date and time information.

Engineering departments may have checkers. These people examine the drawings before they are released (meaning that other parts of the company would be looking at them or working from them). When this is done, the name of the person checking can be shown in the CHECKED BY space. Often, designers check each other's work, so just another designer's name would appear in this part of the Title Block.

Beyond the checking function, a person in responsible charge (an engineer or supervisor) approves the drawing. The name of this person along with the date is shown in the APPROVED BY space.

4.5.11 Additional Approvals

Usually the content of the drawing will be relevant to a specific engineering function (such as electrical systems, hydraulics, structural integrity, etc.). Often the performance characteristics of the design needs to be approved. For instance, electrical schematics need to be functional, and structural components must be able to withstand expected stresses, etc. This sort of specialized approval is usually in a separate area of the Title Block (farthest to the left). Space for these approvals are under the heading of ADDITIONAL APPROVALS. These approvals can be vital legal information since the engineer's name and license number

4.6 THE REVISION BLOCK

In addition to the Title Block, a standard drawing format includes a separate section of the drawing to indicate revision information. As mentioned earlier, revisions are a very important mechanism for maintaining control over the state of a design. It is more than just changing something on a drawing; it needs to be coordinated throughout the company and perhaps into the field where the product is maintained or repaired. It is important for anyone using drawings to know the revision status of the drawings. The formal process (which should exist) for changing the design goes by a number of acronyms. Some of these are ECR (Engineering Change Record), ECN (Engineering Change Notice), or ECO (Engineering Change Order).

As mentioned earlier, the current revision level of the drawing is indicated in the Title Block. The Revision Block, then, is used to record some detail of the "history" of the drawing and the design it is supposed to document. Figure 4.3 shows a typical Revision Block.

The first column of the Revision Block shows the revision level. Obviously if the drawing is at its first version (say a revision level of 0), there is nothing

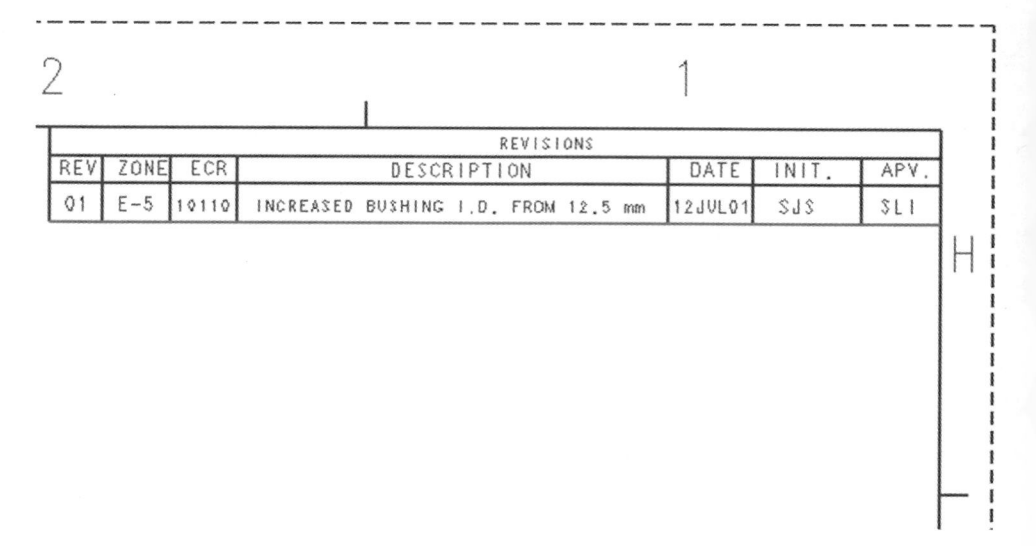

REV	ZONE	ECR	DESCRIPTION	DATE	INIT.	APV.
01	E-5	10110	INCREASED BUSHING I.D. FROM 12.5 mm	12JUL01	SJS	SLI

FIGURE 4.3 Standard revision block.

Drawings and 2-D Design

shown in the Revision Block. But, when the drawing is revised for the first time, an entry is made in the Revision Block starting with a number 1 in the REV column. The next revision would then have a row with 2 in the REV column, etc. If letters are used then REV would show A, B, C, etc. at the start of each row or entry in the Revision Block.

The second column of the Revision Block shows ZONE, the geographic region of the drawing that has been modified since the previous revision level. This is a number and letter combination that indicates a horizontal and vertical slice of the drawing itself. These numbers and letters are shown in the margin of the drawing format.

The third column shown for the Revision Block is ECR (Engineering Change Record), usually a number or character string that is used to track a specific set of changes to the design. A specific ECR number would be shown in the Revision Block of all the drawings that were changed "under" that particular ECR number or process. All these changed drawings together are often referred to as a package.

The next column of the Revision Block is DESCRIPTION, where the designer or user can place text indicating some detail of the changes made to the drawing. For instance, "dimension changed value," or "holes removed from a plate," etc. Obviously, there is very limited space for this description, so it needs to be very concisely written. Some CAD systems allow "memo fields" to be used for expanded description of the history. These would be stored with the drawing in the CAD database, but would probably not be shown on a hardcopy of the drawing itself; it would be printed separately.

The next column in the Revision Block is DATE, when the drawing was changed, or when the drawing's ECR was completed and/or released.

The next column in the Revision Block is for INITIALS, the initials of the person's name that made the revisions to the drawing.

Finally, the last column in the Revision Block is for APPROVAL. This is usually the "initials" of the person that is approving the changes to the drawing.

4.7 TYPES OF MECHANICAL DRAWINGS

There are a few basic types of mechanical drawings. Some CAD software may be better suited to specific types, but in general, they will all handle them to one degree or another. Table 4.3 lists the types and a basic description. These drawings are shown in an approximation order of "creation" during a design project. For example, an engineering sketch would often be for preproduction or prototypes only (i.e. not intended to be manufactured and sold in a final product); while the word drawing would probably only be needed for actual production processes.

TABLE 4.3 Basic Types of Mechanical Drawings

Type of drawing	Released?	Description
Engineering sketches	No	A simple drawing to document new concepts or to prepare for laboratory analysis.
Layout	No	A drawing for overall design system configuration. Detail drawings are made to fit into the context of the layout.
Detail	Yes	A drawing of a single component or part.
Assembly or Installation	Yes	A drawing that shows how various parts (or details) are put together. The Bill of Material is usually used in conjunction with the assembly drawing to specify which specific items are put together.
General Arrangement or Final Assembly	Yes	A drawing that shows the most general design information for the end product of the design. It tries to show the whole design as it appears in the field. Also called a G.A.
Schematic	Yes	A drawing that documents designed circuits used within a product; it may not be geometrically correct, but it is functionally correct.
Word	Yes	A drawing that contains no geometry of a design; it only contains text information (notes) with the drawing format. It is generally employed to officially record some text-based design or manufacturing information.

The kind of drawings that are used for "official" production and design control are often referred to as "released" drawings. This implies that it is available throughout the company and/or its suppliers. Table 4.3 indicates whether the type of drawing is of this type.

4.7.1 Engineering Sketches

An engineering sketch is assumed to be less than a normal drawing. Instead of documenting a design for production and sale, this drawing would be used to document a prototype or basic parameters for a potential product. It may not have a format at all. In many cases, the sketch is made by an engineer (often primarily responsible for calculations); while official drawings are made by designers (often primarily responsible for commercialization of a new product).

Drawings and 2-D Design

4.7.2 Layouts

A layout drawing is often also not a normal drawing. It is a special drawing that often is a master of the released drawings. This is where components are shown in the context of the entire design, to basically figure out how things fit. For a car, the layout drawing would actually look like many views of the car, and the position of the engine, doors, etc. would be determined from this drawing. This can still be done with 2-D CAD software, but much more typically this type of activity is done with 3-D models instead.

Often the layout drawing would be controlled by a head designer for a product. This drawing would then be used by other designers in the more detailed design of "lesser" assemblies. Some CAD software uses data management methods to continue this paradigm with 2-D CAD data shown overlaid or as layers. The master designer can change the layout, but the lesser assembly designers are not allowed by the software to alter the overall layout. Again, 3-D CAD generally does an even better job with this project control function.

4.7.3 Details

A "detail" drawing is really a drawing of a single component within a larger design. It is a drawing of a single part. It usually has an assigned number called the part number. The detail drawing should be completely sufficient to have someone manufacture the part. Using the car analogy, the button one would press to open a door on a car would be a single part. This button would have a drawing of its own, and the button would have a part number and be manufactured as a discreet item. A drawing showing just this button could then be called a detail (or a detail drawing).

4.7.4 Assembly or Installation Drawings

An assembly drawing shows how various parts (details) are put together. For the car door example, the button would need to be shown connected to the handle and latches, etc. that form a door handle; the assembly drawing shows this relationship and how to put the parts together. Keep in mind, that the Bill of Material is usually used in conjunction with the assembly drawing to show or call out which specific items are to be put together.

Also, an assembly like a door handle may then need to be shown being installed on the car door. This type of drawing (which simply shows a larger context of assembly function) can be referred to as an installation drawing.

As with the detail drawing, the assembly drawing should be sufficient to assemble the parts together to meet the requirements of the product being designed.

4.7.5 General Arrangements

Eventually the assembly and installation drawings lead to a drawing that shows the complete final product. This is the general arrangement drawing. It shows the most general configuration for the product. It may not be necessary for actually manufacturing the product. However, if this a top level drawing for an assembled product (such as the car in previous examples) this drawing shows how the last steps are taken for the assembly. Therefore, this type of drawing may be referred to as the final assembly drawing or just the final drawing.

4.7.6 Schematics

A schematic is a graphical tool for showing circuits of various types. Usually they would be for electrical, hydraulic, or pneumatic (air pressure) systems. Schematics are essential elements for documenting the design of products that use these types of systems. Therefore, they need to be carefully documented and controlled in conjunction with the overall product design state.

In order to apply revision control and information release control, the schematics can be shown on the typical mechanical drawing format (with Title Block, Revision Block, etc.).

4.7.7 Word drawings

A word drawing does not contain any geometric information that graphically shows the component documented by the drawing. Instead, it is specified or described in written text (notes) shown on the drawing.

Using the car door analogy, there may be certain lubricants that need to be carefully specified. These lubricants may have a special formulation or chemical characteristics. The lubricant characteristics drawing document could be indicated as an item that is shown or called out on the door assembly drawing. Then using the part number for the call out, one could then find the word drawing for that part number. Then when one looks at the word drawing, the lubricant characteristics can be determined.

4.8 DRAWING STANDARDS

There are a number of industry standards that can be referred to for drawing standards. Often these form the basis of a company's drawing procedures, although companies also usually need to have some customization applied to these standards. Table 4.4 lists some of the more important standards from groups such as ASME International (mechanical engineering professional society) and ISO (an international standards body).

Drawings and 2-D Design

TABLE 4.4 Some Important Standards for Drawings

Standard (recent edition)	Description
ANSI Z210.1 (1973)	Rules for numerical rounding
ASME Y14.1 (1995)	Standard for the drawing formats and sizes (U.S. or inch drawings)
ASME Y14.1M (1995)	Standard for the drawing formats and sizes (Metric or mm drawings)
ASME Y14.2M (1992)	Standard for line conventions and lettering
ASME Y14.3M (1994)	Standard for how to show views and projections
ASME Y14.5M (1994)	Standard for dimensions and tolerancing
ASME Y14.36M (1996)	Standard for notation that indicates surface texture (relates to how closely manufactured parts approximate totally smooth or the planar geometry that is shown in the drawing)
ASME Y14.8M (1996)	Standard relating to information for castings and forgings
ASME Y14.100M (1998)	Standard for drawing practices
DIN 6 Part 1 (1986)	Projections in first angle and views
DIN 6771 Part 1	Title Blocks
ISO 128	Technical drawings—general principles of presentation
ISO 1302 (1992)	Technical drawings—method of indicating surface texture
ISO 2768 (1989)	General tolerances
ISO 5457 (1980)	Technical drawings—Sizes and layout of drawing sheets

4.9 VIEWS

At this point, the basic outline of mechanical drawings has been presented. This has included the sizes and formats for standard drawings without much discussion of the graphical data that represents the product design. Much of the rest of this chapter will be concerned with the geometric part of drawings. The most important concept with respect to this part of the drawing is called views or multi-view projections or perhaps just "projections."

As the name implies, a view relates to what an object looks like from a particular view or viewing angle (refer to Figure 4.4). For instance a Front View is what the object looks like in the front (front being determined by the person making the drawing); then the Right View is what the object looks like from the right direction; the Top View is then from the top. The basic idea here is to fully document in a 2-D medium the complete definition of an actual 3-D object by looking at it from a set of more or less standard directions. The person reading the drawing is then expected to be able to reconstruct or mentally visualize the object's true 3-D configuration.

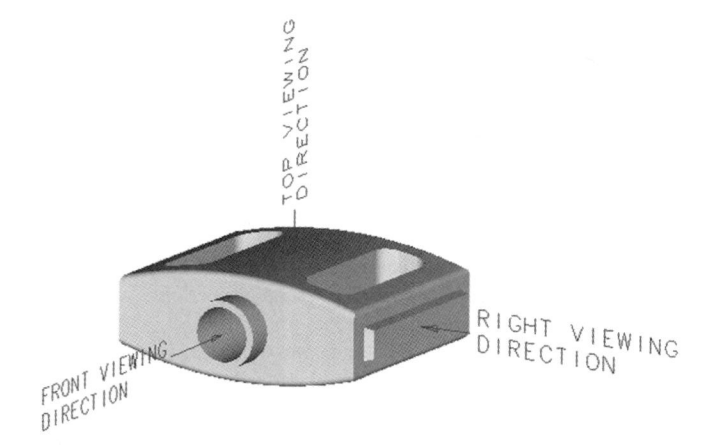

FIGURE 4.4 Viewing directions or angles for an object.

One needs to realize that when an object is shown in a drawing it is generally implied that is from a particular view. Referring to Figure 4.1, one can see that there are 4 views shown. These views are all for the same object; it is not showing 4 separate objects. The lower left view is assumed to be the Front View, the lower right view is the Right View, the top left view is the Top View, and the top right view is called the Isometric View (which is explained later).

4.9.1 Standardization of Views (Third Angle vs. First Angle)

Unfortunately, there are two competing standardizations for the presentation of views on a drawing. The sample drawing shown in Figure 4.1 uses what is known as "third angle projection." This is the standard in the United States. In this case, the front view is selected first, and the view on the right is what the object looks like from the right.

Many other countries (continental European countries, in particular) use "first angle projection." In this case, the view of the object from the right (the Right View) is placed to the LEFT of the front view; the Top View is placed BELOW the front view, etc. The idea here is that one imagines that the view from the right projects onto a piece of paper held up on the left side of the part (like the part's view is projected onto the paper). This is in contrast to the idea of the "third angle projection" where one imagines that the paper is "wrapped" onto the part and then unfolded to reveal what is seen in the drawing. In either case, the graphical information shown in the views is the same, the only difference is where the views are placed on the drawing.

Drawings and 2-D Design

It is important to know which method was being used by the person making the drawing. Therefore, there is a standardized note in the drawing format that indicates first or third angle projection, and a small cone-like object is shown with its Right View projection (the Right View being either to the left or right of the Front View). Refer to Figure 4.5.

4.9.2 "True" Views

An important concept of the views on a drawing is that the angle at which the object is viewed is "true." This implies that the object is viewed head on or not at a skewed angle. This is important since the drawing needs to show dimensions in the views, and these dimensions should all be in a plane that coincides with the face of the object being viewed.

The first three standard views already mentioned (Front, Right, and Top), which are identified merely by their location on the drawing, are all going to be assumed to be true views. If the face of the object shown in these views are not parallel to the viewing plane, then other kinds of views need to be put on the drawing to show accurate dimensions.

Although architectural or A/E/C drawings are not a major focus of this book, these types of drawings will use views called *plan* and *elevation*. The plan view is like a Top View; it views the construction from above (showing the plan or layout of the site or project). The elevation view is like a Front View; it views the construction from the side (as a person would face it), and it obviously would show how high or elevated the objects on the site would be.

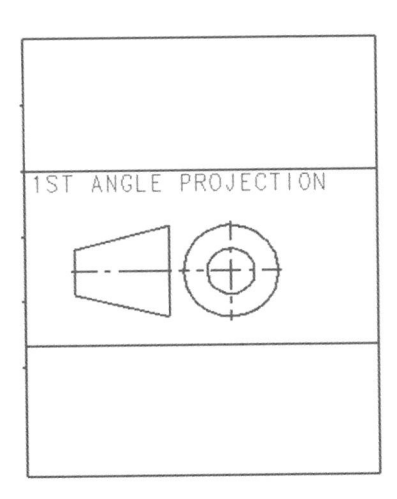

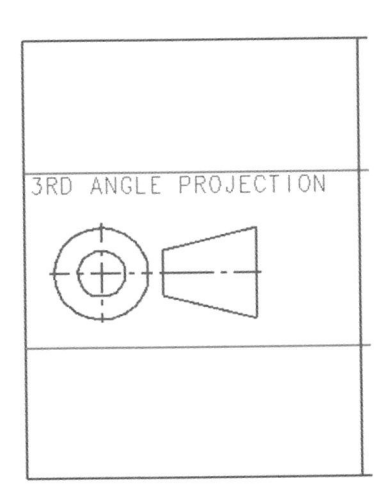

FIGURE 4.5 First angle and third angle projection symbols.

4.9.3 Auxiliary Views

Probably the simplest type of view for showing part of an object at a nonstandard angle is the auxiliary view. As shown in Figure 4.6, the auxiliary view is shown as a projection or at a viewing angle that directly faces the part of the object that needs to be defined. As mentioned earlier, usually the most important characteristic of the view is that dimensions can be accurately shown for that part of the object in the drawing. In this case, the auxiliary view is a "true" view (although the standard views are often referred to as the "true views").

4.9.4 Section Views

Another common type of view for more accurately defining the geometry of an object is the section view. The section view is like a cutaway of the internal volume of a part or object. As can be seen in Figure 4.7, the section view uses crosshatching (the tightly spaced lines at an angle that fills a area of the view) to indicate material of the part that has been cut into.

Sometimes the standard views (such as Front, Top, and Right) will have cutaway sections (with crosshatching shown in these views), but generally they are instead used as the "basis" for the section views. When they are used as the basis for the section views, there are special lines drawn with letter identifiers and arrows. In the Front View shown in Figure 4.7, the thick line with the letter A at the beginning and end indicates what the section view is cutting. Arrowheads at the beginning and end of the line show the direction the cut section is to be

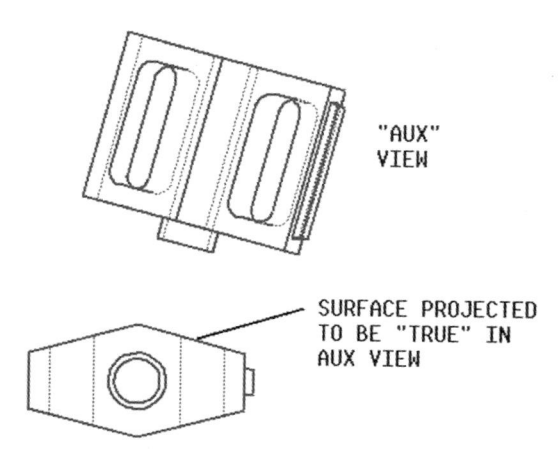

FIGURE 4.6 Sample auxiliary view on a drawing.

Drawings and 2-D Design

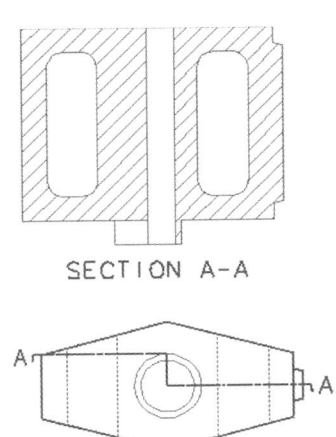

FIGURE 4.7 Sample section view on a drawing.

viewed. The section view, in this case, is called Section A-A, and this is clearly indicated by a note beneath the section view.

4.9.5 Detail Views

Another type of view for documenting a part in a drawing is the detail view. The detail view is generally a magnification or clarification of a somewhat small feature. As shown in Figure 4.8, the area of interest is usually based on one of the other views already in the drawing, and again, it is usually identified with a letter.

4.9.6 Isometric View

The last type of view presented for helping to define more complicated parts is the isometric view. This view is expected to show the object more or less realistically. It shows the part at a more "arbitrary" viewing angle so that one is not limited to only seeing one particular surface or face at a time. The view in the upper right corner of the drawing shown in Figure 4-1 is an isometric view.

The isometric view is generally not expected to show official dimensional information. Instead, it is expected to allow the reader of the drawing to more quickly visualize the object in the drawing. If the reader has just a basic idea of what is in the drawing based on the isometric view, interpreting the remaining views of the drawing is much more efficient. For very simple parts this not a very significant advantage for the reader, but for more complicated parts it can be a very large advantage.

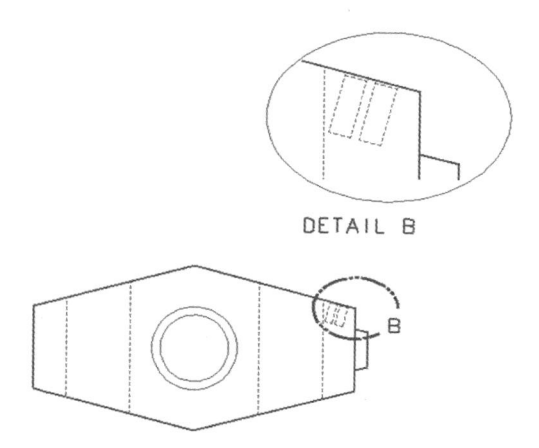

FIGURE 4.8 Sample detail view on a drawing.

Before the large scale adoption of the 3-D CAD systems, the isometric view would rarely be expected on mechanical drawings used in basic manufacturing. It would only be found on drawing connected with service manuals, parts manuals, training materials, etc. The use of the isometric view was limited because the creation of this view was very time consuming (particularly if no CAD system at all is used). However, with 3-D models generally available for parts and/or assemblies, the isometric view can usually be created automatically. These isometric views can also have "hidden line removal" so that the part or object looks even more realistic.

4.9.7 View Scale

An important concept with respect to views is scale. The scale of a view (any view) is the relationship between the image or geometry shown on the drawing and the actual object being documented. For example, if a plate is 100 mm long, and that part's lengthwise edge is shown on a view in the drawing that is at half scale, then the line drawn of the plate shown on the drawing would be 50 mm long. This 50 mm is the length of the line one would measure directly from the hardcopy or print of the drawing (assuming it is printed according to the paper size shown in the Title Block). In other words, the view scale is like making a model of the object using the paper drawing and indicating how big or small that model would be with respect to the real object.

It should be clear, then, that a large object (such as an airplane) would be shown in a drawing with a very small scale, for example, 1/32 or one thirty-second scale. In this case, one inch on the drawing would represent 32 inches on the

Drawings and 2-D Design

plane. For a mm drawing, perhaps a scale of 1/50 or one-fiftieth scale would be used, and 1 mm on the drawing would represent 50 mm on the plane. Another way of indicating view scale is in decimal form. For the 1/32 inch drawing, the scale would 0.03125. For the 1/50 mm drawing, this would be 0.02. Yet another way of indicating scale is to indicate a ratio using a colon (:). In this case, the two scales in the examples would be 1:32 and 1:50 respectively.

If all the views in a drawing are at the same scale (which is somewhat typical), then only a single value of scale needs to be indicated on the drawing, and this value can be put into the Title Block. Often though, section views, detail views, etc. are shown at a different value of scale (usually larger scale to show more detail). In this case, there should be a note below these special views to indicate the scale for these views. In these cases, the scale shown in the Title Block becomes more like a predominant scale; meaning that most of the views in the drawing are at that scale, but not necessarily all of them.

Keep in mind that the dimensions shown in the drawing are not really related to the view scale (as far as the drawing reader is concerned). The dimension on the drawing for the 100 mm plate example should be shown as 100 mm, because that is how long the real plate is supposed to be. As discussed later, however, for CAD users it is important to know the relationship between the view scale and the dimensional values.

In terms of what value of view scale should be used, there are no specific rules. It depends on the complexity of the object between documented and the paper sizes available. If only smaller-sized drawings can be printed, then perhaps the smaller drawing sizes (B and C for inch; A2 and A3 for mm) should be used. In this case, smaller scales would be used to "shrink" the objects image enough to fit on the paper. If larger drawings can be used, then perhaps larger drawing sizes could then be used. Of course, if the drawing contains a great number of edges, lines, circles, notes, etc. then a smaller paper size with the small scale will simply not work (the drawing will not be readable). In this case, the larger paper size, or perhaps using multiple sheets of the larger scale will be required.

4.10 "OBJECT" LINES

Obviously the object (part, assembly, product) documented by the drawing will be shown in the views of the drawings. The lines drawn in the views that are showing the object can be called object lines. They may also be referred to as line work, geometry or visible and dashed lines.

As shown in Figure 4.1, there are solid lines (not broken at all) and dashed lines (broken). The solid lines are usually edges of the object that are visible (from the viewing angle of the view). The dashed lines are edges of the object that are hidden from sight (again from the viewing angle of that particular view). These dashed lines are then called hidden lines. In CAD systems, these appearances of lines (solid, dashed, etc.) are usually referred to as fonts or line fonts. The solid font indicates a solid line type; the hidden font indicates the hidden or dashed line type, etc. Keep in mind, however, that some fonts are not actually for the object, but rather for indicating the center of a line (this is the centerline font).

The main task for the object lines is to help the reader visualize the view of the physical object. Beyond the line type being visible or hidden, the next visual cue is called line weight. Line weight is a term that indicates how thick and dark the line appears. A heavy line weight indicates that the line is drawn thick and dark. A lighter line weight indicates a thinner and lighter line. Often the use of line weight is limited to 2 or 3 levels such as light, medium, and heavy; or they may be referred to as thin, medium, and thick.

The solid lines (indicating visible edges) are generally thick or heavy lines. They are supposed to stand out clearly. Then the hidden lines and other lesser indications are generally thin or light line weights. As mentioned earlier, the point is to assist the reader in visualizing the object.

It is important to realize that the standard drawing is somewhat of a simplification versus the actual 3-D object. For instance, the Front View in a drawing shows some lines that are more than one edge. The object may have an edge facing the reader which is shown as a solid line; but, then there is another edge at that line that is coincident or on top of the first line. This is not really an issue for the manual creation of drawings, but when the drawing is created electronically, and particularly from a 3-D model, this can be an important consideration. The CAD system may regard them as the same line, or the CAD system may regard them as two distinct entities. If the CAD system does treat them as a single line, then usually there is some tolerance or error to indicate how close the edges must be before the line is merged. Unfortunately, this means that a very small rotation of the 3-D model can be incorrectly documented in the drawing if the tolerance is not acceptable.

4.11 DIMENSIONS

Referring to Figure 4.9, some parts of the views show numbers with arrows. These are called dimensions. They are meant to indicate the actual size of the object being documented by the drawing.

Notice that the views in Figure 4.9 do not show all of the possible dimensions for the object in every view. Instead, in each view, certain dimensions are shown that make the most sense for that particular view. For instance, the Front View shows the width and height of the object, while the Top View shows the depth. Of course, the depth could also be shown as a dimension in the Right View, but there is no need to show this parameter more than once in the drawing (repeating it would be considered confusing and a source of potential redundancy and error).

Drawings and 2-D Design 95

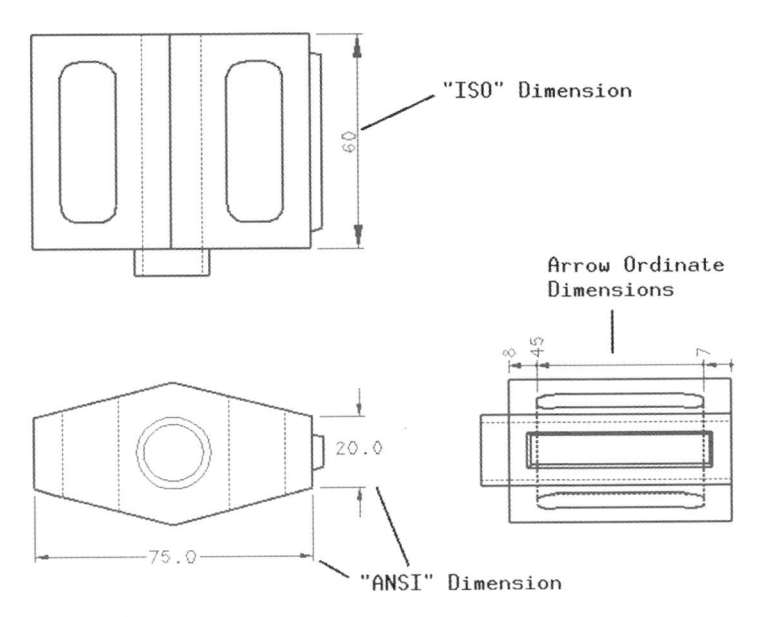

FIGURE 4.9 Sample dimensions.

The most important aspect of dimensions is that they are supposed to unambiguously define the geometric characteristics of the object documented by the drawing. In other words, if a manufacturer is given the drawing, the manufacturer should be able to create the object exactly as intended by the author of the drawing. This seems simple enough, but in reality, this can be rather difficult for some objects.

Figure 4.9 shows the details of some simple dimensions that could be found on a typical drawing. The dimension starts with the extension lines near the object. These are the lines that indicate what exact part of the object is being dimensioned. They may "indicate to" edges of the part, the center of a hole, etc. Next the extension line is met by the arrowhead of the dimension. The arrowhead may point in either direction. The arrowhead is connected to the dimension line. The dimension line extends for the length of the dimension, and then there is an arrowhead at the other extension line.

Note that the American (ANSI) standard for dimensions breaks the dimension line to put the numerical value for the dimension in the "middle" of the dimension. However, the international standard (ISO) for dimensions does not break the dimension line. Instead the numerical value is shown above the dimension line. These styles are shown in Figure 4.9.

4.11.1 Dimension Values and Scale

Obviously the number or values shown in the dimension are very important. They indicate the size required for the object. It is also important to understand the concept of view scale for dimensions since the view scale indicates the relationship between the object or geometry drawn on the drawing and the value shown in the dimension. For instance, if the real object is 100 mm long, and the view scale is 0.1, then the person making the drawing would make a line for showing the object 10 mm long (since the scale is 1/10 or one-tenth). However, the dimension number shown on the drawing must still say 100.

When someone made a manual or "board" drawing (not using a CAD system), they would use a scale. This was a special kind of ruler that had the markings at different scales (1/2, 1/4, 1/10, etc.). Then the person making the drawing could just use these devices to make the drawing at a particular scale. The scale would make sure that the object was being "compressed" or "expanded" correctly. Then the draftsman would just draw the numbers on the dimension. For the example mentioned above, the person could write the dimension as "100" even if the line drawn for the object was only 9.9 mm long (nearly 1/10 scale).

However, for CAD systems, this situation becomes more complicated. The CAD software should use exact values for the lines that are drawn. The user can indicate that a 100 mm object line is needed (at whatever scale). Then the user can ask for a dimension to be drawn for that line, and it will automatically show up as "100." This will only work if the object is drawn very carefully and if the user understands what the scale really means.

4.11.2 Dimension Values and Decimal Places

Besides showing the designer's intention for the size of an object, the dimensions on a drawing can also indicate information about the accuracy needed for the object's manufacture. For instance, a hole in a plate may be dimensioned at a certain location with respect to the bottom and side edges of the plate. However, the hole must be manufactured by some means, and the manufacturing process will have some error with respect to the ultimate value shown in the drawing. Furthermore, some processes will be more accurate than others; therefore the tolerance needs to be indicated in the drawing so that an appropriate manufacturing process can be selected.

If the hole is a critical item (perhaps the hole is for a shaft, and that shaft has to line up with gears on a machine), then the hole might be drilled. Drilling is generally a precise operation. If this hole is critical, then the dimension showing the location of the whole would be shown with additional decimal places. The decimal places are the number of digits shown after the decimal point (or comma). If a dimension is for an integer value, such as 12 inches or 100 mm, then greater accuracy is implied by showing these numbers in dimensions as 12.0,

Drawings and 2-D Design

12.00, or 12.000 inches or 100.0 mm. If a dimension already has decimal values since they are fractional, then added zeros may or may not be needed. For instance, 1.250 and 1.125 each indicate a tolerance to the thousandths place. The accuracy being implied by the decimal places in the drawing is usually shown in a special note near the Title Block (for example, "3 Decimal Places equals +/– 0.001 inch").

This approach to indicating the needed accuracy of dimensions is only a basic tolerance methodology. It simply indicates a general tolerance to the dimensions shown. There are, however, other types of critical geometric characteristics. For example, 2 holes may need to be drilled in a accurately parallel fashion (perhaps the holes are for 2 shafts that have gears that need to mesh). This more sophisticated approach to controlling the accuracy of a part is an issue for what is known as "GD&T" or Geometric Dimensioning and Tolerancing. This is mentioned in a little more detail below, but it is really beyond the scope of this work.

The use of decimal places for indicating basic accuracy creates somewhat of a problem for CAD systems. As mentioned above, the values for dimensions can actually be automatically calculated by the CAD system based on the precise geometry that has been drawn with the software. If a rectangle is drawn by the user to be 100.5 mm by 50.75 mm, then the CAD system is able to show the dimension as 100.500 by 50.750 mm. However, this indicates a very accurate tolerance ("to 3 decimal places"). Then the designer may have the CAD system reduce the decimal places to a more typical tolerance of 1 decimal place. The 100.5 should be no problem, but the 50.75 has to be rounded. With the dimension shown as 50.8, the original intent of the dimension is lost. The original rectangle was 50.75, and someone may now assume that the original CAD-created rectangle was 50.8 (when in fact 50.75 was used). Whether a value is rounded up or rounded down should be done according to a standard such as ANSI Z210.1 (1973).

4.11.3 Arc and Angular Dimensions

So far, linear dimensions (dimensions that indicate lengths) have been discussed. There are other kinds of dimensions, however. The most common would be for circular geometry and angles.

Figure 4.10 shows examples of the circular and angular dimensions. Note that a special symbol that looks like a zero with a line through it indicates a value to be a diameter of a circular feature. Also, note that an R is used to indicate a radial dimension. This is common for features known as fillets (pronounced "fill-its"). Fillets are regions where two intersecting planes of material are blended together. They may be the result of the mold used to make a single part; they may be the result of machining; or, they may be the result of welding two segments together.

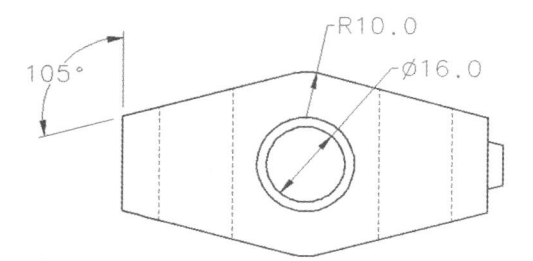

FIGURE 4.10 Examples of arc and angular dimensions.

Also, note that there are lines that connect from the dimension values to the dimension lines. This extra type of line is known as a "leader line" or just a "leader."

4.11.4 GD&T

As mentioned above, there are sophisticated approaches to indicating the allowable errors or tolerance to the object being documented by the drawing. This is called GD&T or Geometric Dimensioning and Tolerancing or just Geometric Tolerancing. GD&T involves an entirely new set of standard symbols and notes for the dimensions and data shown on the drawing. Only a few examples are shown here; consult a standard such as ASME Y14.5 or www.asme.org for more information.

The most obvious indicator of the presence of GD&T on a drawing is a Feature Control Symbol or FCS. As shown in Figure 4.11, the FCS is a set of symbols and values in a box that is connected to or "associate" with a dimension. The letters such as -A-, -B-, or -C- are used to specify a datum. The datum is like the reference or grounding point for the dimensions. They refer to surfaces of the part that are considered a base for the remaining dimensions.

Besides the GD&T symbols, tolerances of dimensions can be indicated by over/under conditions. In this case, the dimension value is shown at a maximum permissible high and low value. Figure 4.11 shows some dimensions with this approach as well.

In general, drawings are not supposed to be physically measured (using the hardcopy). The system of dimensions shown on the drawing is to solely indicate the size of the object being documented (so measuring is unnecessary). To reinforce this idea, a note on the drawing will generally state "NOT TO SCALE" (even if the drawing is quite accurate in terms of scale). This "never measure" approach is more problematic with the CAD data. If a user has carefully created a

Drawings and 2-D Design

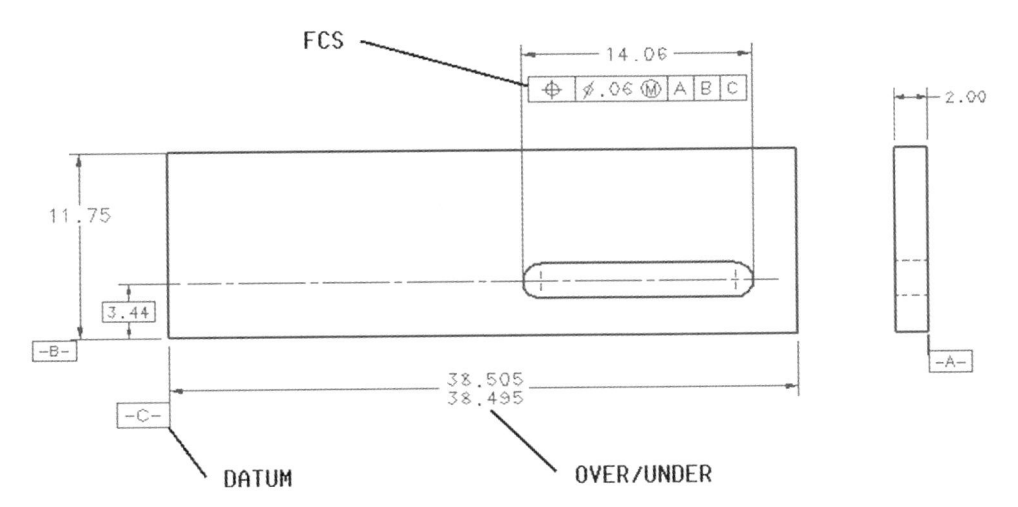

FIGURE 4.11 Simple examples of GD&T symbols and over/under conditions.

2-D CAD model (or a 3-D model for that matter), then there really should be no problem with measuring the model (using the CAD system) to obtain some extra geometric properties or the designer's intent. However, this measuring of the model will not indicate any tolerance information; at this point, only the GD&T symbols, the decimal places, or other notation can really reflect the designer's intentions for tolerancing.

4.12 NOTES

An important source of information on drawings is notes. Notes are just instances of written text on the drawing. Some notes are very general, such as "NOT TO SCALE" or "SPECIAL PROJECT." Other notes may be very specific to the part or assembly shown on the drawing. For instance, "USE LUBRICANT X345 WHEN ALIGNING ITEMS 22 AND 23." Notes can be short like these examples, or they may even fill the entire view of the drawing (i.e. a "word").

Notes are often standardized as a numbered list in the upper left hand corner of the drawing.

4.13 BALLOONS

Another important indication on some types of drawings is balloons or bubbles or item numbers. As shown in Figure 4.12, balloons are numbers in a circle with an

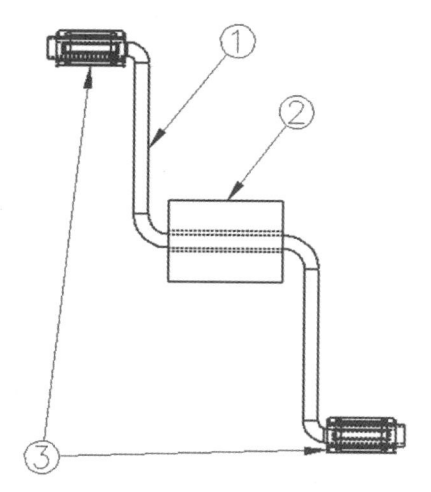

FIGURE 4.12 Balloons in an assembly drawing.

arrow pointing to a part in the drawing. This arrow that points from the number to the object is known as a leader.

The numbers in the balloons usually correspond to an item number in a list that appears on an assembly type of drawing. The assignment of specific numbers is found in a parts list (shown on the drawing) or in a Bill of Material (BOM). These items numbers could also be shown in the drawing with underlines (instead of balloons)

4.14 CENTERLINES

A special line type is used for various circular features such as holes. In this case, a centerline is used to locate the center of the feature. As shown in Figure 4.13, the centerline is a font using thin line weight that has a long solid section with occasional breaks to a shorter line. The centerline is useful in quickly identifying that some lines are associated with a hole or other circular feature. Figure 4.13 also shows a sort of "bull's-eye" that locates the center of a hole within its "true view" (viewed directly along the axis of the hole). The lines of this type of centerline are often used as extension lines to a dimension to locate the hole.

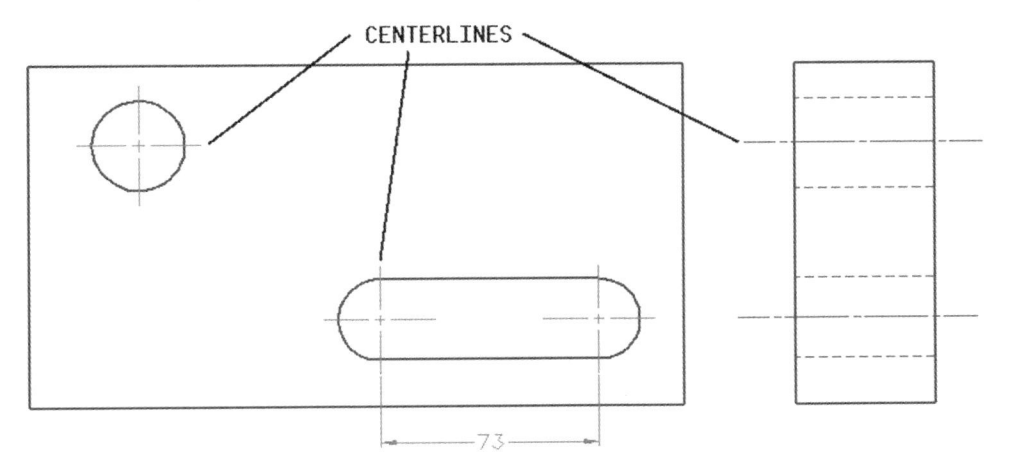

FIGURE 4.13 Centerlines in a typical drawing.

4.15 CROSSHATCHING

Referring again to Figure 4.7, one can see that the Section View has areas of the object geometry that has a pattern of diagonal lines. This indicates that this part of the object is actually internal and solid. The crosshatching shows the imaginary geometry that would be apparent if the object was actually cut open.

The sample in Figure 4.7 shows a pattern of 45 degree lines evenly spaced for the crosshatching. This is the typical pattern. However, there are various types of crosshatching patterns that could be used. The different patterns can be used to indicate different types of materials (such as iron, aluminum, nonmetallics, etc.) Refer to standard ANSI Y14.2M. Standard ISO 128-1982 (E) paragraph 4.2 indicates that the meaning of the hatching shall be clearly defined.

4.16 DESIGN METHODOLOGIES USING DRAWINGS

The use of drawings for design activity lends itself to a certain methodology. This methodology involves a sort of hierarchy of tasks, and each level of this hierarchy usually works with a certain type of drawing. For some companies or design activities, all these levels may not actually be present, but for a large company or a system integration design activity these levels are most likely present. Some common examples of the methodology would be in the automotive, aerospace, or engineered equipment sectors. Virtually all these companies are using 3-D processes now, though, so this drawing methodology may no longer be relevant. However the same principles can be applied to 3-D design methodologies (this is discussed in later chapters).

4.16.1 Specifications

The top level of the 2-D design methodology is the General Arrangement (GA) drawing. As mentioned earlier in this chapter, this drawing basically shows the end product being designed. This is the drawing that would be used to form marketing assessments and the engineering specification. The engineering specification or "spec" is a document (often referring to the General Arrangement drawing) that is supposed to specify the requirements that the design is supposed to meet. This specification may also be vital to the ISO 9001 quality system for the engineering department.

Obviously, then, the General Arrangement drawing is often created first (before any other drawings for design). This stage of the design process is often called Conceptual Design. At this stage, the basic concepts of the design are being assessed and adjusted to meet the various market requirements. It is very important to be as accurate as possible in this stage of the process. If a major problem with design is discovered later in the project (after 100s or 1000s of drawings have been made), then it will be very costly to correct. If problems can be anticipated at this stage, there is only the General Arrangement drawing to be corrected.

The General Arrangement drawing would be typically controlled by a project engineer (although this person may or may not create the drawing). The project engineer is usually assumed to work on the project over its entire life cycle. As the design proceeds to a level or greater and greater detail, problems with the original assumptions of the design will arise, and the General Arrangement drawing will then be revised to reflect the evolution of the design. Because the General Arrangement drawing is at such a high level, it is rarely used in the actual manufacturing of the product. However, it is often used in consultations with prospective and actual customers. This drawing is also likely to be used in post-sales situations such as retrofits, product support and maintenance, technical manuals, and product liability.

4.16.2 Layouts and Assembly

Assuming the Conceptual Design phase is successfully completed (the General Arrangement drawing being finalized as much as possible), the next phase is usually Preliminary Design. The type of drawings used at this stage are called layouts. These drawings are used to actually start sizing and fitting together components,

Drawings and 2-D Design 103

parts, and systems. As with General Arrangement drawings, layouts are not really used in the manufacture of the product, and they would generally not be "released" (meaning they are not used outside of the engineering department).

Layouts are generally going to start from the graphical information available in the General Arrangement. That is, the views of the General Arrangement would be copied and probably scaled to larger size. The layouts will often also concentrate on specific areas of the product being designed. For an automotive example, the group responsible for the engine would work on an engine layout, and one designer would be given the main control over the engine layout drawing. The layout drawing is also probably going to have more views that most other drawings; it may show the Front View, Right View, Left View, Top View, Bottom View, etc.

The person doing the layout would then try to figure out how individual components and parts would fit into the overall design. This usually means that a specific component is being selected by the designers (and whether it is going to be purchased or manufactured). Based on that decision, the component would be drawn in all the views necessary on the layout. This often involves creating an image of the component in one or two of the views, and then using a method known as projection to figure out what it would look like in the other views. This was done with specific mechanical drawing methods prior to CAD systems, but now the CAD system should be able to do this for the designer. It is very important that a proper understanding of scale be maintained in the layout drawing. Components will not fit properly in the eventual physical product if the components are not properly scaled to the view scale being used in the layout drawing.

Once the detail design process described below has made sufficient process, the layout drawing would often be the basis for an assembly drawing. The assembly drawing would show how to assemble components. The assembly drawings would be the drawings that have the "balloons" or "bubbles" mentioned earlier.

4.16.3 Detailed Design and Drafting

Eventually, the layouts progress to the point that the geometry or shape of specific components have been determined, and the layout designer feels that the components have been positioned and sized properly to fit and work properly. Based on the views of the component in the layout drawing (say Front View, Right View, and Top View), a "detailer" or detail designer can create a drawing that can be used by manufacturing for that single component. This type of drawing is called a detail drawing or just a detail, and the individual component in the design is often just called a detail. In most cases, a 3-D CAD system would refer to the detail as a Part or a Part Model.

Since detail drawings are used to dictate or guide the manufacturing process, these drawings are the ones that would show most of the dimensions, tolerances, GD&T, etc. Since these drawings are used in processes outside of the engineering department (such as manufacturing or purchasing), these drawings are usually the type that need to be released. They will need the appropriate Title Block with approvals and revision levels. The detailer needs to know the proper way to document the individual components; it is best that they also understand the manufacturing processes that are likely to be used to create the component. However, a detailer may not need to know anything about the overall system design and performance.

4.16.4 Drawing Package

With the detail drawings completed, a full package could be created that included all the information assumed to be needed to build the entire product. The full package would include the General Arrangement drawing, the assembly or installation drawings, the detail drawings, and all the Bills of Material or parts lists. Prior to the adoption of CAD systems, all this information would typically be captured on "miniature" media such as microfilm (photographic film type of media with different frames on a roll showing drawings or lists), microfiche (a photographic film card that may contain dozens of images of drawings), or aperture cards (mainframe-style punch cards with a small photographic film image embedded in the card). This miniature media could be then be filed and archived for maintenance and future reference. Although these types of systems may still be used for legacy information, most drawings and packages are now controlled via electronic imaging systems.

4.17 CONCLUSION

This chapter has presented basic information on mechanical drawings. There are many more features of these drawings, but other resources can be consulted for more detailed information. A few of these additional features would be surface finish symbols (generally indicating how rough or smooth surfaces need to be), weld symbols (indicating how parts are to be welded together), screw and thread representations and notes, and "broken lines" (where objects are too long to be shown completely). However, what has been presented are the most essential elements to allow someone to become basically familiar with drawings.

4.18 CHAPTER EXERCISES

1. Try to locate a board drawing (drawing created by hand) from your campus or company. Record the date it was created, the project is was made for,

Drawings and 2-D Design

the drawing size, the revision level, and how it was approved. In a campus situation, check with the on-campus machine shop, plant engineering, or plant services department if needed.

2. Create a hand sketch drawing for a simple object using a Front, Right, and Top View. If you can find them, devices such as a drawing board, T-square, triangle, compass, and scale could be used.

3. Study a copy of an ASME Y14 or ISO 128 standard. Record whether the drawing in Exercise 1 (or any other drawing from your campus or company) conforms to the standard.

4. Study copies of a corporate drawing or drafting standard. Record whether the sample drawing or drawings conforms to the corporate standard.

4.19 CHAPTER REVIEW

1. In the First Angle projection system, the Right View for an object is placed on the right or the left of the Front View?
2. Which projection system is standardized in the United States?
3. If a view scale is set to 0.5, and a line on a drawing is drawn to a length of 20 mm, what is the value that needs to be shown in the dimension for the length of that line?
4. If a face of a part is oblique and not directly viewed by the Front, Top, or Right Views in a drawing, what type of view should be created to document and dimension features in that face—an Auxiliary View or a Detail View?
5. In an ISO standard dimension, is the value shown above the dimension line or in the middle of the dimension line?
6. What is a datum?

5

Two-Dimensional CAD

5.1 INTRODUCTION

This chapter provides information on CAD systems relevant to 2-D design and drawings. The previous chapter explained what drawings really are and what requirements they are expected to meet. That chapter also discussed how drawings (CAD or manual) are used in implementing a 2-D design methodology. Hopefully the information in this chapter will help the reader (a user, manager, or administrator) now understand how the CAD system can implement these methods.

5.2 BACKGROUND

CAD systems have developed over a number of decades, and they are continuing to become more powerful. Throughout the 1970s, these systems became more and more capable, but they were quite expensive and generally ran on mainframe computers. As the minicomputer (such as the former Digital Equipment Corporation's VAX) became available in the early 1980s, the CAD systems started to be-

2-D CAD

come less expensive. A long trend toward lower cost per seat continued with the advent of the personal computer. So, over these decades (and certainly by 1990), CAD became the standard method of producing drawings.

5.3 "SMART PAPER"

Probably the best way to think of doing 2-D CAD design and drawings is to think of the system as "smart paper." Although the geometric entities (lines, arcs, etc.) being created on the screen could also be manually drawn on a piece of paper, the CAD system has the ability to provide important information and intelligence about that geometry. The key to this intelligence is the simple mathematical model that supports the entities.

If one works with a drawing as a piece of paper, and one wants to draw a circle at a specific location on the paper with a specific radius to size the circle, then there is no choice but to approximate the circle. The CAD system, however, has the ability to idealize that circle and essentially make it perfect. The CAD system can be "told" to make a circle that has a 10 mm radius and lies at a point 100 mm from the bottom and side edge of the paper. Although the CAD system will eventually make a hardcopy that is (like the manual drawing) only an estimation of the circle, internally the CAD system can remember the exact dimensions that the CAD user had in mind. It can store the 10 mm and 100 mm values as actual numbers. Although the CAD system also has a finite accuracy of perhaps 8 significant digits, and the 10 mm ideal value may actually be stored as 9.999999 mm, this is well within the accuracy of virtually all manufacturing processes.

There are a number of advantages that can be exploited from the "smart paper" analogy or the idea that the CAD system stores a mathematical model of the design. Table 5.1 lists a number of these advantages. It is important to note, however, that these functions will not work properly if the CAD system is not used accurately. For instance, whenever possible the actual numerical values for the geometry (i.e., "object lines") should be entered to the CAD system (as opposed to just guessing based on the appearance). If a circle is supposed to be 15/16 of an inch in diameter, then the value of 0.9375 should be typed into the CAD system somehow. Even if the eventual dimension shown on the drawing might be 0.93 (to reflect the tolerances associated with decimal places), it makes the most sense to have the CAD system automatically round the number instead of just making the circle only 0.93" in diameter (which will degrade the mathematical model).

TABLE 5.1 Advantages of the "Smart Paper" Concept in 2-D CAD

"Smart Paper" Advantage	Description
Dynamic measurement	Although one can calculate distances, areas, angles, etc. by manually writing out formulae and solving them, basically all these values can be quickly and easily calculated and displayed by the CAD system. It can continue to provide these measurements throughout the design process.
Projections	If the CAD system tracks views, their scale, and their viewing angle (Front, Top, Right, etc.), then it should be able to create projections. This option allows the user to create geometry in one view and have some of it automatically created in other views with the correct orientation. With the advent of 3-D models however, this is a much less used capability.
Scaling	Geometry that is created properly can be scaled to larger and smaller sizes and create accurate new geometries based on existing ones. Some systems may also allow the scaling to not be uniform in the X- and Y-directions. For instance, a part could be stretched to be longer in the X-direction, but not changed in the Y-direction.
Accurate moves and rotations	Geometry that is created properly can be reoriented in a number of ways, for example, shifting or moving what has been drawn by a specific amount (say move all holes to the right by 10.523 mm). It can be much more accurate than following methods that only look right.
Automatic dimensions	Geometry that is created properly can have the CAD system determine the values of dimensions easily and automatically (even if a variety of drawing view scales are used in the drawing). In this case, after the design is drawn, the user can just pick the lines, edges, points, etc., and the system will figure out the proper distance between them and create a dimension that shows the appropriate value.

2-D CAD

5.4 PAPER SPACE AND MODEL SPACE

Given the many advantages of the "smart paper" concept, it becomes necessary to understand exactly how the CAD system is implementing the mathematical model. The first part of this understanding is to figure out how the CAD system uses *paper space* and/or *model space*. These "spaces" are like a grid or a field of possible data or numbers associated with the drawing geometry (lines, arcs, etc.). For example, a line is usually going to be defined by two sets of X and Y values (one set for the beginning of the line; the second set for the end of the line). The actual numbers used for these X and Y values are going to be dictated by the "space" implemented by the CAD system.

5.4.1 Paper Space

A CAD system that uses a paper space approach essentially "knows" the size of the drawing (refer to drawing sizes in the previous chapter). Furthermore, the drawing size can indicate the unit system (inches, millimeters, etc.) and the scale needed to enter the geometric data for the drawing. For example, if an E-size inch drawing is being used, then the basic paper space is going to go from 0 to 44 in. in the horizontal direction and from 0 to 34 in. in the vertical direction. So, the paper space approach not only indicates the boundaries of the drawings, but it also sets the units. If paper space is for an "inch size" (such as E-size just mentioned), then obviously the units are inches. If a millimeter size such as A0 is used, then obviously the drawing is going to use millimeters.

Once the user is forced to work within the given set of values based on the paper size in the paper space system, a CAD system of this type is probably also going to offer the option of working with viewports. As shown in Figure 5.1, viewports are a common computer graphics technique for segregating and manipulating different parts of the computer monitor. Viewports have distinct boundaries, and the graphics programming usually does not plot anything beyond the boundaries (this is known as "clipping"). Most importantly viewports have their own mathematically defined space with a minimum X- and Y-value and a maximum X- and Y-value. It turns out the concept of computer graphics viewports corresponds very well with the 2-D drawing concept of views (such as Front View, Right View, etc. presented in the previous chapter). So viewports are easily applied to become drawing views.

Once the paper space–type of CAD drawing has the viewports equated to drawing views, then the viewports can easily have a mathematical scale factor that corresponds to drawing view scale as well as its own origin. Now each of the views can have its own value of scale, and any size design can be worked on within the context of the size drawing selected. The user can be released from being to forced into just the paper's boundaries (such as the limits of 44 in. and 34 in. mentioned above for E-size). For example, if the object being designed is

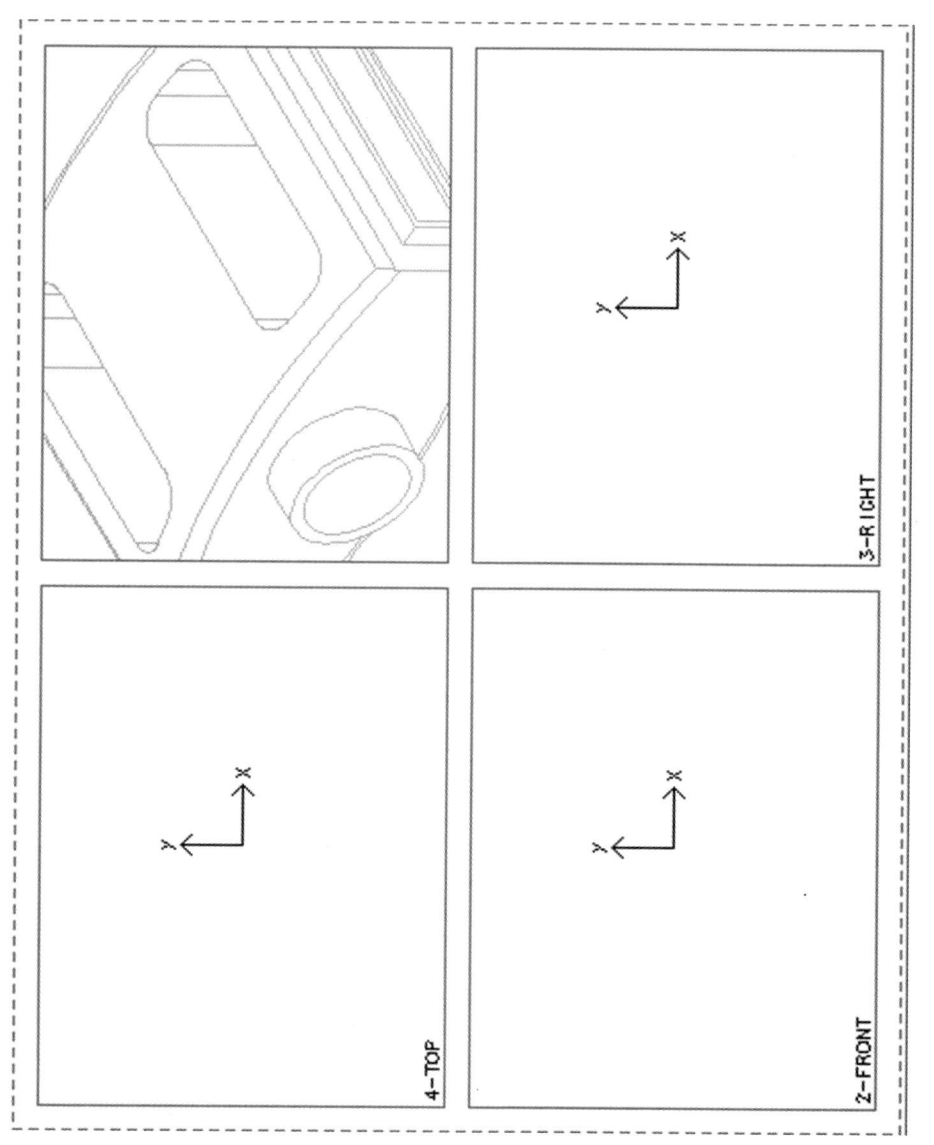

FIGURE 5.1 Viewports (solid border) and paper space (dashed border).

2-D CAD

very large, then view scales such as 1/32 can be used and as the user enters the length of line as 320 inches to the CAD system, the CAD system can automatically apply the 1/32 scale factor and put a line 10 inches long onto the CAD drawing. Furthermore, if the user measures the length of the line in the CAD drawing, the CAD system can automatically "undo" the effects of the view scale (thus indicating the length of the line just mentioned as the product design's "real" 320 inches), even though a hardcopy of the drawing would only have the line 10 inches long on the paper.

The main concept to keep in mind, then, for the paper space approach is that there is a predetermined relationship between the actual, physical object and the object's appearance on an actual size of drawing.

5.4.2 Model Space

Some CAD systems do not use the paper space approach. Instead, they use an approach that can be called *model space*. In the model space approach, the creation of geometry is not concerned with the drawing sizes while the drawing is being created. The CAD system simply considers all of the information to be drawn in an unbounded two-dimensional field or plane.

This type of CAD system needs to indicate the origin or the center of the model space (this is where the X- and Y-values are zero). Since there is no mathematical consideration for the drawing size, the coordinates of the geometric entities can stretch to any values, and without concern for the units being used (inches, millimeters, etc). For instance, the edge of the physical object may be 250 mm long. The CAD user can then create a line in model space that is 250 units long (regardless of inches or millimeters). At this point, there is no necessity to add a scale factor, since the drawing does not need to be shrunk to fit on the printed drawing. One can consider this approach as at "full scale" or "1 to 1." When the drawing is completed, and all the pertinent geometry has been shown in the model space, "analyzing" or measuring distances and other geometric properties is rather simple. It can all be done with respect to the single mathematical field.

Of course, the drawing also needs to be plotted. Now the model space must be converted to fit some paper space. This is easily done by overlaying or mapping the drawing size to the geometric entities in the drawing. This can involve scaling (shrinking or expanding) either drawing size or the geometric entities. Once the geometric entities fit into the border of the selected drawing size, then the drawing can be printed. Often the user has a drawing size in mind while creating the drawing (instead of waiting until the hardcopy is needed) so that it is not overly difficult to fit the information into the drawing size. To prevent difficulties with this, it is often possible to show the overlay of the drawing size border or format while the drawing is being created.

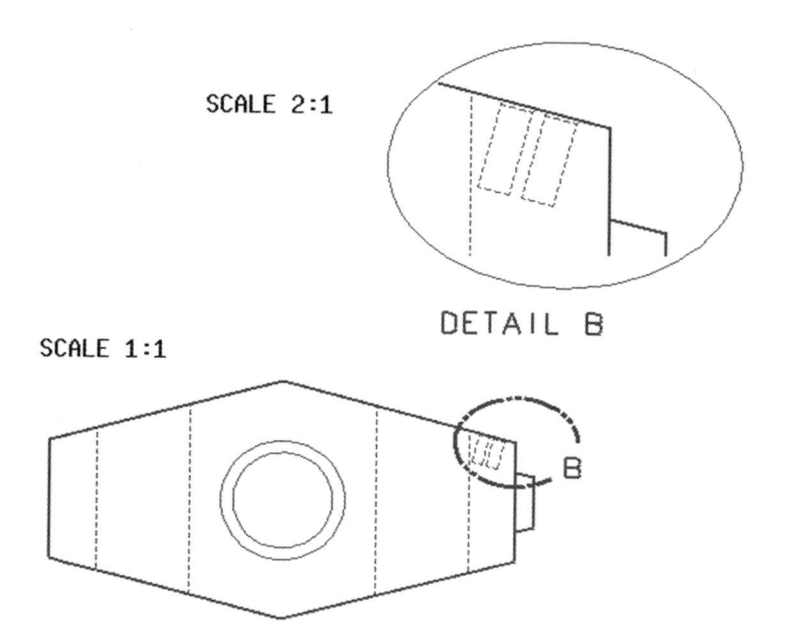

FIGURE 5.2 Drawing with views at different scale.

It is also possible to work with a scale in the model space approach. For some CAD systems, however, there will be just a single scale applied to the model space (as opposed to a different scale for each view shown on the drawing). Figure 5.2 shows a drawing where a Detail View B is showing a "close up" of a part of the object in the drawing. This Detail View B is at a different scale than the "main" view. This can be a problem for CAD systems using the model space approach without viewports. This is why many drawing "translations" from one CAD system to another shows parts of the drawing blown up or shrunk. One CAD system is probably using viewports and the paper space approach with a mixture of view scales, but the other CAD system can not handle this.

5.5 DIMENSIONS

Considering the concept of using a 2-D CAD system as "smart paper," where the CAD system is used to create an accurate but limited mathematical model of the object, the methods for creating dimensions becomes extremely important. It is the dimensions that bridge the mathematical model and useful information that needs to be shown to the eventual reader or customer of the drawing. Also, if the geometric entities in the drawing are created accurately enough, then the dimen-

2-D CAD

sions can be automatically created by the CAD system. This leads to a significant savings in time and effort.

Dimensions are usually created in the CAD system by selecting, picking, or clicking on the geometric entities that already exist in the drawing (the CAD system should have a method of highlighting or otherwise previewing what is being selected). In Figure 5.3, the center of the hole and the right edge were selected. Subsequently, the CAD system usually creates or previews the dimension for review by the user.

As simple as this dimension seems, there are a number of important issues to be concerned with. First, one needs to be aware of the dimensioning standard. The choices usually are the ANSI or ISO standard. The ANSI standard shows the dimension in the middle of the dimension lines; the ISO standard has the dimension above an unbroken dimension line. The example in Figure 5.3 shows the ISO standard dimension at the top. Usually the CAD system will have the user decide which to use. There could also be a system-wide default that appears normally, but then the user can override it.

Second, one needs to be aware of the number of decimal places shown in the drawing. This is the number of digits shown after the decimal point, and it implies a degree of manufacturing accuracy. As with the dimensioning standard, the user should be able to set a normal number of decimal places, and there could

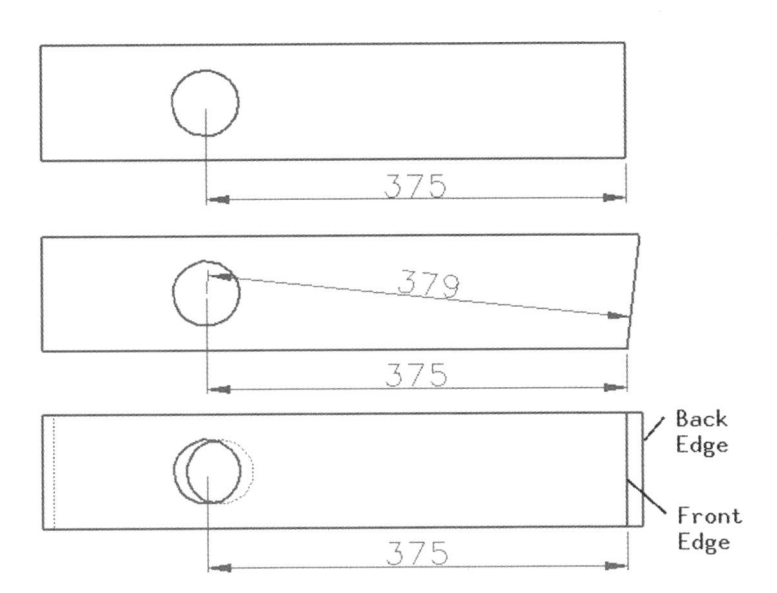

FIGURE 5.3 Creating a dimension with a typical CAD system.

also be a system-wide default. In Figure 5.3, the number of decimal places was set to 0 by the user since there are no digits shown after the decimal point.

Third, one needs to be aware of the round-off inherent in the CAD system. Once geometric entities (such as the hole and the edge in Figure 5.3) are created in the mathematical model sense, and the user picks the entities for dimensioning, the CAD system must then decide how to "round-off" to the number of decimal places selected. Particularly with inch drawings, this can be a significant change in value. For example, if the "exact" mathematical model for the object has a length of 3.875 in. (3–7/8 in.), and the user has selected 2 decimal places for the dimension on this length, then the CAD system is actually going to show 3.88 in. as the dimension. Of course, most readers of this dimension (at least readers accustomed to inch drawings) will figure that the designer really drew the object to 3.875 in. But it would be wrong to show the dimension as 3.875 in. instead of 3.88 in. on the drawing, since 3.875 would be 3 decimal places, and that would mean that a higher level of precision would be expected in manufacturing, and an unnecessarily expensive manufacturing process would have to be used.

Fourth, one needs to be aware of the application of any view scale in the dimension. If the mathematical model is accurate, and the scale has been specified for the view that the dimension is placed in, the CAD system should automatically account for the view scale. Unfortunately, not all geometric entities are accurate in the mathematical model. For instance, if a company needs to show something on a drawing that is actually designed by another company (such as a supplier), then a drawing translation may have occurred. In this case, the supplier's CAD system creates data that is "read into" the user's drawing. Since this geometry may lose some accuracy in the translation process, or worse yet there is a conflict between a paper space and a model space approach in the two CAD systems, a dimension's value created by letting the CAD system work with the mathematical model may be wrong. At this point, the user needs to either correct the mathematical model (by totally recreating the geometric entities) or create an "out of scale" dimension. The "out of scale" option would be telling the CAD system to ignore the mathematical model, and force it to show a desired value for the dimension. This is a rather dangerous option, and it should be avoided whenever possible. Although the originating user may understand that this has been done for a dimension, future users may not. Some CAD systems do make some appearance change in the dimension to warn users that this has been done (perhaps underlining or changing colors of the dimension).

Fifth, it is important to realize that there are geometric assumptions being made by the creation of most dimensions. The dimension shown in the top of Figure 5.3 obviously just shows the distance of the hole from the edge of the part. However, the reader of the drawing may assume that the dimension is along a line that is perpendicular to the edge. This is a good assumption, but what if that edge

2-D CAD

is not a perfectly vertical line in the mathematical model? Perhaps that edge leans to the right; now the dimension shown in the middle of Figure 5.3 may mean a perpendicular distance from the edge, or it might mean that the dimension is along a perfectly horizontal line (and not perpendicular to the edge). Some CAD systems will allow the user to make the distinction between the perpendicularity and the horizontal orientation, but the user needs to realize that further dimensions need to be created to carefully define how that edge is leaning.

Lastly, the CAD user needs to be aware that dimensions can have *associativity*, a term that was apparently invented by CAD software vendors to indicate that one drawing or 3-D model can have a relationship to other entities. In the case of 2-D CAD systems, associativity is likely to arise between a dimension and the geometry that is selected for the dimension (such as the edge and the hole in Figure 5.3). If this dimension is associative, then if the hole is moved to be closer to or farther from the edge, then the value shown in the dimension will automatically reflect that move. Often this level of associativity requires that the 2-D CAD data be based on a 3-D model of the object shown in the drawing. Further complicating matters, keep in mind that if a 3-D model is the basis for the drawing, selecting a line (such as the edge shown in the bottom of Figure 5.3) may be making more geometric assumptions. There may be an edge at the front and back of the part at the line's location and by selecting one or the other edge, different dimensions or associativity may result.

Considering all these caveats for creating a simple dimension between a hole and an line, it is clear that properly creating and documenting a design with 2-D CAD can still be demanding. However, if one spends the time needed to create accurate geometry in the mathematical modeling sense, then there should be no problem getting accurate and productive results when creating the dimensions. And, users should not consider getting a complete understanding of the 2-D CAD system as waste of time when moving to a 3-D CAD system. The 3-D CAD systems still require that concepts (such as dimensions in a 2-D plane) be used to create 3-D geometry.

Finally, dimensions in the 2-D CAD system may be able to take advantage of GD&T. Usually this takes the form of adding Feature Control Symbols (FCS) to dimensions. These added notations and symbols expand the meaning of the dimension by showing such things as datums (basically flat surfaces on the part that dimensions are expected to start from), tolerances and tolerance zones (how accurate a dimension must be in a more accurate sense than the decimal places mentioned earlier), and more. If a dimension is drawn to a particular hole, and the hole needs to be more accurate in the horizontal direction as opposed to the vertical direction, then the user needs to indicate that by having the CAD system create the proper symbols. Note that the concept of associativity also needs to be applied to this system (in both the 2-D CAD only sense as well as the 3-D model to 2-D drawing sense).

5.6 VIEWS AND VIEWPORTS

As already mentioned, views are a very important concept in drawings. They provide a standardized means of interpreting the three-dimensional character of an object through two-dimensional information. Furthermore, some CAD systems use viewports to handle these views. This section covers some important implications of views for the CAD system. Some of these concepts will not be relevant if the viewports are not used.

5.6.1 Viewport Clipping

Probably the most obvious ramification of the use of viewports in a CAD system is the idea of clipping. Clipping means that even though the user has selected mathematical values for geometric entities (such as lines or arcs) that can fit on the drawing's overall size, the entities (or parts of them) are not shown. They are cut off or clipped at the border of the viewport. Figure 5.1 shows some geometry that is clipped. The solid lines forming rectangles are the borders or edges of the views.

Although clipping may seem detrimental, it can be of value if used properly. Clipping can assure that the different views are properly "contained" and segregated (which is usually how the drawing needs to appear anyway). If a Detail View is created (a view that "zooms in" on a small area to show greater detail), then the user can just select a segment of a larger view and the CAD system will automatically clip all the unneeded geometry. Figure 5.2 shows this situation.

5.6.2 Viewing Angles

If the CAD system can create projections, then the views need to have defined viewing angles. These angles specify the precise direction at which someone is looking at the object shown by the drawing. The angles indicate the direction by three rotations (about the X, Y, and Z axis), or the components of a 3-D vector. The angles are standard values for the standard views (such as Front View is 0,0,0 or no rotation; Top View is 90,0,0 or rotate 90 degrees about the X axis to see the top, and Right View is 0,-90,0 or rotate 90 degrees about the Y-axis to see the right side of the object; the Isometric View would be more general such as 45,-35,-30).

If an arbitrary view is created for a drawing (one that looks at the object in some oblique or ad hoc direction), then viewing angles may be entered for these views in some CAD systems. Of course, the viewing angle is important for viewing faces of objects that are "true." In this case, if the viewing angle is true, then dimensions created automatically will be correct. If the viewing angles are not true, and do not view a face directly (or along a vector normal to the surface),

2-D CAD

then dimensions created by the CAD system will be skewed and the values may not be what is desired.

5.6.3 View Origins

In addition to specifying the boundaries, scale, and viewing angles of views, users need to be careful to specify and manage the origin of views. An origin is the location where X is 0, and Y is 0. For a CAD system that does not use viewports, then there is probably just one origin for the entire drawing, so there would probably be little problem in managing it. When there are viewports, then there can be an origin for each view, and more care needs to be taken with them.

It is important to keep the viewport origins aligned. In Figure 5.1, these origins are indicated by the X-Y symbols. Notice that they are in alignment vertically and horizontally. This alignment is best understood with respect to the real object being shown in the drawing. The real object in 3-D space has only one origin where X and Y and Z are all zero (refer to Figure 5.4 where the origin is shown as an X-Y-Z coordinate system). When one views the object from the Front direction (along the Z axis) and the Right direction (along the X axis), one can see that origin for the Y axis (where Y is zero) is going to be at the same height in both directions. Therefore, the 2-D drawing views should show the Y direction starting from the same location (i.e. the 2-D view origins should be

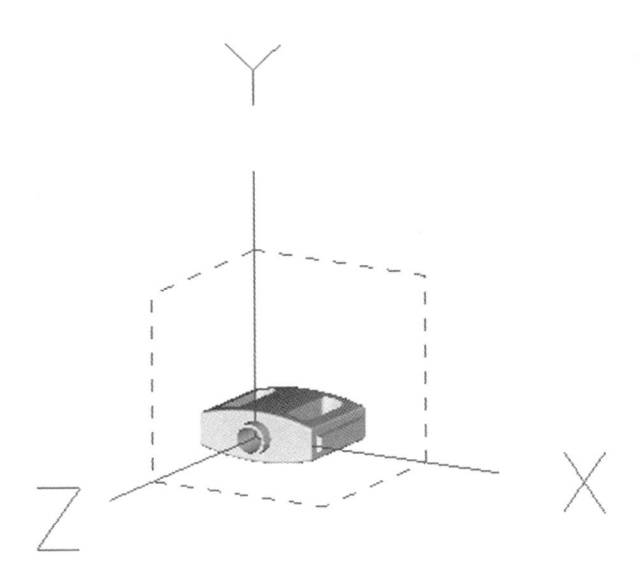

FIGURE 5.4 An object's real 3-D "origin."

aligned in that direction); this is done by making sure that the origin in both the Front and Right views are at the same vertical location (refer to Figure 5.1).

5.6.4 Viewport Calculations

It is important to note that the careful specification (alignment, scale, etc.) allows one to calculate the exact location of a point with respect to the entire drawing based on the characteristics of the view. Calculating this normalized location involves a transformation or mapping. It transforms local view coordinates to the global coordinates for the entire drawing. It is often expressed in terms of a position vector such as **R**. The calculation looks like the following in vector form:

$$\mathbf{R}_{global(X,Y)} = \mathbf{R}_{origin(X,Y)} + scale * \mathbf{R}_{local(x,y)}$$

where:

$\mathbf{R}_{global(X,Y)}$ is a vector in global X,Y coordinates (such as model space) that points from the global origin to the desired point within the viewport. Its components would be X_{global} and Y_{global}.

$\mathbf{R}_{origin(X,Y)}$ is a vector in global X,Y coordinates that points from the global origin to the origin of the viewport (i.e., local origin). Its components would be X_{origin} and Y_{origin}.

$\mathbf{R}_{local(x,y)}$ is a vector in local x,y coordinates that points from the local origin to the desired point within the viewport. Its components would be x_{local} and y_{local}.

And, scale is the scale factor for the viewport (such as 0.5 for half scale).

When vectors are added, their separate components can be added in "scalar" form. Thus, the calculation for the global X and Y for a point in a viewport with a scale factor becomes the following:

$$X_{global} = X_{origin} + scale * x_{local}$$

$$Y_{global} = Y_{origin} + scale * y_{local}$$

Understanding this transformation calculation can be important if the CAD system allows geometric entities to be created in either the local or the global coordinates. In this case, 2 points located in 2 different views (even if they use different scales) can be related to each other accurately (by using the global coordinates).

5.7 LAYERS

Another method of organizing the graphical data displayed on a CAD system is through the use of layers. Although different CAD systems will use layers in dif-

2-D CAD

ferent ways, this feature primarily allows for the selective viewing, editing, and management of the entities in the drawing. For instance, a CAD drawing may have a layer for dimensions, a layer for hidden lines, a layer for notes, etc. The layers can be thought of as clear paper that can be overlaid onto the geometric entities.

Layers can be used in a variety of ways to improve working with the drawing. If a drawing is too cluttered, and the user needs to no longer view data temporarily (such as all notes or dimensions), then the user can hide these items by turning off or hiding the entire layer. Or, some attributes of the data on the layer may need to be changed. For instance, if all the lines on a layer are for holes drilled into a part, and the user wants to change the color of the circles for the holes all at once, then the user should be able to select the layer and then make the attribute change for them all.

In some cases, the user is responsible for making sure that the particular types of items are placed on independent layers as a company standard. If so, users should determine if this is the case before working on a drawing. A company may also need to establish a standard naming convention for the layers since the CAD system may allow user defined names or numbers for them.

Some CAD systems may have the layer capability, but layers may not be the only means of filtering or selectively viewing or editing the different entity types. These systems may offer entity filtering instead. In this case, all the dimensions or notes in a drawing could simply be selected automatically by a single command. In this case, the CAD system actually "knows" the different entity types in the drawings data structure instead of relying on the user to change layers as they create the different types of entities for the drawing.

If the entity type filtering option is available, then the layers could be used for other functionality. For instance, if the designer of a component was exploring different possible configurations of the design, then the layers could be used to contain different geometry for the different designs. Then, by managing the layers, the different options or configurations could continue to be refined until the final design is achieved. The layout type of drawing (discussed in the previous chapter), in particular, can take advantage of this approach.

There may be many more applications for layers within the context of a particular CAD system or design task. The user will need to refer to the CAD system documentation or experts for more detailed information. The intention here was simply to explain what layers are and what they can accomplish in general.

5.8 VECTOR GRAPHICS

An important ramification of the "smart paper" analogy for CAD systems is that the CAD drawing (particularly the geometric entities such as lines and arcs) is independent of the computer devices used. In other words, the drawing is correct regardless of the size or resolution of the computer monitor or the hardcopy device used to print the drawing. This is a result of the fact that a mathematical model has been created for the drawing.

In the mathematical model, a line has a beginning X and Y value and an ending X and Y value. These 4 numbers represent the mathematical model for the line. Of course, in order to see this line on the computer monitor, the mathematical model must be "mapped" or converted to the screen. This mapping process is automatically performed by the CAD system and the computer hardware. It takes the X and Y values the user entered and converts them into specific instructions for the computer to change the color of specific pixels on the screen to show the line. Of course, this shown line is an approximation of the mathematical model, and "better" graphics hardware will show the line more clearly and precisely, but the exact numbers (the X, and Y in the mathematical model) are not affected by the computer hardware.

This important concept of model vs. computer graphics for the CAD system can be expressed by the term *vector graphics*. A vector is a specific mathematical concept that is helpful in managing graphics and different viewports (such as the equation shown earlier for finding the relationship between the local and global X- and Y-values). Any CAD system (or any program actually useful for design work) must be based on vector graphics.

The alternative (which could not really be called a CAD system) would be to use *bitmap graphics*. In this case, there is no mathematical model. The software (such as a "paint" program) would not always intelligently recognize any particular geometric entities. What is seen on the screen is just a collection of dots.

Appreciating the difference between vector graphics and bitmaps is a major benefit for CAD users and CAD administrators alike. For instance, one can imagine that it is a simple matter to convert vector graphics to a bitmap. Indeed, the mapping process to display the drawing on the computer monitor does just that. The screen has a finite number of pixels, so what one sees on the screen is a bitmap itself. But, it is extremely difficult to convert a bitmap to vector graphics (particularly for drawings). How would a computer program tell the difference between a set of vertical lines in a drawing (some being the object and others being dimensions), or how would it tell the difference between a string of number ones, and a string of lower case letter Ls? This process is known as "raster-to-vector" conversion, and it is rarely extremely successful with drawings. It is similar to object character recognition (OCR) for documents, but at least in a document, the OCR program only needs to find numbers, letters, etc. In a drawing, there is a much wider variety of content to figure out.

2-D CAD

Another instance of appreciating the difference between vector graphics and bitmaps is that one can not really import or insert many popular types of files into CAD systems. Files such as GIF, TIFF, and JPEG are bitmap files. Therefore, they have no mathematical model that can be readily used in the CAD system. The data in these type of files can be projected or overlaid onto the screen, but they cannot have any useful connection to the mathematical model. One example of this would be a "decal" shown in a CAD drawing. The decal can be moved and resized (as a bitmap), but one would not be able to connect a line from what is shown in the decal to the rest of the normal geometric entities in the CAD drawing. One solution to this is for software to allow a user to trace the geometry from the bitmap. In this case, the user can manually create a mathematical definition for the geometry. The results of this tracing process can be acceptable for design purposes, but there will be a loss of some accuracy versus the original design's mathematical model (the image can only be as good as the bitmap's finite resolution).

5.9 GEOMETRIC ENTITIES

Naturally an important feature of the CAD drawing is the geometric entities that represent the object being documented by the drawing. There are a variety of geometric entities that are usually provided for by the CAD system. Each of them will eventually be useful to an experienced CAD user. This section will give a very brief overview of these entity types as well as some discussion on how geometry is recognized and/or selected in the CAD system.

5.9.1 Snapping

One of the most common assisting techniques in CAD systems is *snapping*. Snapping allows the user to make certain that only specific points on the screen will be recognized for selection. These points form a grid on the screen (refer to Figure 5.5). Usually the user can indicate or customize the interval for the grid.

Snapping is one way to ensure that various elements or geometry are aligned to each or with respect to an origin or other X,Y position that is meaningful to the design. However, the grid can also be a hindrance. If the grid interval is set to 1 mm, and a geometric entity needs to be located at a fractional location of a millimeter (such as 1.5 mm), then there will be a problem. The snapping is going to only allow the selection of whole number millimeters. This technique is probably called snapping since the mouse is only going to recognize the grid points, previewed entities appear to jump around or "snap" from grid point to grid point.

FIGURE 5.5 An example of snapping grid.

5.9.2 Geometric Relationship Recognition and Utilization

Some CAD systems are not restricted to snapping to the imaginary grid locations. These systems can recognize relationships in the existing geometric entities. For instance, if two existing lines cross at a point, then a new line can be drawn such that it starts at this intersection point. This would be the recognition of an intersection. Some other possibilities include starting point of a line, ending point of a line, coincidence with a line, the center of circles, the center of arcs and fillets, tangency to circular entities, midpoints of lines, etc.

Beyond just recognizing the geometric relationships for particular points in the model or paper space, some CAD systems may also permit the relationships to be utilized in the creation of new entities. For instance, after allowing a specific point on a line to be selected for the starting point, the CAD system may then recognize a relationship such as perpendicular or parallel or tangent to that original line or other surrounding geometric entities. Figure 5.6 shows a case

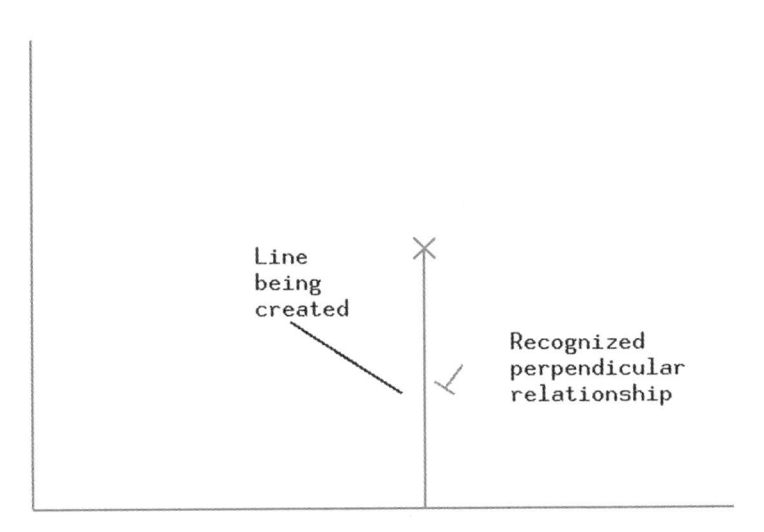

FIGURE 5.6 An example of recognition of a geometric relationship.

where the previewed direction of a line about to be drawn is at a perpendicular relationship to another line. Notice that when this condition is recognized the CAD system needs to let the user know that this has happened (such as showing the perpendicularity symbol in Figure 5.6). This allows the user to judge whether this is an intended condition.

5.9.3 Odometer

Another feature of some CAD systems is a dynamic indicator of the current position of the mouse (or other pointing device) with respect to the model space or viewport coordinates. This is often called an odometer since it shows how far the mouse has moved from the origin. As the mouse moves, the numbers in the odometer part of the screen are constantly changing. This is helpful for deciding where to draw the geometry (particularly in terms of positive or negative coordinates); however, this technique probably should not be relied upon to get precise geometry. If a point needs to be located at a specific X, Y coordinate, then these numbers should be entered directly from keyboard (using whatever input data method is used by the particular CAD system). Refer to Figure 5.7.

5.9.4 Points

Points are probably the simplest geometric entity to create in the CAD system. All that is required are the X- and Y-positions of the point. Of course, points are

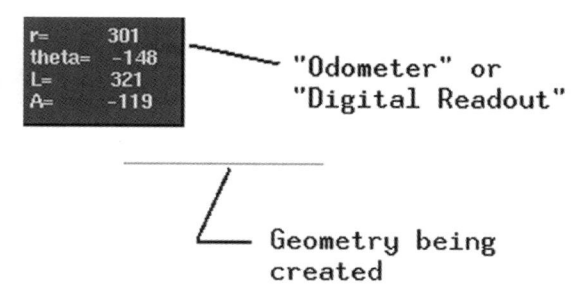

```
r=       301
theta=  -148
L=       321
A=      -119
```
"Odometer" or
"Digital Readout"

Geometry being created

FIGURE 5.7 An example of an odometer.

of little use in creating geometry for the object of the drawing (which really only arise from the edges and surfaces of the object or part). Points are typically used to assist with the overall design and drawing procedure.

For example, if a designer is designing a device such as a motor casing, the center of rotation of the motor's rotating assembly is of vital importance. However, this imaginary location is not actually part of the casing; the point is in the middle of the air for the physical object. In order to help with the design, then, the CAD user may create a point in the drawing to represent this location. It may then be easier to measure from or analyze the motor casing. If nothing else, it will help as a visual reference to the location. Refer to Table 5.2.

5.9.5 Lines

Lines are probably the most prevalent and important geometric entity in the CAD system. They are used for many functions. Of course, they are used for the object of the drawing. These object or "geometry" lines will be for straight edges of the

TABLE 5.2 Point Creation Information

Geometric entity	Point
Geometry defined by	1 X, Y coordinate in paper space or model space
Possible methods for creation	Enter the X and Y values directly
in the CAD system	Snap to existing geometry with recognition of: intersections, midpoints, centerpoints, etc.

2-D CAD

TABLE 5.3 Line Segment (i.e. a "line" in the drawing) Creation Information

Geometric entity	Line segment
Geometry defined by	2 X,Y coordinates (start and end)
	1 X,Y coordinate with a direction angle, heading, or "slope" and length
	Y = mX + B (m = slope, B is Y-intercept; infinite line)
Possible methods for creation in the CAD system	Enter the X and Y values directly for start and end points.
	Snap to existing geometry for the points with recognition of: intersections, midpoints, centerpoints, etc.
	Snap to or enter X and Y for the start point, then enter an angle or slope value or indicate a direction for the line with recognition of parallelism or perpendicularity from existing geometry.
	Polyline: snap the X and Y for the start point of a line segment and then continuously snap or enter X and Y values for a "chain" of line segments (each new segment begins where the previous one ends).

part, and for locations where a curved surface meets a flat surface ("seams"). Lines are also going to be used for centerlines (the imaginary central axis for cylindrical features such as holes), and in layout drawings, they will be used to create "construction" geometry that helps the design process.

Generally in the context of a CAD system, a line is straight, and it is actually a line segment. In geometric terminology, a line is infinitely long, and what one sees is just a segment of that infinite span. This infinite character is the result of the mathematics that describe lines. In the context of 2-D CAD, it can be assumed that these lines are going to just lie in a plane (such as the paper space mentioned earlier). Refer to Table 5.3.

5.9.6 Circular Entities

The next geometric entity is circular features. This would include circles for holes, shafts, etc. It would also include arcs (parts of a circle) and the special arcs called fillets.

There are a few special considerations to keep in mind for the circular entities when using a CAD system. First, there is likely to be quite a variety of commands or methods for creating these types of entities. Naturally, a circle can be created by selecting the center of the circle and then specifying the radius. But it can also be created by selecting the center and one point on the edge of the circle,

TABLE 5.4 Circular Feature Creation Information

Geometric entity	Circular
Geometry defined by	X,Y coordinate for center (X_c, Y_c) and a radius (R) $(X - X_c)^2 + (Y - Y_c)^2 = R^2$
Possible methods for creation in the CAD system	Enter the X and Y values directly for the center and a radius (or diameter)
	Snap to existing geometry for the center point with recognition of: intersections, midpoints, etc. and also snap to a point on the edge of the circle.
	Snap to existing geometry for 3 points on the edge of the circle.
	Create fillet geometry by selecting 2 existing lines (not necessarily perpendicular to each other), entering a radius, and then selecting from among ambiguous choices.
	Select 3 existing lines that have tangency with the desired circle.

specifying 3 points on the edge of the circle, specifying 2 points on the edge and then a radius, etc. Secondly, some circular feature methods have redundant choices for the geometry. Although the 2 points on the edge and a radius specifies a size of circle, there are actually two possible circles in two different locations (see Figure 5.8) for this method. When this situation arises, the CAD system should indicate this redundancy and permit the user to make a selection of the desired circle. Refer to Table 5.4.

Finally, there are issues that arise with respect to tangency. The CAD system will generally offer the capability to create circular entities that are tangent to lines (a circle's edge is tangent with a line when the circle and the line "just touch" and have a single point or X,Y value that is part of both entities). Creating the circular feature with tangency is especially important for fillets. Fillets represent a blending of two surfaces of the object or part shown in the drawing. Since the fillet blends with two different surfaces, it is supposed to be tangent to two lines simultaneously. Fortunately, the CAD system can quickly calculate the fillet to meet this geometric constraint; however, as mentioned earlier, there may be redundancies. Indeed, two lines (mathematically) have 4 possible fillets, but usually only one is what is desired. Refer to Figure 5.8.

5.9.7 Ellipses

Another fairly common geometric entity is the ellipse. The ellipse is basically oval-shaped. It is usually described by a major axis and a minor axis. As the

2-D CAD

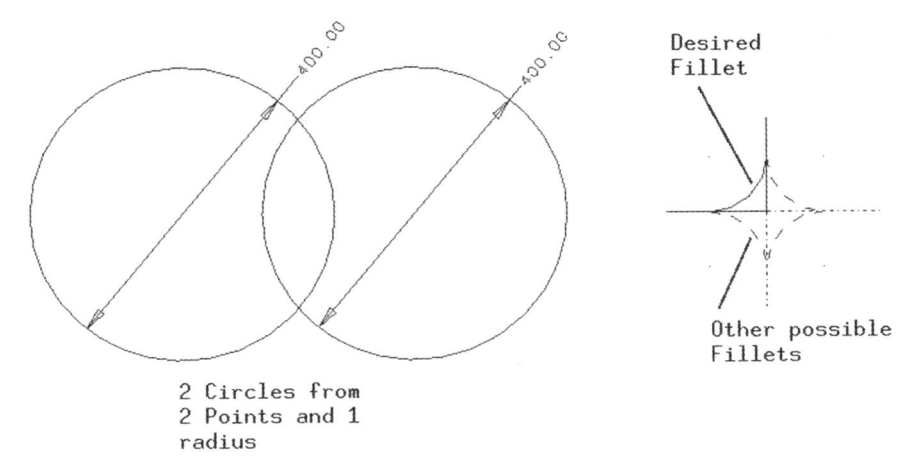

FIGURE 5.8 Examples of redundant geometry creation, tangency, and fillets.

names imply, the major axis is the longer or stretched distance, and the minor axis is the shorter or compressed distance. Refer to Table 5.5.

Ellipses are most often created in CAD drawings for cases where circular features of the object or part are seen at an angle in a given view. Commonly these would include shafts, holes, fittings, etc. Manufacturing a section of a part to be elliptical might be rather unusual, but the CAD system will handle this quite

TABLE 5.5 Elliptical Feature Creation Information

Geometric entity	Elliptical
Geometry defined by	X,Y coordinate for center (X_c, Y_c), direction and length (a) of X-axis direction of ellipse, and length (b) of Y-axis direction of ellipse $(X - X_c)^2/a^2 + (Y - Y_c)^2/b^2 = 1$
Possible methods for creation in the CAD system	Enter the X- and Y-values directly for the center and axis lengths. Snap to existing geometry for the center point with recognition of: intersections, midpoints, etc. and select a length for the major and minor axis. Snap to existing geometry for points that lie at the corner of a rectangle that would bound or contain the ellipse.

easily. The center and major and minor axes can be entered, and the CAD system will create the geometry.

The ellipse is a conic section. That is, the mathematics of the shape is derived from intersecting a cone with a plane at a specific range of angles. Some CAD systems contain commands for creating other conic sections such as a parabola's and a hyperbola's. These may be useful geometric entities in some cases.

5.9.8 Splines

The last geometric entity presented here for a typical CAD system is the spline. The spline is modeled after a bendable strip of plastic material that was weighted down at specific points on a table. Then a free-form curve was drawn along the strip. The spline is able to meander through an arbitrary set of points and create a smooth curve between them. Splines would be used for creating geometry in the drawing for such things as hoses, pipes, wires, or free-form surface items such as body panels. Splines are also interesting because they can crossover themselves. Refer to Figure 5.9 for examples.

The spline is usually created in the CAD system by having the user click on or select specific points that lie on the spline that needs to be created. For instance, the user would select clamping locations where a hose must pass through. Then the CAD system would create the actual spline curve that passes through all those points. However, the spline's mathematical formulation is actually based on something known as a Bezier Curve, and this curve uses control points. These are points that actually lie some distance off the final spline between the start and end

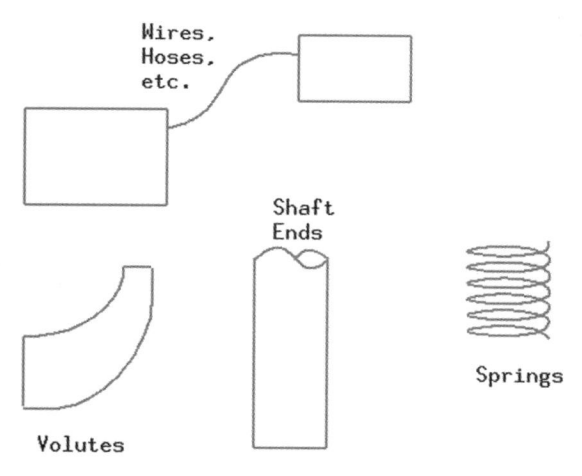

FIGURE 5.9 Examples of splines in a CAD drawing.

2-D CAD

TABLE 5.6 Spline Creation Information

Geometric entity	Spline (or cubic spline)
Geometry defined by	X,Y coordinates for at least 4 points. On the spline or "control" points depending on the system.
	Bezier-curve formulation (control points approach):
	$x(t) = a_x t^3 + b_x t^2 + c_x t + x_0$
	$y(t) = a_y t^3 + b_y t^2 + c_y t + y_0$
	t is the distance along the spline
	$a_x, b_x, c_x, a_y, b_y, c_y$ are solved for based on the 4 points.
Possible methods for creation in the CAD system	Enter the X- and Y-values directly for the through points and/or control points.
	Snap to existing geometry for the points.

of the spline. They indicate an influence or a direction that the spline must head toward, but the spline does not touch them. Some CAD systems will allow these control points to be used to create and/or modify splines. Also, due to the mathematics of the spline, there must be at least 4 points indicated for the spline to be created (either points on the spline or control points).

5.10 SPECIAL TECHNIQUES FOR 2-D CAD

Although the fundamentals of creating mechanical drawings explains much of the functioning of a 2-D CAD system, there are many procedures, techniques, or short cuts that the CAD system can provide that have no real analogy with the manual drawing process. These special techniques are essential for making optimal use of the CAD system.

5.10.1 Groupings

Probably the most common special technique in 2-D CAD systems is the logical grouping of geometric entities. These groupings can then be manipulated as if they were a single entity. These groupings may be called blocks, symbols, clumps, or detail page items. A typical example of one of these groupings would be an engine shown in a vehicle layout drawing. One advantage of this grouping is that the engine can be moved to different locations on the drawing as a unit. This is far more productive than having to select and/or move all the individual entities in the engine (and their might be hundreds of lines, arcs, and splines). This grouping can also be rotated as a unit to help position it within the 2-D design. Other common examples of these groupings would be drawing formats, standard or catalog items such as nuts and bolts, and schematic symbols (hydraulic, electrical, etc.).

These groupings offer many advantages for users, but usually CAD users need to first understand the concept of instance or instancing (for groupings as well as even 3-D CAD techniques). An instance of something in a CAD system is merely an intelligent copy of something that the software can track (if desired). The instance is basically a copy made from some master item. If the master item changes, then the CAD system can have all the instances change or update as well. In many cases, the use of groupings is an exercise in using instances. The CAD user may create a grouping and make it a master. Then the CAD user can create many copies of that master so that they are all the same. This would be a common activity for groupings used for standard or catalog items such as nuts and bolts. Each copy of the grouping could be called an instance of the master.

What happens if the user no longer wants to treat the grouping as a unit? In this case, the CAD system should provide a procedure or command to disassemble or break or smash an instance. After this is done, the individual geometry entities in the grouping can then be manipulated; the CAD system will no longer consider it an instance of the master. If the instance needs to changed, but not broken, then the CAD system should also provide a procedure to edit or modify the grouping. Also, there may be an option to create instances of groupings that are independent and not affected by future changes to the master geometry.

Some CAD systems will permit the groupings to be nested. A grouping is nested if it is part of yet another grouping. Considering the engine example again, there might be parts of the engine that, in turn, might be helpful to consider a grouping as well (such as the cooling fan in front of the engine). If the CAD system allows the engine grouping to use subgroupings such as the cooling fan, then that system permits nesting. If the engine symbol is smashed, it would then expose its subgroupings. Then these subgroupings could again be smashed to eventually get back to the individual geometric entities (such as the lines, arcs, splines, etc.).

Another important issue with respect to groupings concerns dimensions that may be related to a grouping. Specifically, when groupings are scaled (compressed or expanded to be smaller or larger), should the dimensions be scaled as well? This usually depends on why the user is scaling the grouping. If the grouping is being scaled because the physical object is changing size, then usually the user wants the dimension to show a new value. If the "smart paper" approach has been followed properly, and dimensions are associated with the geometry in the drawing properly, then when a grouping changes size, then the dimensions will automatically change to reflect the new size of geometry. However, putting the actual dimension in the grouping should be avoided since the grouping will have a scale factor, and the view the grouping is in may also have a scale; now one has a scale of a scale, and this can become difficult to manage.

In another situation, a grouping may be scaled just because the geometric entities do not fit on the selected paper size. In this case, the physical object rep-

2-D CAD

resented in the drawing is not actually changing size. The value shown by the dimensions should not change. The height of the letters and numbers in the dimension shown in the grouping may also need to be scaled to fit the new paper size, as well.

Another case of scaling a grouping would be to change unit systems. Perhaps geometric entities and a grouping were originally created using the inch system (obviously in a CAD system that uses the "paper space" system mentioned earlier). Now, when an instance of this grouping is created in a drawing using mm, the instance will be much too small. Instead of something being 10 inches long, it will show up as only 10 mm long. To correct for this, the grouping can be scaled by 25.4 either in a separate operation or as the grouping instance is actually being created. In this case, the geometric entities in the grouping need to change by the factor 25.4, and the values shown in the dimensions should also be scaled (i.e. the number 10 in a dimension meaning 10 inches should change to 254 for 254 mm).

Finally, CAD systems may allow for groupings to be imported and/or exported. Many computer software systems use these terms when information is taken from one system to another. In this case, a grouping may be created in a CAD drawing, but it may be useful in other CAD drawings. In order to get this grouping to the other drawing, it may need to be exported using the CAD system's data management functions. Once the grouping is exported, it can then be imported into the other CAD drawing. A set of standard groupings may also be inherited as part of a template or prototype drawing that is referenced each time a new drawing is created. Or, a entire customized library or catalog may be created across the entire network. This functionality will often be accompanied by a set of administrative functions that will allow someone to manage the library and control revisions to the groupings included in the library.

5.10.2 Projections

The next special technique for 2-D CAD systems is called projections or projecting or 3-Viewing. This technique has to do with creating geometric entities in different Views based on an originating view. For example, in Figure 5.10 there is a Front View, a Top View, and a Right View that shows an object from different viewing angles. Note that the Front View shows a square hole that has been added to the object. This hole is shown in the Top View, but now is being added to the Right and Isometric Views as hidden (using the hidden font which is a short dashed line).

The projecting special technique assists with creating the lines in the other views by the CAD system analyzing one view and using the known viewing angles (how the object is supposed to be viewed in 3-D space) for other views. Looking at Figure 5.10, since the Right View is looking at the object from the

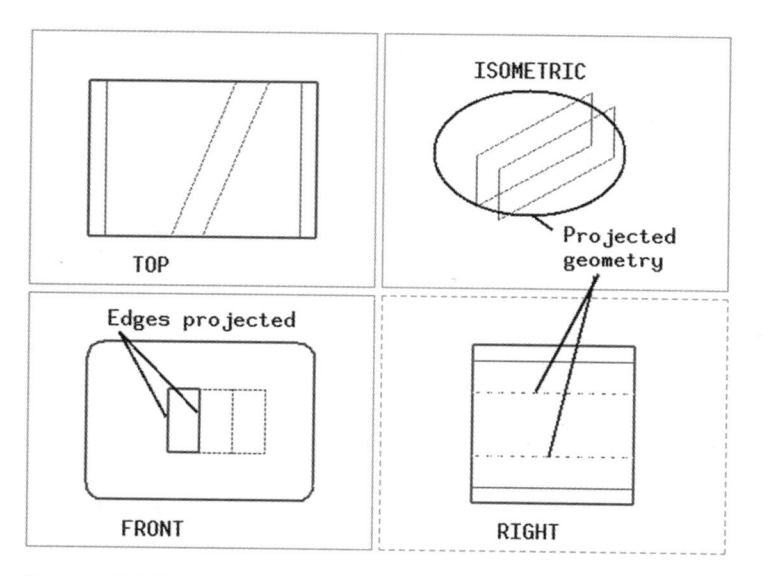

FIGURE 5.10 A "3-View" Drawing with a new feature in the Front View being "projected" to another view.

right side of the object (shown to the right of Front View using the U.S. standard Third Angle projection system), then the location of the edges for the hole can be properly located in the Right View using the CAD system's projection capability. The Front View does not indicate the depth of the hole, so the Top View had to be used to tell the system where the hole starts and stops (a beginning and ending projection plane). Sometimes projections are done more simply as infinite lines. In this case, the projected geometry would be based on a point in the Front View, and then a line that completely crosses the other view or views would be shown. The CAD user may then "relimit" the line in the Right View manually to the proper depth.

Keep in mind that the projection special technique may require that the CAD system creates and manipulates views. Also, the CAD user will need to make sure that the "smart paper" approach to creating entities is being followed. The origins (where X=0 and Y=0 for the views) will probably have to be carefully defined for each view so that there is a consistent calculation between the views. The user also needs to consider the impact of different view scales between the views (although the 3 most standard views—Front, Top, Right—would be expected to be at the same scale).

5.10.3 User-Defined Options

Most CAD systems have a number of special techniques or capabilities related to user-defined options. These are options that allow the user to enhance, customize, or even override the CAD system's capabilities. Three of these options are mentioned here; some CAD systems will offer others. Also, in the next chapter more information on CAD system customization is presented.

Most CAD systems will offer the ability to create user-defined colors, fonts, and line weights. This allows the user to create a drawing that reflects specific company policies or just makes a more complicated drawing easier to create and interpret. In terms of colors, the user will probably be given a chance to set RGB levels or scales. R stands for red, G stands for green, and B stands for blue (unlike pigments that use red, yellow, and blue as the primary colors; light sources use red, green, and blue for primary colors). The scale for each of these primary colors may go from 0 to 100 percent, or they may go from binary-based values such as 0 to 15, or 0 to 255. Depending on the amount of each, a new color can be created (most CAD systems should offer up to 255 × 255 × 255 or about 16 million possible colors). Colors can be useful in visualizing information in a drawing since different items can be color-coded (for instance, different details in an assembly drawing can be different colors). Also, different entities such as dimensions, notes, balloons, etc. could be color-specific. Again, this could be quite helpful in reading a drawing (at least if the electronic CAD drawing is viewed). Obviously, one problem with using colors is that they generally will not be seen by the reader or customer of the drawing. Most drawings are either printed or scanned in a black-and-white format. Even if it is scanned with a color-enabled file, most often a drawing will eventually be copied or printed in hardcopy form; in this case the colors will be lost. Therefore, it is common practice to make sure that a drawing is completely readable in the black-and-white format.

The next option that might be user-defined is a font. This is not necessarily as useful to the CAD user as colors, but they may help with improved hardcopy quality. For instance, if the standard or default appearance of the hidden line font (should be short dashes) is not clear enough (perhaps the dashes are too close together), then a better hidden line font could be created by the user. Fonts are usually defined by a bit pattern. For instance, the CAD system may allow a repeating pattern using 16 ones and zeros. In this type of system, the hidden line font might be represented by 1111000011110000. The system may also require that this be specified in hexadecimal. The example bit pattern would be F0F0 in hexadecimal. Keep in mind that the pattern is repeated over and over in a line, so one needs to use a pattern that is continuing or symmetrical as desired.

The next option that might be user-defined is line weight. This also may not be necessarily useful to the CAD user as a design is being created, but may be used to enhance the appearance of the hardcopy. The only information that needs to be supplied is the how thick or "heavy" a line needs to be when printed. This thickness will probably be specified by a measurement in inches or mm. At one time, line weights may have been controlled by selecting different pens that a pen plotter could use in making the lines. Some CAD systems may still refer to pen mapping where different line weights (such as thin, medium, and thick) are given a "virtual pen" number (such as pen 1, 2, and 3). This gives a CAD administrator the opportunity to actually define the line weight by adjusting the printer (using commands that the printer receives independently of the CAD system). In any case, it is important that line weights make the drawings as readable as possible, and it may be necessary to create a user-defined line weight to accomplish this.

5.10.4 Hiding and Showing

The next special technique can be referred to as hiding and showing. This is the CAD system's ability to temporarily remove parts of the drawing, and then restore them again at a later time. For example, if changes to a design are being contemplated, but the previous design iteration still needs to be preserved, then the old design information (geometric entities, dimensions, notes, etc.) can be hidden. The user will usually just select the items to be hidden and then select the command that hides the data.

The hidden data should be retained by the CAD drawing, even though it will not appear on the monitor or in the hardcopy. Another example would be having too much information shown on the drawing as the user is trying to design a specific component or part. In order to see certain geometry more clearly, less important entities in the drawing might be hidden temporarily. When the user wishes to restore the information, the CAD system should have a method for indicating what has been hidden already and then selecting the entities or information that needs to be restored. Having the information re-appear (i.e., showing) can also be considered an undo function for hiding.

One issue to keep in mind with hiding entities is that if the drawing is exported to another CAD system (through a neutral file such as DXF or IGES), the neutral file may or may not handle this special technique. Even if the neutral file is able to store the fact that entities are hidden, the other CAD system may not handle it anyway. If there is uncertainty of whether this will cause problems, then most likely the hidden entities should be removed before the export or translation is performed. This may be referred to as a cleanup procedure for the drawing.

5.10.5 Selecting Techniques

Another important advantage of a CAD system is its ability to automatically recognize and alter a variety of items in the CAD drawing. Examples of these types of items would be geometric entities vs. dimensions, notes vs. balloons, etc. An example of the manipulation would be to change all the heights of all the letters in notes (but not change the heights in balloons). However, this is usually accom-

2-D CAD

plished by first selecting or trapping all the particular items of interest, and then making the manipulation. Therefore, it is important to understand and master the special techniques for selecting the items.

One special technique would be intelligent recognition of the types of items. In other words, if the user wants to select all dimensions, then the CAD system may be able to simply recognize them all at once. This type of selection can be referred to as filtering. For example, the user would select the Dimension filter, and then have the system select everything that passes the filter. Some CAD systems even allow a logical mixture of filters (i.e., select all Circles and Splines). Another way to accomplish this type of selection may be to use a layer. The user may put all dimensions on a dimension layer, and then the user would have the system select everything on that layer (and thus select all dimensions for alteration or manipulation).

Another special technique would be various uses of windowing. Windowing relates to how a pointing device (such as the mouse) is used to select items. Most CAD systems would have the ability to have the device click in 2 locations to form a box. And, usually, the system will select anything that is completely in the box. This can be referred to as "trapping in." Some CAD systems may offer other options, such as using a circle instead of a box, or allowing the user to sketch an arbitrary shape that eventually closes to form a selection area. Another possibility is to allow selecting items that are only partially in the box area. Yet another possibility is to select all the items that are outside of the box (i.e., "trapping out"). If viewports are part of the CAD system, this allows for more possibilities or sophistication in the select process. For instance, filtering and select boxes can be combined with selecting items within the different views.

5.10.6 Undo and Redo

Another special technique for CAD systems is being able to undo and redo operations with a CAD drawing. The undo simply undoes the change or command of the user. The redo redoes the operation. This capability is generally provided due to a journal file. This file keeps track of all the operations that the user is making. The CAD system reads this file to know what to undo and/or redo.

Keep in mind, however, that CAD systems may have limitations on this journal process. For instance, the journal may be "cleared" every time the user makes certain operations (such as saving the file, deleting view ports, etc.). In other systems, there may be a limit on the number of steps that will be journaled, and thus the number of steps that can be undone or redone may be limited.

5.10.7 Related Geometry Creation

Once some geometric entities (such as lines, arcs, etc.) are created, the CAD system can create new entities based on these existing entities. This is possible since

TABLE 5.7 Special Techniques for Creating Related Geometry

Special technique	Description
Trimming + extending	This involves altering geometry so that they touch. If a line is supposed to end at another line or a circle, then the system can find the intersection. If the line in question overshoots the other line, then it is trimmed to the intersection. If the line in undershoots, then it is "extended" to end at the intersection.
Dividing	If a line is too long, and it needs to stop at a specific point or intersection, but the user also wants the "remainder" of the line left as well, then this technique will perform this operation by making a single entity into 2 separate new pieces (using the intersection as the point of division).
Offsetting	This technique creates new geometry at a specific distance away from existing geometry. This usually can make use of lines, arcs, circles, etc. After existing geometry is selected, the user needs to enter the offset distance, how many copies, as well as the direction to offset (there will be two directions away from the existing geometry).
Merging	This technique would be the opposite of "Dividing" mentioned above. In this case, 2 separate geometric entities are merged into a single new entity. Of course, the entities must already be "just touching" in some manner. For instance, if 2 lines are going to merge, they should already have the same line mathematical definition (having the same slope and sharing a point).
Making corner	This technique does a "Trim/Extend" type of operation mentioned above, but it is done to 2 lines, and the result is a corner. One can imagine 2 lines that intersect in the shape of a cross, and then the system cuts off 2 pieces to leave the corner.
Filleting	This technique creates a new arc where 2 lines intersect. This is a very, very common operation in mechanical drawing since most parts with features that join have some sort of transition from one feature to the next (they aren't really a totally sharp corner). See Figure 5.8. The user will usually just select the 2 lines and then enter the radius for the arc that is then tangent to both lines.
Patterns	This technique would create a pattern or array of a specific entity or entities. For instance, a bolt circle is a common mechanical design technique where 2 components (such as a pipe and a flange) are held together with a set of bolts in a circular pattern. The system creates a bolt circle by having the user select one bolt that is in the correct location and size, and then picking the origin of the bolt circle and the number of bolts needed. The system would then draw a series of bolts whose center lies on the bolt circle's center. Another common pattern arrangement would be a rectangular pattern.

2-D CAD

the CAD system can perform mathematical operations on the entities. For instance, the CAD system can quickly determine the intersection of 2 straight lines (assuming they are not parallel). Then this intersection point can be used to trim or extend lines to make them meet at this exact point. So the user can create 2 lines that are supposed to touch, but only do it roughly, and then have the CAD system make the precise correction. Other possible special techniques would be taking 2 lines that cross and cutting them so that a corner is formed, creating new lines or arcs that are "offset" a given distance from existing lines or arcs, or making patterns or arrays of existing geometry. Table 5.7 lists some of these special techniques.

Keep in mind that these techniques will not work well if the "smart paper" paradigm is not carefully followed. It is the "underlying" mathematics of the geometry that makes these special techniques work. So, if lines are not drawn in the proper scale, in the proper view, at the proper angles, etc., then these special techniques may not work.

5.11 ENABLING 2-D DESIGN METHODOLOGIES

In the previous chapter, an overall philosophy or design methodology was presented based on drawings. This involved a hierarchy of designers and types of drawings (such as layouts, detail drawings, assembly drawings, etc.). The 2-D CAD system can be a powerful enabler or enhancer for this methodology.

For instance, the detail drawings can be automatically extracted from layouts. Recall that the layout drawing has the overall locations of items in the design. The layout designer determines the basic envelope or region of space that a specific smaller part (or detail) is supposed to fit into. So, if the layout designer exports this envelope or boundary to a detail designer (perhaps as a grouping symbol, block, clump, etc.), then this detailer can be pretty certain that his or her new drawing's part will fit into the overall design (assuming the layout designer does not alter the overall layout without relaying the change to the detailer).

Another enabler for the 2-D design methodology is using full-scale CAD drawings. Although it would be quite impractical to design a building using paper that is at full scale (since the paper would be as big as the building), the CAD system has no problem having the user design in a mathematical space as big as a building. This approach has the advantage that any time a geometric entity is created or an entity is measured, the result is correct in the context of the overall design. If the CAD user needs a beam that is 10 m long, then a line can be entered with a length of 10,000 mm and the resulting line will accurately reflect the length of the real beam. Of course, if the CAD system does not support viewports and view scale, then this "full scale" approach is always available (since everything in the CAD drawing would be at "full scale" in a mathematical sense). For the CAD systems that use viewports and view scale, then the user needs to be more careful in interpreting the geometric data from the drawing.

A few CAD systems actually provide techniques to "overlay" multiple CAD drawings so that they can all be seen at the same time. This capability is quite valuable in the context of the 2-D design methodology. Now the top level designer's drawing (that shows the full product) can be seen on top of other designer's layouts. In this case, it can be quite easy to make sure that the detail drawings will correctly document the needed parts and items. These CAD systems will typically also use the viewport and projections techniques mentioned earlier so that the drawing's definition in the Front View, Top View, Right View, etc. can all be made consistent, and hopefully, the final design will have fewer interference problems (where parts do not work or fit properly or be easily assembled). Although this overlay technique is pretty powerful, it is somewhat irrelevant given that the 3-D design methods provide all these benefits and many more. In a sense, the overlay 2-D method with projections is a "poor man's 3-D design." But, given that 3-D design has become relatively affordable, it makes little sense not to just use 3-D design methods instead.

5.12 TWO-DIMENSIONAL CAE BENEFITS

Most, if not all, analytical functions performed by engineers have been affected by the use of computer software. These types of computer programs are usually classified as CAE or Computer Aided Engineering. Probably the most common of these programs is called FEA or Finite Element Analysis. These programs attempt to predict the behavior of solid materials (such as metals) under various loads or forces. Another common type of program is called CFD or Computational Fluid Dynamics. These programs attempt to predict the behavior of fluids. In each of these cases, the CAE software needs to use a fairly large amount of geometric information about the components being analyzed.

Assuming that a 2-D CAD system has been used properly (i.e. following the "smart paper" approach), the CAD drawing can be useful for engineering analysis. If the geometric entities are scaled and drawn accurately, then the data about these geometric entities probably can be transferred to the CAE program directly. Usually this involves the use of a neutral file format such as IGES. The desired result is to have lines or line segments from the CAD program to be recognized as lines or line segments in the CAE program. And, when 2 line segments meet at a corner, for instance, the line segments must both really end at the intersection. Although a small gap between these lines would still look correct and would have no detrimental effect on the designer's drawing, such a gap could cause difficulties in the CAE program. Many of these geometric difficulties stem from the generation of something in the CAE program called a mesh. A mesh is a grid or framework of points within the confines of the part's geometry that con-

2-D CAD

trols where calculations can be performed (refer to Figure 5.11). The creation of meshes often involves the application of automated geometric calculations, and issues like lines in a corner not touching will often cause errors.

Another aspect of engineering analysis with 2-D CAD data will involve the 2-D geometric properties. One obvious calculation would be area. If a part of a drawing has geometric entities that form a closed section, then the CAD system should be able to calculate the area of that section. Another common requirement for calculation in a CAD system is section properties such as inertia or moment of inertia. The inertia gives an indication of a cross section's ability to resist being deformed based on a specified axis; it weights the distribution of the area with respect to an axis. This calculation is also useful for finding a CG or Center of Gravity (generally these are misnomers for a mathematical centroid or center of area, but the relation to a real center of gravity is too strong and so the term CG is used anyway). The CG would be a point at the geometric center of any arbitrary shape. Both the inertia and the CG are very often applied in formulae for analyzing mechanical parts and assemblies.

5.13 CAD DATA FILES

Most of the issues covered in this chapter are primarily for the CAD user. Most of the items in the next chapter are most relevant to the CAD administrator or manager. But, one issue that is important to both groups is how the CAD data is actually stored on the computer system or network. This is done with something called a file or a CAD data file. The files will be somewhat automatically managed by the computer system, and the CAD system may hide some aspects of the file management, but it is still essential that CAD users understand as much as possible about the files to make optimal use of the system.

5.13.1 File Types and Names

Files are often identified by their extension or type. Most users would recognize that a file that ended with DOC would be a word processing document file, and a file that ends with DBF would likely be a database file. The same applies to 2-D CAD, but the file extensions may or may not be associated with a specific CAD system vendor. The most common extension would be DWG. This extension is used by a number of CAD systems. On a PC-based or Windows system, the DWG file would be in upper case. On a unix system, the file extension would often be lower case (such as dwg). When looking at the computer system, then, files will appear with names such as Bracket.DWG or bracket.dwg.

It is important to know the way the computer system stores these names, since there may be restrictions or limitations on the names. Often the file extension is limited to 3 characters, so when creating a CAD data file for a CAD draw-

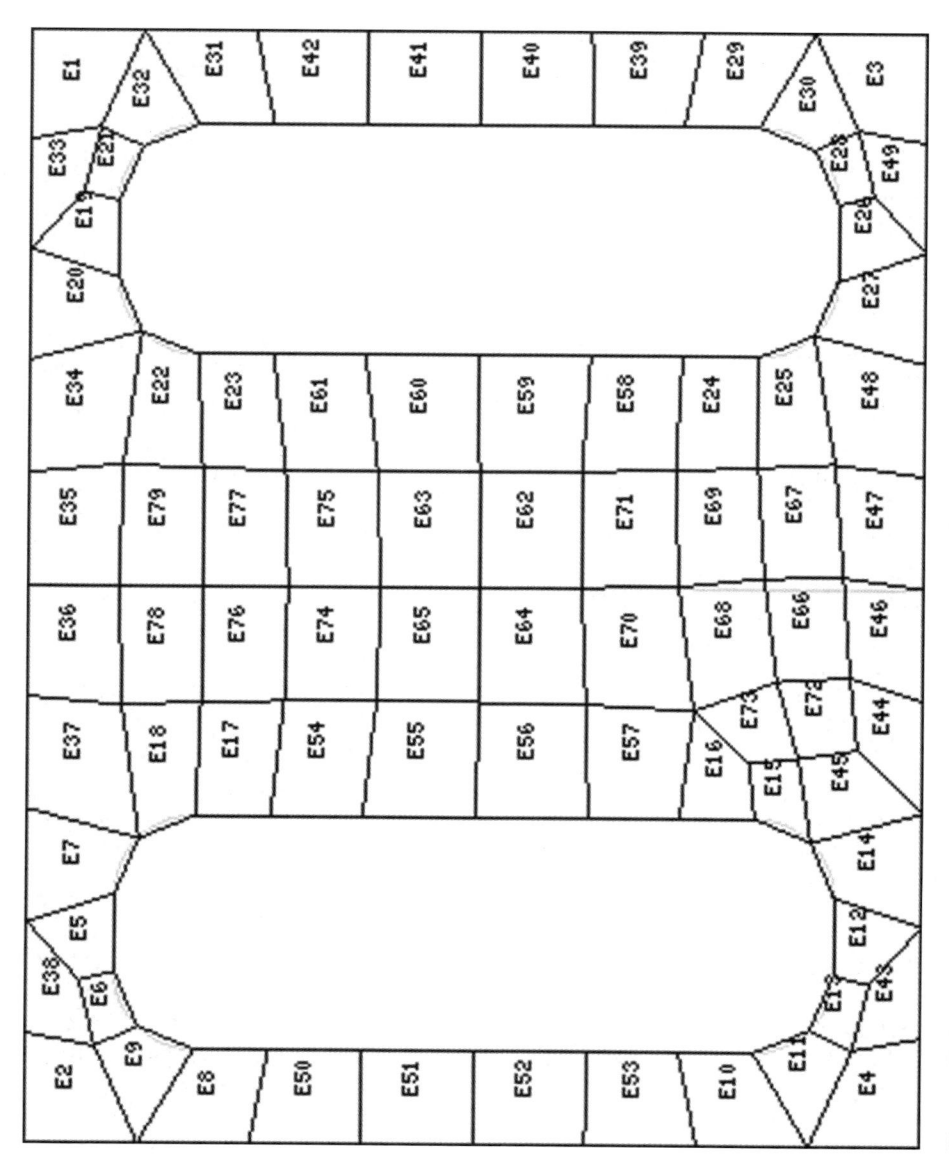

FIGURE 5.11 A sample 2-D mesh.

2-D CAD

ing, generally always limit the file extension to 3 characters (the CAD system may automatically apply the extension). Another issue is that the period (.) is often used to separate the main file name from the file extension; therefore, the period should be avoided as part of the drawing file name. Depending on the system used, the delimiter for directories would often be restricted in the CAD data file naming. For the "Windows" systems, this would be the backslash (\); for the Unix systems, this would be the forward slash (/).

In terms of the main file name, it can be descriptive (such as bracket.dwg), but usually companies have a part numbering system that manages all the product data for the company. This is often a proper choice for the CAD data file name. A company may produce many drawings that would say "BRACKET" in the Title Block, but there will only be one drawing that would use a given drawing number or part number. Of course, considering the 2-D design methodology presented in the previous chapter, there were some types of drawings in the hierarchy that were not usually released (such as layouts); they may not have part numbers. These nonreleased drawings often can benefit from descriptive file names (such as engine_trans_layout.dwg), instead of the part number file names.

Some more sophisticated CAD systems allow the user to create "normal" names and numbers for drawings (such as "ENGINE TRANSMISSION LAYOUT 44202") independently of the CAD data file name. Systems like this would probably allow the use of things like the period or the slash in the drawing name and/or number. This metadata-based type of CAD system uses a special database or lookup table to cross-reference the real names to the data file names.

5.13.2 File Format

Most, if not all, the CAD file types are considered binary data files. This means that one can not directly view and manipulate the data in the file. Files which one can directly view are known as ASCII files. Some CAD systems will have the ability to read and write both formats.

One implication of the file format (both its structure and specific data) is that you usually have to read, write, and view the file with the exact CAD system that created it. This is a serious limitation in general, but this proprietary data file approach does allow the CAD system vendor to accelerate performance of the CAD system. The CAD data file is custom-made for the particular CAD program.

However, care has to be taken when copying or moving the binary data files from one type of system to another. Not all computer systems use the same byte order (the 1s and 0s in the binary file go in different directions). In this case, the CAD system may be forced to use an ASCII format of the CAD data file instead (the ASCII file does not have this byte order problem). This is also why "neutral files" are almost always in ASCII format.

Another issue for file format is translating CAD data files from one CAD system to another. Getting CAD drawings from one CAD system to another is known as a translation. The translators usually use a neutral file. Usually, the sending CAD system creates the neutral file from their proprietary format. Then the neutral file is copied or moved to the other CAD system. Finally, the neutral file is translated into the receiving CAD system. The most common neutral files are called DXF and IGES. Refer to the next chapter on 2-D CAD management for more information on drawing translations.

5.13.3 File Size

The size of CAD data files can be an important issue. Although most computer systems have a great deal of available disk space at a relatively low cost, it is generally important not to waste the space. Also, most company's CAD systems use a file server. The file server's disk space is going to be a resource that needs to be carefully monitored and managed, and it also should not be wasted.

Probably the biggest impact on the size of the CAD data file is the file format (as discussed above). The binary file format is much more compact than the ASCII file. Therefore, the binary file format should be used as much as possible (unless the drawings need to be quite portable between systems).

The next biggest impact on the size of the CAD data file is likely to be the number of entities in the CAD drawing. This may generally be related to the complexity or busyness of the drawing. A drawing with many views of a complicated or intricate object is going to take many more lines, arcs, splines, etc. to describe, and the CAD data file is going to be larger accordingly. However, there are some issues that can aggravate this situation. For instance, if a drawing has free-form geometry (such as wires, hoses, wavy curves), and the curves are actually being stored in the CAD data file as 100s or 1000s of tiny line segments, then the file will be unnecessarily large. This situation of having many tiny lines is usually not done by the CAD user directly; instead, a drawing may be translated from another CAD system through a neutral file, and that neutral file does not support entities such as splines. The neutral file breaks the curves down into the many small line segments. Although the CAD drawing looks correct, it is actually using a large amount of disk space.

Another issue that may be a problem is off-screen entities. In this situation, entities are drawn outside the borders of the drawing or of individual viewports (if the CAD system uses viewports). These more or less invisible entities will not appear on the hardcopy of the drawing, and if the user forgets that they put these entities in the drawing, there may be a large amount of wasted space in the CAD data file. CAD users may use this technique to store design information that is relevant to the design, but is not supposed to be shown.

2-D CAD

Some CAD systems will have limits on the CAD data file size or at least the number of entities that can be stored in the file. The older mainframe-based systems were notorious for their limitations in this regard. Most of the CAD systems developed since that time are less likely to have problems.

In terms of a typical CAD data file size, it can range from 50–100 kilobytes for detail drawings to 5–10 megabytes for a very complicated assembly drawing. Also, note that the neutral files used for translations will usually expand the file size by perhaps by a factor of 2 or 3. Therefore, a large drawing could wind up at 15–30 megabytes as an IGES file.

5.14 CHAPTER EXERCISES

1. Use the 2-D capability of your CAD system (not using 3-D at all, if possible) to create a drawing of the object that was sketched in the Chapter 4 Exercises. Use the same views, scale, etc.
2. Record whether the CAD system use the "paper space" approach or the "model space" approach.
3. Change the scale of the views so that the drawing fits on a larger paper size.
4. Determine the formula of a line in the drawing. Use this formula to extrapolate the X and Y values for a point on the line. Use the CAD system to measure or analyze this location on the line. Record how closely the values agree.
5. Try using a grid snap option, if available.
6. Try using an automatic geometric relationship detection, if available.
7. Try to export the drawing in a neutral file, and then try to import the drawing into an engineering analysis program.
8. After saving the drawing to a file, record the size of the file in bytes. Make changes to the drawing and resave the file. Record any change in the file size.
9. Try to create a user-defined line font (such as long-short dashes) or a user-defined color (specifying the percentage of red, green, and blue).

5.15 CHAPTER REVIEW

1. Explain the difference between using a CAD system as "smart paper" versus a graphics program to create a drawing that creates an image. What are some of the advantages of the "smart paper" approach?
2. If a line is needed on a CAD drawing that models the edge of a block that is 11.55 mm long, but the tolerance for the dimension showing the length of the line is only to 1 decimal place, what X and Y values

could be entered into the CAD system to create the line that preserves the model?
3. Should the user or the CAD system round off the dimension values? Why?
4. Are viewports and viewing angles necessary to project geometry automatically between views? Why?

6

Managing Two-Dimensional CAD

6.1 INTRODUCTION

This chapter provides vital information to someone managing a 2-D CAD environment or a CAD user interested in the administrative aspects of CAD. Depending on company size, the management function may be the responsibility of a few users, a single person, or even an entire department. It could be controlled by the engineering department or the IT (Information Technology) department. Regardless of the situation, it is probably best that those administering the CAD system be somewhat familiar with design and engineering functions. The previous chapters should provide sufficient knowledge for IT professionals.

Some of the information in this chapter would be relevant to the 3-D CAD environment also. A later chapter covers management for 3-D CAD. Issues addressed in this chapter include plotting, drawing management, translations, customizations, and system administration.

6.2 PLOTTING

Since the end product of the 2-D CAD system is going to be drawings, an essential aspect of managing the 2-D CAD system is going to be plotting. Plotting is meant to refer to all kinds of hardcopy of the drawings. However, it is generally called plotting since the original machines that produced all the hardcopy from CAD systems were plotters (pen plotters, bed plotters, electrostatics, etc.). These machines could trace out lines as a pen was mechanically moved across the paper. For a long time, this would look very different from what computer printers could do. Computer printers either printed just characters (numbers, letters, etc.), or they produced a low resolution image of dots, and they usually only worked with smaller sizes of paper (such as A or B size or metric A4 or A3).

Now, computer printing technologies (such as laser printers, jet printers, etc.) use the large paper sizes, and they have resolutions high enough to look as good as the plotters. Therefore, plotting may now be done with printers as well as plotters. In fact, actual pen plotters are probably only rarely used. But, one needs to realize that the concepts and terminology from pen plotters are still relevant, and that many printers still "emulate" the pen plotter operation.

Another important development is the use of on-line drawings and other documentation. In this case, there is no physical hardcopy from the CAD system directly. Instead, the electronic data of the CAD drawing is sent to a computer system file server, (or virtual vault or cabinet). Instead of just viewing hardcopy prints, end-users of the drawings search, view, and print via a computer system. This is discussed in more detail later.

6.2.1 The Plotting Process

Figure 6.1 shows an overall plotting process. This process needs to coordinate the actions of the CAD system (the application program), the computer's operating system software, the plotting device itself, and perhaps a computer network. With so many functions to coordinate, setting up a plotting system can be somewhat challenging. Each of the parts of the process must be properly configured.

The first step is solely a function of the CAD system. This program contains the actual data or CAD drawing, so obviously it must start the process. The demand for the hardcopy may be driven by the user, or it might be generated by an automatic system that creates and/or updates drawings. The CAD system may also have access to a number of different print queues (a special computer program and resources that is always running in the background waiting and controlling the sending of data). So, at some point, the user may need to not only select the drawing to be plotted, but also indicate which queue is going to be used. Often CAD users have access to smaller hardcopy systems for check prints (used by people verifying that drawings are acceptable), and then they would also have access to a larger system plotter or plot room system for full size prints.

Once the request for the specific hardcopy has been initiated, the CAD system and/or operating system must then create data that can be used by the specific plotter device. The CAD system's proprietary data file (such as a DWG file)

Managing 2-D CAD

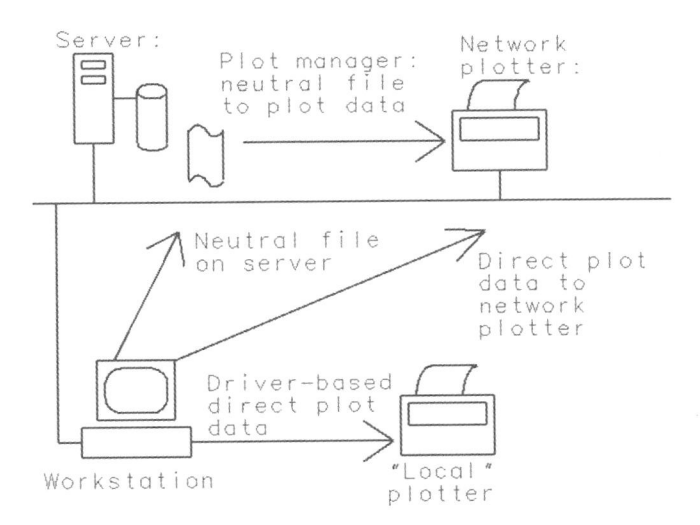

FIGURE 6.1 Schematic for plotting process.

is not appropriate for the plotter since the plotters are sold to companies with many different CAD systems. In order to enable the plotters to work with various systems, printers have their own language or data they are able to receive. Therefore, at some point in the process, the CAD data must be converted to the printer's language.

In Figure 6.1, there are few paths shown for getting this plotter data to the plotting device (or actually its queue). In the simpler path, the user's computer (which has the CAD system running on it) has a plotter attached directly to it. In this case, the CAD system would create the exact data instructions to get the drawing produced on the plotter. This process is often done via a special, device-specific programming called a driver. In this simple setup, getting the correct driver is often the biggest problem. Some of the formats that plotting devices use would be PostScript®, HPGL, or PCL. So, these systems would need a PostScript driver, an HPGL driver, etc. In this arrangement, the data is probably going to a queue that is running on the same computer as the CAD system. This queue will allow multiple drawings to be sent to the plotting device, and it allows the user to continue work with the CAD system or other activities while the copies are being printed. Figure 6.1 also shows the CAD system on a workstation sending the instructions to a network printer.

The more complicated path shown in Figure 6.1 uses a sort of off-line approach. In this case, the CAD system is not communicating directly with the plotter. Instead, some sort of independent plotter file is created by the CAD system and stored on a disk drive somewhere. This could be a proprietary plot file for the CAD system; this could be a "neutral" file of some kind (such as a CGM file); this could also be a plotter device file (such as an HPGL file) that contains the specific instructions to create the hardcopy. In this case, the file is going to be used by a program that is independent of the CAD system. Although this plot manager program may be written and supported by the CAD system vendor, it could also be a third party program that accepts the independent files and manages where, when, and how they are plotted.

The plot manager program now becomes the program that will need the proper driver and access to the proper print queues. The plot manager may also be a program that operates across an entire network, and thus can print to plotter devices located throughout a building, a campus, or even across an entire enterprise with offices across the country.

6.2.2 Plotter Setup

Assuming that a process is in place to send the plot data from the CAD system to an actual device, the next issue to consider is the setup of the plotter device. This procedure has to be coordinated with the CAD system's or the plot manager program's requirements. Indeed, plotting devices should not be obtained until one has checked that the device is compatible with the particular CAD system. For instance, if a CAD system only generates HPGL data for plotting, then the device is going to need to support HPGL (or the plot manager program is going to have to do a translation). Once the device is obtained, then the device will probably have to be "set" to accept HPGL. Fortunately, many plotter devices now have automatic format detection where the device looks at the data stream coming to it, and it then automatically changes to accept that format as needed.

Another important setup issue to consider, which is probably unique to CAD systems, is line weight. Line weight is a term that refers to how "heavy/thick" or "light/thin" the lines are on the drawing. Different parts of the drawing traditionally are made with lines that are given standardized weights (refer to Section 4.10). This helps the reader of the drawing more easily visualize the object in the drawing documents (e.g. visible edges of the part are heavier; hidden edges are lighter). When the plotting device was a pen plotter that used mechanical pens moving across the paper, it was possible to have the plotter select different pens that produced the needed line weight. Of course, the user must have somehow added the information in the CAD system to indicate which lines are heavy, medium, or light weight. This might be shown as different colors in the CAD drawing by selecting all the appropriate entities and entering a command that the CAD system would save as a tag in the data file for those entities.

Managing 2-D CAD 149

Support for line weights becomes more complicated when the plotting devices no longer have pens. Now there are a few choices for handling line weight. One approach is to have the CAD system actually show the line weight as part of the image on the monitor (instead of colors or tags as mentioned earlier); then the CAD system uses that information to create the proper bitmap data to send to the plotting device.

Another approach uses pen mapping in the CAD system or plot manager with a plotting device that can emulate the pens. In this case, there are no mechanical pens on the plotting device, but the device can behave as if they existed. The plotting device would allow the administrator to define a certain number of virtual pens. For each virtual pen, the device would allow the administrator to indicate the weight of the pen. The weight would be a distance for the thickness of the line (e.g., in inches, mm, picas, or points, etc.). For example, Pen 1 would be 0.5 mm wide, Pen 2 would be 1.0 mm, and Pen 3 would be 1.5 mm. There are a few issues to be addressed with this approach.

First, the file format must be able to handle this "pen" type of information. HPGL would be an example of a format that handles pens. Secondly, the CAD system must know which pen number is mapped to which tag in the CAD system (this pen map would generally be established by an administrator for the system). Now, when the CAD user creates a line and a medium weight is desired for that line, the user uses the command that tags that line's weight as medium. The CAD data file saves this information. When the CAD system needs to make a hardcopy of this CAD data file, the CAD system translates the request for medium weight to a specific virtual pen number (that has been pre-established for this weight). The plot data file eventually gets to the plotting device, and the request for the specific pen number is interpreted by the device to switch to the desired line thickness (switching virtual pens).

Lastly, the data files could actually define their own line weight information or use its own "overstrike" algorithm (repeating the line with a slight offset to get thicker lines). In this case, the plotter data file or the neutral plot file (such as CGM mentioned earlier) actually contains the information on how thick each line in the file is expected to be. This has a few drawbacks. First of all, the plot files will become larger since each file will contain information on line weights that could have been stored in the plotter device. Secondly, the plotting device will have to spend more time interpreting the data file since it must determine what is real plot data and what is line weight data.

Table 6.1 summarize issues to consider for setting up the plotter device.

6.2.3 Drawing Print Release

An important concept with respect to the hardcopy drawings is the drawing or print release. This is simply an event which is the clear delineation of when the

TABLE 6.1 General Plotter Setup

Issue	Comments
Pen mapping	If a pen plotter format (such as HPGL) is chosen for the CAD system and plotter, then a map may be developed and saved in the system to indicate what pen information on the device is equivalent to the entity information in the CAD system (color, etc.).
Clipping	Many plotting devices can have a maximum plot size. This would involve clipping or removing unwanted data from beyond these sizes. This is often an issue for CAD system that uses the "model space" approach explained in Chapter 5.
Nesting	More sophisticated plotting devices handle the optimization of paper use by making various plots of various sizes fit together on the paper size (perhaps even rotating the images). This is referred to as "nesting." Normally paper is fed from a roll.
Rotating	Many plotting devices or plot managers will allow the plots to be rotated. This may help to fit more plots on a given sheet or roll.
Scaling	Many plotting devices or plot managers will allow the plots to be scaled. In this case, the plots can be made larger or smaller as needed. This scaling can often be done in the CAD system as well as the plotter device, so administrators should be careful to avoid getting "scale of a scale" problems.

drawing is complete enough to have those beyond the engineering department access the drawing. Completion of the drawing would include being checked and approved as well as having the appearance of the drawing completed. Those beyond the engineering department could include manufacturing, purchasing, marketing, technical publications, and field support functions within the company (with manufacturing also meaning to include suppliers or vendors). In the case of engineering being done as a contractor to another company or for a government project, release could be to the end-user or customer, as well.

With paper drawings being used as the sole end-product, release of the drawing would mean physically making the drawing available to the other groups. The released or official drawing would be filed in an appropriate vault. When different parties need to access the drawing, then they would request a copy of the drawing, and then it would be sent to them according to a documented procedure.

The actual release procedure goes by many different acronyms. The term generally used in this work for drawings is ECR (meaning Engineering Change Record or Release). Other terms are ECO (Engineering Change Order) or ECN (Engineering Change Notice or Number). Even though these terms indicate a change, this process is assumed to apply to the release of brand new drawings as

Managing 2-D CAD   **151**

well as revisions to existing drawings (or a combination of new and revised drawings). One vital characteristic of the ECR is that they would be cataloged or tracked via a number for each release (ECR number, ECN number, etc.). This allows records keeping of what drawings are affected by a particular release activity. The ECR process would typically include assigning the unique ECR number (or number/letter combination), listing the drawings (by drawing or part number) that are part of the release, having the package approved for release, recording the date and time of the release, etc.

One effect of the release process on the hardcopies of the drawings is that a date/time stamp may be placed on the hardcopies. This may be done manually (with an actual stamp), or it may be automatically printed on the drawing by the hardcopy device. In either case, the master hardcopy of the drawings would be given a marking that indicates the official date and time that the copy was made.

When drawings are duplicated or copied for use by other departments, there should also be consideration given to putting an "UNCONTROLLED" or "FOR REFERENCE ONLY" stamp on the drawing (again, either manually or via a customization to the plotting process and hardcopy device). Clearly once a paper copy of the master drawing in a vault has been made, the CAD system is not going to be able to automatically inform those receiving the copies when there is a change to the master drawing in the vault (at least if a totally paper-driven process is relied upon). In order to inform anyone that receives the copy that it may or may not be the latest revision level, these other stamps are put on the drawings. This is an important task in having a quality system comply with standards such as ISO 9001.

This section has somewhat oversimplified the drawing release process, but the important aspects and its relevance to the CAD system are clear (particularly with respect to file management or data management discussed later in this chapter). CAD users and administrators will need to work with their own companies to determine the actual details of their process and how it affects the plotting process.

6.2.4 "On-Line" Imaging Systems

The computer systems or networks that run the CAD system have also developed means for digitally storing and retrieving drawings without any hardcopy at all. This type of system can be referred to as an on-line system. In order to see the drawings, personnel (in the engineering department or otherwise) must use a computer as well. Of course, at that point, the person should be able to make a hardcopy on their own printer; but, companies must be clear to consider these hardcopies as uncontrolled.

A major distinction among the on-line systems is whether they use a native CAD data file, a neutral CAD data file, or an image or bitmap file. If the on-line system uses the native CAD data file, then not only is a computer required to even see the drawings, but one also needs to have a copy of the CAD system software, or a special viewer program. This can be a significant problem if people outside of the engineering department, or outside of the company, or outside of the country need to see the drawings. One advantage of this approach, though, is that the drawings can be analyzed, red-lined, or even revised with little difficulty by the wider audience (assuming they have the software/and have been trained to use it).

In order to still use a CAD type of data file, but make the drawings more accessible, a neutral CAD data file could be used instead of the native file. Examples, would be DXF or IGES files. These files still contain CAD vector data (so they can be analyzed and edited), and many different CAD systems can recognize and view these files. Unfortunately, not all CAD systems will translate them 100% correctly, so there is a significant potential for error unless a carefully approved list of tested programs are enforced for the wider audience.

In order to break free from the CAD system (so that anyone, anywhere could see them), many companies have taken the image file approach. This is a very different type of file than the CAD file. As mentioned above, it is a bitmap. This means that the program showing the drawing to the user cannot really recognize any of the entities in the drawing (there are just individual pixels shown). An image file viewer program (many of which are provided at no cost) is generally not be able to distinguish between the number 1 in a dimension and the small lines that might be a segment of the geometric definition of the part documented by the drawing. Or, this program would not be able to distinguish between the upper case letter O and the number zero. Typical file types for an image file would be TIFF, GIF, JPEG, BMP, and PCX (with TIFF probably being the most common). An advantage to this approach (besides being CAD system independent) is that there are many programs that catalog and manage these image files. And, there are many other types of software (besides CAD) that can read these image files (notably Web browsers). Therefore, by using the image file approach, these packages would be able to easily work with the CAD images.

6.2.5 Batch Plotting

In the course of releasing packages of drawings, it may be necessary to create hardcopies (or image files) for an entire list of drawings. This is usually referred to as batch plotting. In many cases, this operation can be done automatically. Usually, a company can obtain computer programs that can work from a data file with a list of drawings (by file name, drawing number, revision level, etc.). These programs then retrieve the images or the CAD data files and produce the hardcopies (or image files). The list of drawings required may be generated by an ECR process, a manufacturing job request, or in conjunction with a purchasing request for quotation/outsourcing.

6.3 FILE OR DATA MANAGEMENT

As many drawings are created, released, and stored in the CAD system, it becomes necessary to implement a file management or data management strategy for the drawings. This strategy is meant to organize the information so that it can be protected (from intentional or accidental destruction), controlled, and made easily available for future revisions. Beyond the inherent benefits of organization, a documented procedure that implements this strategy can be a requirement for the company quality system to comply with standards such as ISO 9001.

There are a few typical strategies for this organization. One strategy is to use the operating system's file management alone. Obviously operating systems are not written to assist CAD management, so this is often not the best choice. Another strategy is to use a database and database programming. This database would be populated with various metadata for drawings (drawing or part numbers, revision levels, etc.). At that point, users can intelligently search, sort, retrieve, and revise the drawings. If the database and database programming is implemented within the CAD system itself, then this whole approach can be invisible to the user. This also means that the CAD user could be insulated entirely from the operating system. This could also mean that a mixed or heterogeneous set of platforms (unix workstations and PCs) could be used in the same CAD system at the same time.

6.3.1 Drawing Metadata

An important concept with respect to CAD data management is metadata. This is a computer systems term that means "data about data"; this term is used by this work a number of times to help explain data management. In the context of the CAD system, this is information in a computer file (such as a database) that is relevant to the overall organization of the many drawing files.

As an example, it has been mentioned already that drawing hardcopies can be placed into vaults. In an actual vault, there were flat file cabinets where the drawings were laid onto large horizontal drawers. These cabinets were based on standard drawing sizes (as discussed in Section 4.4). The C-size drawings, for instance, would be in a C drawer; D-size drawings would be in the D-size drawer, etc.. A catalog of all the drawings in the vault might have entries such as D081462 or E0604062.

The size letter would be an example of something relevant to the "metadata" for those paper drawings. It is a tag or "attribute" that would be needed along with the drawing or part number in order to actually find the drawing in the vault. One needed to know which set of drawers to look in to find a particular drawing, so the drawing size letter is important in that file management system. Other examples of information that would be relevant to drawing management would be the ECR number, revision level, the person that approved the drawing, and the title of the drawing.

Although metadata can be created and used in a non-CAD drawing environment, it is certainly easier to create, track, and utilize in the computer-based system.

6.3.2 Database-Driven CAD Management

When CAD systems were first implemented, the CAD system probably had no drastic impact on the vaulting and release process. The CAD system merely generated the hardcopy that was placed into the vault. However, the computer that runs a CAD system is quite capable of creating and maintaining the catalog of drawings, so naturally the cataloging function should eventually be taken over by a computer system. Something like a catalog is usually called a database for any type of computerized system. A database uses and maintains many fields or specially designated chunks of information that can be sorted, searched, and printed, etc.

With a CAD database managing the drawing data, it becomes possible to have the CAD system use and maintain many more tags or attributes for the drawings by using the metadata items as the database fields. Now, users can obtain valuable information such as reports that list and sort drawings by part number, ECR number, date of release, etc. Users can also rely on the CAD database to help find drawings and research the various impacts of a change to a drawing. As CAD systems become more complex (particularly having to handle 3-D models), metadata as database fields becomes a necessity for managing the very large amounts of data. Therefore, this capability should be used whenever possible.

A good database-driven system allows users to use standard database functionality such as relational data base linking, record-locking, purging, etc. This system should also allow CAD administrator to setup a scheme for organizing the drawings according to designer and management needs. For instance, drawings may be stored in a system of virtual vaults, cabinets, libraries, or group/user based on an important field such as drawing number. Or, drawings may be stored in the virtual cabinets based on product type or design team. Obviously, if these cabinets are arranged and sorted by drawing number, then it will be easy to locate a specific drawing by drawing number alone. If the cabinets are arranged and sorted by product line, then it will be easy to locate a specific product item alone. Hopefully, the choice of part number or product line for cabinets will not be a large problem, however. The database manager software should allow the user to search by product line despite the primary key or field being by drawing number, and vice versa.

Although one of the fields in the database would probably be the operating system file name, it is not necessary to have that field shown to the user. In this

Managing 2-D CAD

case, the user may be indicating a drawing title or drawing number to a field that makes sense to them, while the CAD system itself would then automatically assign a file name for the operating system. Now the operating system level can no longer be of concern to the user. In terms of retrieval, the user can simply tell the CAD system what the drawing or part number is, what the revision level is, etc., and the CAD system can then file the name accordingly. Furthermore, this allows files to be stored in different directories, folders, file systems, or network file servers even though the drawings all may appear to the user to be in the same cabinet or library. This allows the system management functions to be independent of the CAD data management.

Building on the database driven CAD data management scheme, a CAD system may also offer the capability to add a layer of security on top of the automated file management. In this case, not only does the CAD database manager software control the finer details of how the data is stored, it may be used to restrict and grant access to the drawings at different stages of the development process. This type of CAD system would keep track of the legitimate users for the entire system, and then a CAD administrator could determine which of these users would be able to see, mark up, revise, and/or delete drawings. The need for this level of security is not necessarily driven by a need to stop malicious users (such as hacking). Instead, this level of security simply allows a complex process (such as product design) to be managed effectively. It is not unusual for two different designers working on two different design projects to find themselves conflicting with each other. They may need to revise the same drawing at the same time (since common parts may be shared across products for manufacturing efficiency). A good security system in the CAD database will make sure that only one user will gain revision control of the drawing at a time. While one user is granted access to modify the drawing, other users would be locked out.

This layer of security also allows different types of users to be granted different privileges as a group. Perhaps only designers can revise drawings, only managers can mark up drawings, and only CAD administrators can delete drawings. This database driven CAD management can make sure that is what actually happens.

Going beyond the definition of the different classes of users, a CAD system may also be able to keep track of the stages of a project. So, when a drawing is intended only as a concept or a preliminary layout (and not to drive actual manufacturing of a component or assembly), the CAD system may be able to make sure that drawings are not allowed to be revised by anyone or released to manufacturing. Then, when the drawing is released and available for revisions, then the CAD system can allow this to occur. This highest level of data management may be referred to as PDM (Product Data Management); it may be part of a separate computer software system called a PDM system.

The larger the company or enterprise, or the more complex the final product, the more important it is to manage the drawings as flawlessly as possible. Poor business and even life-threatening situations for the public at large can result if the wrong drawing is used at the wrong time. Just one change in a note on a drawing that indicates a stronger material to be used to make a single part can be disastrous. Drawings must be well-organized and controlled through all revisions for the life cycle of the product. A CAD system with a good database system will make this task much easier.

6.3.3 File-Driven CAD Management

Having extolled the benefits of the database manager approach to the CAD system, some companies may still rely on a pure file management strategy. In this case, it is left to the CAD administrator and users to make use of the files and folders provided by the operating system. This is really not data management, but just file management. In this case, it is up to the users to maintain and organize the drawings in a somewhat manual process. Users are being relied upon to name the files correctly, put them in the correct directories, and probably delete them when they are no longer needed.

The drawing must be identified by the file name alone. If someone needs to find a specific drawing they will have to be able to identify the proper folder needed (folders based on drawing number or product line, no doubt), or they will be forced to search all the available directories based on a file name matching method of some kind. Of course, if folders are based on drawing number, there is no way to look in folders based on product line instead, and vice versa. Without the CAD data base and the metadata fields, one will not be able to search for drawings based on attributes such as who created the drawing, when it was released, the ECR number, revision level, and so on. One can only search based on the file name, and the file name is going to have a very limited amount of information within it. Obvious choices for inclusion in the file name would be drawing number (assuming no characters that violate operating system conventions are used in the company's drawing number system), revision level, sheet number, and the title from the Title Block.

Another problem with simple file management is revision control. As mentioned earlier, often there are conflicting needs for drawing revision (with more than one user needing control of the drawing at the same time). In order to control which user actually has the control over revisions using just file management, schemes such as moving CAD data files between different kinds of folders have to be used. However, this file management is not automatic, and the scheme is unlikely to be able to consider the stage of the product development process (i.e. prototype versus existing product) or the state of the drawing (i.e. new drawing versus revision).

Managing 2-D CAD

It may be quite difficult to prevent intentional or accidental deletion of files in the file management approach. Practically by definition, users will have direct access to the operating system and files (since they will be copying, moving, renaming files, etc.). If users are able to have read/write access to the directories or folders (so that they can be revised or moved), then these users are often going to get delete privileges as well. One solution to this problem may be to have CAD administrator users that have the privilege to move the drawings between folders; they would then be responsible for the file management on behalf of the general population of users. However, this is clearly a function that can be automated with the database management system, and if the database approach is used, then this would allow CAD users to spend more time on design functions and less time on administrative tasks.

6.3.4 Multisheet Files

Some CAD systems allow multiple sheets of a drawing to be stored within a single operating system file. Generally, as drawings become larger (since they are documenting something large or complex), the designer will put the drawing on larger and larger drawing sizes. Eventually, the largest size of drawing (E for inch or A0 for mm for instance) is reached. At that point, designers will make use of multiple sheets or pages to document the design under the single drawing number. Recall that Sheet is an attribute of the standard Title Block.

Using the file manager (i.e. no "metadata") approach, one is forced to use a file naming scheme to record the sheet number. In a database manager CAD system, the sheet number may be a metadata attribute. If the CAD system offers this option, the multisheet file is likely to be a better approach than a separate file for each sheet. It can be quite cumbersome for a designer to have to open one file to see one sheet, and then open another file for the next sheet (while trying to document the design with all the sheets at once).

6.4 DRAWING TRANSLATIONS

Almost certainly the biggest complaint about CAD systems in general is the inability to exchange drawings between different CAD systems perfectly. As mentioned earlier, most CAD system vendors have chosen to use a proprietary data format for their files (primarily so they can develop their software to do whatever they want). Therefore, the files are not directly interchangeable. Although it would probably be a simple matter to have near perfect interchangeability for 2-D drawings, it is hard to imagine how the software companies could actually agree on what format to use. Instead de facto standards prevail where one vendor becomes so prevalent the others must follow to survive. One can assume that until a single vendor becomes that dominant, there will be competing formats.

In order to deal with this file incompatibility issue, neutral files have been developed. These neutral files may arise from a vendor that decides to make the format public domain to some degree (such as the DXF file from AutoDesk®). The neutral files may come from industry groups such as the National Computer Graphics Association (i.e. the IGES file). Or, they may come from government agencies such as National Institute for Science and Technology (NIST). These files are like a third party for the CAD data. The originating CAD system creates the neutral file. The neutral file is exchanged between the two computer systems. Finally, the receiving CAD system reads or imports the neutral file.

6.4.1 Errors in Translation

In the process of using the neutral file, errors or problems are generally incurred. If one CAD system uses viewports and the paper space technique, but the other system does not, then the data moved between CAD systems may be stripped of the view scale information. In this case, some of the geometry will be either much larger or much smaller than the rest. Or, the geometry may appear to be the correct relative size, but dimensions show incorrect values. In the latter case, a dimension from a Front View with a scale of 0.5 that indicates a value of 100 mm may show a value of 50 mm after the view scale intelligence is lost (although the line would be 50 mm long if one measured the hardcopy of the drawing, one needs to look at the view scale to know that the real, physical part would really be 100 mm long as indicated by the original CAD system's dimension value). This is a very serious problem. Users of the 2 CAD systems simply need to be aware of the problem and verify that the translation issues are resolved. Fortunately, once two companies have determined the problems and exchanged a few drawings successfully, then most future translations should be acceptable (at least as long as both companies remain on the same versions of the same two CAD systems). Figure 6.2 shows some typical results from this translation problem.

Another typical problem for the CAD data exchange is the use of hidden or no-show entities. Many CAD systems allow the user to hide entities (such as lines, notes, dimensions, etc.) so that they do not appear in the hardcopy of the drawing. The CAD system may permit groupings (symbols, clumps, blocks, etc.) to be defined, but never used or instanced. However, this is information that is stored in the CAD data file that can be restored or shown at a future date. One needs to be aware that the neutral file may, or may not, transfer this data and/or leave this information hidden. As with the viewport problem, testing must be performed to determine what happens for the two CAD systems in question and deal with the situation accordingly.

Managing 2-D CAD 159

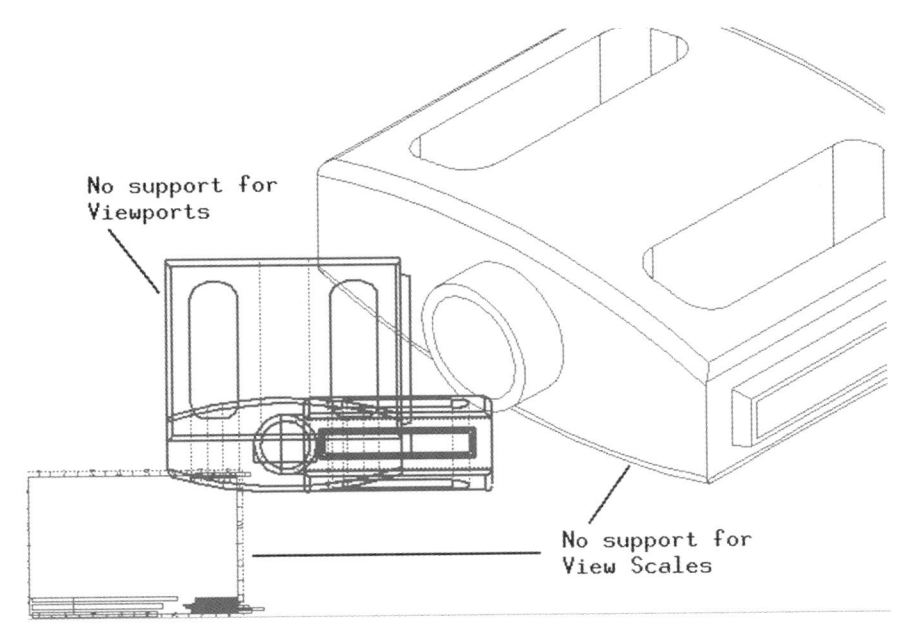

FIGURE 6.2 Bad results of CAD data exchange with dissimilar CAD systems.

6.4.2 Vector vs. Raster Data

One concept that the CAD administrator needs to understand clearly is the difference between vector and raster data. This has a direct bearing on the feasibility of drawing translations. Vector data in a file means that there is mathematical definition to the parts of the drawing that show the object or part. There are real X and Y values for the geometric entities. This means that the CAD system can create and calculate whatever it needs to function properly. Files such as DXF, DWG, and IGES contain vector data.

Raster data can be thought of as a bitmap. It is just a set of bits that indicate the color (or just the black/white) of dots on a monitor or a paper copy. There is no distinction between dark dots that are part of a dimension or note or the dots that are part of a line segment for a geometric entity (such as a line or circle). There are no X- and Y-values in the raster data that says where a line starts and where it ends. It should be clear that it is basically impossible to use this type of data to exchange CAD data between CAD systems. There are raster-to-vector conversion programs, but they are of extremely limited value. They may find that certain dots form a line, but they will not be able to recognize the difference between a line segment that is geometric, and a line that is part of a note or dimension (at least not virtually every time). Furthermore, the X- and Y-values that would be extrapolated by a raster-to-vector program are going to be of limited precision; they must follow the pattern of dots, but a line segment may not mathematically end at a point of a dot. Files such as TIFF, GIF, JPEG, PCX, and BMP contain raster data, and they would not be appropriate for direct CAD data translations.

6.5 TEXT FONTS

Another unique problem for CAD data administration is the fonts used for text (or character data). This is not to be confused with line fonts that are used for geometric entities in the drawing (things like solid, hidden, phantom, center line fonts). For common computer software such as word processing, text has fonts that indicate the appearance of the letters, numbers, etc. Examples are Courier, System, and Arial, etc. However, these fonts are generally shown as bitmap data. It does not lend itself to the kind of data that CAD data files use (i.e., vector data).

Therefore, CAD systems generally have their own fonts defined for the text based on a stroke table. This is a set of instructions that can create the various letters, numbers, and symbols that would be shown on a drawing. For instance, making a number 1 would involve creating 3 lines (a bottom line, the tall vertical line, and the small line at an angle at the top). Clearly this means text shown on a drawing is somewhat primitive, but generally that is desirable. The text should be as crisp, clear, and simple as possible. There is no real advantage to the "prettier" styles available in word processing. Figure 6.3 shows some sample text based on

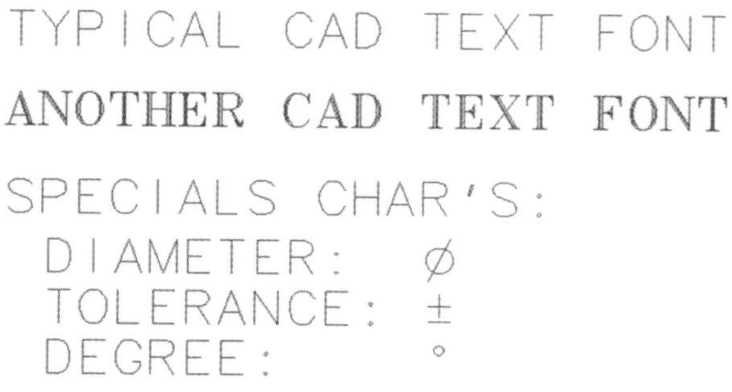

TYPICAL CAD TEXT FONT
ANOTHER CAD TEXT FONT
SPECIALS CHAR'S:
　DIAMETER:　⌀
　TOLERANCE:　±
　DEGREE:　　°

FIGURE 6.3 Typical text fonts for CAD.

a few standard stroke tables. Some CAD systems will also allow the customization of these stroke tables to create new letter styles.

6.6 CAD SYSTEM CUSTOMIZATION/ AUTOMATION

An important advantage to using a CAD system is the ability to customize and automate processes. The amount of customization and automation varies widely between CAD systems, but the need for this capability also varies widely (usually depending on the size and scope of the company using the CAD system). Examples of customization would be setting system-wide defaults, defining plotting and printing functions, templates, and user interface modifications. Examples of automation would be macros, application programming interfaces (API's), accessing system commands and utilities, or combinations of these.

6.6.1 Defaults

System-wide defaults allow an administrator to define the company-standards for the drawings. For instance, if a company wants all drawings to use a text height of 5 mm (so that all notes are easily readable), then the CAD system often allows this to be done. This information is usually stored in a file on the operating system, and the parameters in this file are usually changed through a graphical user interface program. This type of file is often referred to as a config file. Depending on the operating system, this file may then be secured for access by approved administrators only. This allows the company to enforce the standard text height. Keep in mind that there may be hundreds of parameters that may be customized in this manner. For text height alone, character data may appear in Title Blocks, notes, dimensions, GD&T symbols, decals, balloons, etc. For each of these potential character fields, one may need to set standards for text font, height, slant, gap, line spacing, etc. For many companies, the settings for the default CAD system will be adequate.

Other config files may be used to customize plotting. Considering the potential for complexity in the plotting process (as discussed earlier), most CAD systems have to have the capability to set defaults or standard settings for such options as paper size, plotter and printer queues, multiple copies, color or black-and-white, orientation (such as landscape and portrait), plotting to file, etc. As with the system-wide config file, there will often be a graphical user interface utility to set and manipulate this information.

6.6.2 Templates

Another typical customization option in CAD systems is the template file. This is a standardized CAD drawing that has been saved on the system using desired defaults, setting, configuration, etc. On some systems this is referred to as the prototype drawing. When the user creates a brand new CAD drawing, this file is used as a starting point. Besides being useful for customization settings, this may also be useful for standardizing the Title Block, Revision Block, etc. Since the template drawing should be an actual file on the operating system, it should again be possible to control access to the file so that only appropriate users or administrators are able to change it.

6.6.3 User Interface Alterations

Some CAD systems have the ability to augment or modify the standard user interface. For instance, the graphics and/or text for icons on the CAD system's icon panel or menu bar can be changed. This would allow a CAD administrator to restrict access to certain functions or commands. Or, if a company wanted to have a standard new command added to the CAD system, this option would allow for creation of new icons on the icon panel. Usually this capability is used in conjunction with the customized automation functions. In other words, new icons are added to the interface, and these new icons run new programs that are developed by users and/or administrators of the CAD system.

6.6.4 Macro Programs

Macro programs generally are computer programs that capture the user's actions as if it is recording the user. The user tells the software to start recording, the user performs the functions that need to be captured, and then the user tells the software to stop recording. The user is usually prompted to give a name to this macro as a file. After this process is complete, the user can then run this macro whenever the set of steps is needed. This allows the user to automate some complex functions of the software. The most important aspect of macro programs is that it is assumed that users (not professional programmers) are going to create and use them. CAD systems may offer this capability.

Once the CAD macro file is created, it may be possible to edit this file to augment or alter the automated function. Considering the complexity of functions and commands within a CAD system, this can be a very significant advantage to using a CAD system. For instance, if a company makes a standard component (such as a pressure vessel), and there are formulae or parameters that can indicate the overall geometry of the component, it may be possible to use macro programs to automate the creation of a drawing.

Another important potential source of increased productivity with the CAD system is the use of the operating system via the macro program. Using this capability, the macro program may be able to create and manipulate files, plot or print files, or start other programs or applications running on the computer. For instance, if users need to e-mail electronic copies of a CAD drawing, and this func-

Managing 2-D CAD 163

tion is not found in the standard CAD system's interface, then a macro program could be created that uses the operating system's mail function to send the file. In this case, the user may click on a customized icon or type in a single command on the keyboard that runs an entire series of commands automatically.

Keep in mind however that a macro program needs to be executed by a command processor that is built into the system. This is a separate function in the CAD system that must read each line of the macro program and then execute the commands found in each line. This is transparent to the user, but from a computer programming point of view it is the slowest possible approach. Macro programs should execute faster than a user can enter the commands or click the mouse, but it will be far slower than a real computer program that uses the API approach.

6.6.5 API Programming

A more advanced approach to automation is the API (Application Programming Interface). In this case, the CAD system vendor makes available the object code, libraries, foundation classes, objects, components, routines, or source code, etc. This programming data can then be used by programmers to create new software that uses the CAD systems capabilities but solves problems or performs functions that are beyond the core capability of the original CAD system. It is important to note that this option is generally done by programmers that have been trained for this activity (unlike the macro programs mentioned above).

There are many advantages to this approach. First of all, the new programs created are not using the command processor in the CAD system (as was the macro program). Instead, the API-type of program can be considered compiled. This means that the new program may be able to run as fast and as efficiently as the CAD system itself. This is very important with programs that are going to automate a large amount of geometry creation or manipulation. For instance, in the previous example of creating an entire drawing of a pressure vessel, if the drawing has 1000s of lines, arcs, and dimensions to be created, it could take quite a while to complete if a macro program is used. The same operation for an API-type of program could be finished in just seconds. If a customer is running this program via a Web page to get a standardized drawing, this time difference can be quite important.

Another advantage of the API-type of program is program sophistication. Most macro programming languages are going to be limited to the functions that the command processor can handle. The command processor may not handle even basic computer programming techniques such as subroutines. This means that the macro programs can be quite limited in their scope and readability. The API-type of program, on the other hand, is only going to be limited by the computer language being used. An API-type of program would use languages such as FORTRAN, Lisp, ADA, C, C++, or Java.

6.6.6 Utilizing "Metadata"

An important consideration in the standardization and optimization of a CAD system is the metadata. This would be information such as part numbers, who created the drawing, what revision level it is at, where it is filed on the computer system, the release information, etc.

The metadata is vital to significant automation of functions and optimization of the CAD system. A CAD system is simply not going to come initially configured to search for and then sort all the drawings created by a certain user using a certain ECO number for a part that is purchased instead of manufactured in house. This information is simply too specific to the company that uses the CAD system. Some companies will use numbers for the ECO, others will use letters, or some other technique. Thus, it is going to be left to the automation programs (either macro program or API-type) to provide this functionality.

However, providing such sophisticated automation is not going to be possible unless all the drawings are created consistently and given metadata consistently. The information such as ECO number and purchased/manufactured indicator also may need to be extracted or exported from the drawings. Since this information will need to be found by the automation programs, the drawings must have the data in the correct location in some manner. For best results, the metadata or Title Block information should not be created as a standard note in the drawing, but rather as an attribute or tag operation. The use of templates is also valuable in this regard. The template drawing file can have these tags ready for use.

6.7 SYSTEM ADMINISTRATION ISSUES

There are some issues that must be addressed by the system administration or IT group within a company. These issues are generally standard procedures for IT departments, but some procedures may have unique aspects due to the nature of the CAD systems.

6.7.1 File System Organization

Probably the most important system administration issue to consider carefully and with a clear understanding of the nature of the CAD system software is the file system organization. In particular, CAD system software tends to be quite demanding of computational resources in comparison to common office software. CAD software uses a large amount of RAM, floating point processing, and graphics processing. The CAD software executables and runtime libraries are also pretty enormous for 3-D CAD (1 to 2 GB). Therefore, it is usually considered a requirement that the CAD system software be loaded on a user's individual

Managing 2-D CAD

computer system (i.e. loaded locally or on the client). It is not a good application for access via a file server.

However, CAD users are going to be creating CAD data files that are going to generally be shared. These files will be shared among users during the development of a product design. And, once the project is finished, these files are usually going to be carefully maintained and revised over a fairly long period of time (perhaps decades). This aspect of the CAD system naturally leads to the use of a network and file server for these files. This allows the CAD data files (at least the completed ones) to be controlled, managed, and shared from a central location.

The network architecture, however, needs to consider the size of the files being accessed via the network. Some CAD data files will be 10 or 20 MB, so the system needs to be able to reliably handle this amount of data in as short of a time as possible.

When the CAD user decides to work on a CAD data file, some systems may be configured to copy the file from the file server to a local disk. This means that the network traffic is only generated at the beginning of the process of revising the drawing. While the user is changing the drawing, the file is only stored locally, and performance may improve. Then, when the user is finished, the drawing may be copied back from the local disk to the file server again. In some cases, this sort of process must be created and controlled by the user by simply manipulating files with the operating system. However, some CAD systems will make this process totally automated, and the CAD system may institute a data management scheme that locks out other users from revising a drawing when one user checks out the drawing. If this process is available, it must be exploited if at all possible. It is often very difficult to rely on users to "manually" handle all the aspects of file management.

Figure 2.5 shows a typical configuration of the local and file server data for a large CAD system (2-D or 3-D).

6.7.2 Backup/Restore

Clearly a backup and restore system must be instituted for all CAD system data. The dollar value invested in drawings is rather significant (perhaps 10 or 20 times the annual engineering department budget), and these drawings represent an invaluable repository of intellectual property. Since property needs to be protected from disaster, hardware failure, intentional corruption, etc., it becomes essential to have backup copies of all the drawings, data files, and/or metadata for the CAD system.

Beyond the need to restore data to the CAD system for disaster recovery or accidental deletion problems (which involves restoring files of recent use), some companies may want to permanently archive all drawings or data files. This involves keeping copies and permitting restoration of files that have been superceded by newer revisions. The most important task, in this case, is to have proper listings and databases that allow future users (particularly for field service issues) to find and restore drawings that have not been used for a long time.

For the CAD systems that use metadata or complex data management schemes, one has to consider whether certain files can be restored at all. If the CAD system database information at one point considers a particular drawing to be at a certain revision level, and a user checks out that revision to make a new revision, but then this drawing is accidentally deleted from the database (and this deletion is not noticed for some time), then it won't matter if the drawing data file (DWG) is restored from tape. The CAD database (like a DBF file) already recorded that the drawing revision was removed, and unless there is a way to reset or synchronize the entire database with the operating system file, the revision to the drawing will have to be started over again (using the last version properly recorded in the database). The only way to recover from this situation might be to restore a dummy image of the entire network from the older date and then export a new copy of the drawing with its attributes or relational data intact. This copy would then be imported into the production CAD system.

6.7.3 Licensing

In addition to monitoring the CAD system software, the CAD data files, and customization files, there may also be a need for configuring a license manager. This is a special program that runs on the local system or on a network file server that controls how many times or how many users can be using the CAD system. The license manager is usually provided by the CAD system vendor. License files or password files are often tied to the ethernet address of the computer hardware that is going to run the license daemon.

6.7.4 CAD System Upgrades

One of the most demanding tasks with respect to CAD system administration is upgrading the system. The CAD system and the network it runs on is a rather sophisticated and specialized arrangement. Furthermore, despite its sophistication, it often has to be customized to become fully useful to users (as described above). Therefore, there is a great deal of functionality that needs to be assessed during upgrades. In addition to testing all the new functionality that comes with a new release of the CAD system, one needs to ensure that all the existing functions are still functional. The only way to preserve the production ready status of the system for an upgrade is to test the software. Regardless of what has been indicated by the CAD system vendor, each company should test the upgrade before making it widely available.

Besides the upgrade of the CAD system, administrators should also carefully test new releases of operating systems, computer hardware (particularly

Managing 2-D CAD

graphics adapters), and networking software. These all have the potential to disrupt production, as well.

6.8 CHAPTER EXERCISES

1. Create a hardcopy or print of the CAD drawing created in the Chapter 5 Exercises.
2. Record the plotting process used by the CAD system. Determine if a neutral file is created, and if so, what kind of file. Determine if the data goes directly to the plotting device from the computer system or if it is sent via a network.
3. Record the format or language of the data received by the plotting device.
4. Determine if the CAD system uses a database manager driven data management or if just file names and folders from the operating system are used. If the database manager approach is used, record the metadata that can be used as fields in the database program (i.e. can you search for drawings based on part number, revision level, the person that created it, etc.).
5. Export a CAD neutral file (such as DXF or IGES) for the drawing, and then try to import it back into the CAD system as a brand new drawing. Record what problems appear in the re-imported copy of the drawing.

6.9 CHAPTER REVIEW

1. If a drawing in a released vault is copied for someone to study the part but not to change it, what wording should be stamped on the copy? Why is this important?
2. If a drawing in a released vault is checked out from the vault to be changed what happens to the Revision Level shown in the Title Block of the drawing?
3. What is the purpose of the ECR process?
4. What are some of the problems that typically arise from the translation of drawings between different CAD systems?
5. Whenever drawings are translated for use by another user or customer with a different CAD system, should a print or image file be delivered as well? Why?
6. What are some considerations for whether CAD system software should be installed on a local drive versus a network file server?

7
Three-Dimensional CAD

This chapter covers important concepts that need to be understood before proceeding with detailed information on 3-D CAD. If the reader is already fairly comfortable with 3-D CAD, then one can proceed to following chapters. These later chapters provide details on the subjects of part modeling (creating a 3-D model of a single object or part), surface modeling (a special approach to part modeling), assembly modeling (creating a 3-D model of a collection of parts), and 3-D CAD management. Of course, if a concept or term encountered in the later chapters causes problems for the reader, then refer back to this chapter.

7.1 INTRODUCTION

Up to this point, discussion of CAD systems has focused on the two-dimensional approach. In the 2-D case, work with the CAD system is based on planar mathematics. Users of these CAD systems see only the flat representation of data. The output of the system is just drawings (whether in electronic or paper form).

Although these 2-D CAD systems have basically eliminated all drafting and the manual production of formal mechanical drawings, they still require that the designer (or other type of user) mentally visualize the physical object or design based on flat views. Although this is not a significant problem for simpler parts, it can be an enormous problem when the object or assembly is complex.

3-D CAD   169

For instance, when the drawing is showing how to weld together 30 or 40 interconnected plates at a variety of angles, it can literally take hours of studying to fully understand the design.

Clearly, the process of designing and documenting something geometrically complex would benefit from a CAD system that could actually show and manipulate three-dimensional models. For many years, this was simply not practical for industrial users due to the relatively limited power available from computers (particularly mainframe computers that had to be shared with many other users). However, throughout the 1980s and 1990s, the power of non-mainframe computational systems expanded consistently and aggressively. Eventually (at least by the mid 1990s), the computational power that could be afforded to each designer was sufficient to create and manipulate 3-D models. Now, 3-D methods can be considered the norm for mechanical design.

Table 7.1 lists some of the advantages of the 3-D CAD system. It may seem odd that these need to be listed, but there has generally been great resistance to the adoption of the 3-D CAD system in most companies. It has simply been such a large break with previous systems that the transition was generally disruptive. It has been disruptive to work processes since decades-old paper-driven design methodologies have had to be changed. It has been disruptive to productivity since there is a significant learning curve as designers have to be retrained, but then they slowly gain back and then exceed the 2-D work output. And, it has been disruptive financially since the initial contracts for obtaining the systems can be quite significant (as opposed to existing 2-D CAD systems that are already paid for).

Finally and perhaps most importantly, it has been disruptive at a personal level since the 3-D CAD system is more demanding of the user. They must be comfortable with following the logic and organization of the software's methods. Unlike 2-D CAD, where virtually any particular method of construction could be followed as long as the final product (the drawing) looked right, 3-D CAD users must create a "real" mathematical model. These models are expected to be used, re-used, and tweaked over time. In the 3-D CAD system, more particular methods of construction must be followed. Some methods could be disastrous in the future, while other methods are robust and powerful. The trick, in a sense, is to think like the software, and to maximize its strengths and minimize its weaknesses. Many designers, particularly those who have done 2-D work for a long time, are not prepared for this level of involvement with the software. Hopefully, the following chapters in this work will be helpful in minimizing the risks for these users.

There are some important issues for users to consider at the start. First, 3-D CAD is not really that difficult. Most people able to do design work can learn how to do 3-D CAD effectively, at least if they keep practicing. Second, there is a fair amount of jargon and assumed knowledge hidden in the 3-D CAD systems.

TABLE 7.1 Advantages of 3-D CAD

Visualization	The most basic advantage of 3-D CAD is that the designer can really see the design. There is no need to study the views on a drawing. And, 3-D CAD provides a means of navigating or dynamically viewing 3-D models. This means the designer can view it from any angle and rotate it any direction. Usually, the designer can also make it translucent, cut open the model, etc.
Automated drawings	Since drawings are often still needed to communicate designs with suppliers and customers, 3-D CAD is used to automatically create the geometric entities in drawings. Once a user indicates the viewing angles (Front, plane, datum, etc.), 3-D CAD can view object, remove or process hidden lines, and then create drawing views. 3-D CAD meets or beats times for drawing new geometry created in 2-D CAD.
Geometric properties	Once 3-D model is created, it can be analyzed in many ways that are not possible with 2-D CAD. For instance, the volume of a properly created 3-D model can be easily calculated. Assuming material of part is known, the weight can then be calculated. Even if the 3-D model is an assembly of many different materials, the 3-D CAD system can still account for this and give a very accurate weight. Distances in 3-D models can also be measured in sophisticated ways (3-D point-to-point, points-to-lines, points-to-edges, surfaces-to-points, lines-to-lines, etc.), so designs can be more refined and understood than in 2-D.
Interference checking	Once a 3-D model is created, it is possible to have the 3-D CAD system automatically determine if models or pieces of models interfere.
Improved quality	3-D CAD provides easier transition to the creation of analytical models (for predicting failure). Drawings that are automatically created from 3-D models have fewer chances for geometry errors. Isometric views can be produced easily; an isometric view on a drawing is helpful for the drawing reader to quickly assess what the drawing is attempting to show; reader can search for the proper level of detail and understand the drawing more productively and make fewer errors in interpreting.
"True" design reviews	Design reviews are "high level" meetings that discuss the state of a product design (including designers, engineers, managers, customers, marketing, manufacturing personnel). They are meant to make sure that the design is meeting requirements. Although they can be done with drawings, there is very limited feedback since attendees are forced to imagine the product design state (often non–engineering personnel are not adept at studying drawings). This problem practically evaporates when viewing a projected image of a 3-D model. Feedback is plentiful.

3-D CAD

TABLE 7.1 Continued

Intelligent models	3-D modeling software generally encourages designers to include design intent. This may take the form of creating parametric relationships, entering equations, family tables, constraints, etc. This can allow a generic part model to generate related parts automatically. It might also encode into a 3-D model which dimensions or parameters are most important in the design.
Standardization	Although 2-D CAD may have a means for storing and organizing standard geometry, once it is placed into a drawing, there is typically no tracking of reuse. 3-D CAD can keep a record of the times the standard 3-D part models are used in an assembly.
Associativity	3-D CAD can keep track of individual geometric entity's relationships to each other. Using this associativity feature, when the 3-D model is revised, the geometry and dimensions in the 2-D drawing can automatically change. Relationships can also drive changes in an assembly or analysis model based on a part change.
Integrated product development	With capabilities such as visualization, associativity, and intelligent models, it becomes much more realistic to create interdisciplinary teams for product development. Disciplines that might be represented on such a team would be manufacturing engineering, marketing, process planning, purchasing, system integration, etc. Although these groups can be used on a team with 2-D CAD, in 3-D CAD these groups can very easily integrate the state of the design with their disciplines (just by looking at and manipulating the 3-D models).

These systems have developed over a long period of intense research and development, and often academic terminology and techniques creep into the software. Users that have not been engrossed in the systems over this long time are often blind-sided. Hopefully, the remainder of this chapter will expose some of these pitfalls. Third, 3-D CAD can not be beaten. The intrinsic advantages of a complete geometric model of a product are simply too great to bother using 2-D any more for any serious mechanical design activity. Once a 3-D model exists, it can be interrogated, studied, analyzed, and sliced and diced in a way that not even a physical prototype can equal. So there is no reason to fight the 3-D approach.

7.2 TERMINOLOGY

Although the glossary section of this work provides a list of terms with respect to 3-D CAD, it is important to define some of the most basic terminology at this point.

7.2.1 Model Terminology

Model The most basic item within a 3-D CAD system is the 3-D model (or referred to as just a model). A model is the 3-D computer graphics based object that users create and interact with. It most often refers to a part model, but it can also mean an assembly model, a surface model, or an analytical or finite element model. Keep in mind that it is referred to as a model because it is based upon underlying mathematical equations. Although these equations and their parameters are usually invisible to the user, they dictate what construction methods will or will not work, and whether the model will be robust or fragile. So, it is vital to keep this basic mathematical nature in mind.

Part model As mentioned, this is a model of a single component or part (often referred to as a detail in older drafting systems). A part model is self-contained in that it can be stored in the CAD system by itself, and the underlying mathematical equations are totally sufficient to regenerate and display the part. Figure 7.1 shows a part model.

Feature A feature is a self-contained segment of a part. For example, in Figure 7.1, the circular region that sticks out of the front of the part is a feature. It is a type of feature called a protrusion. The slotted region that cuts through the part from top to bottom is also a feature. It is a type of feature called a cut or cut-out. Sometimes the features are not obvious. The part shown in Figure 7.1 has some rounded or smooth edges; the process of adding this rounding is also a feature of the part. It is an example of a feature called a fillet (pronounced "fill-it").

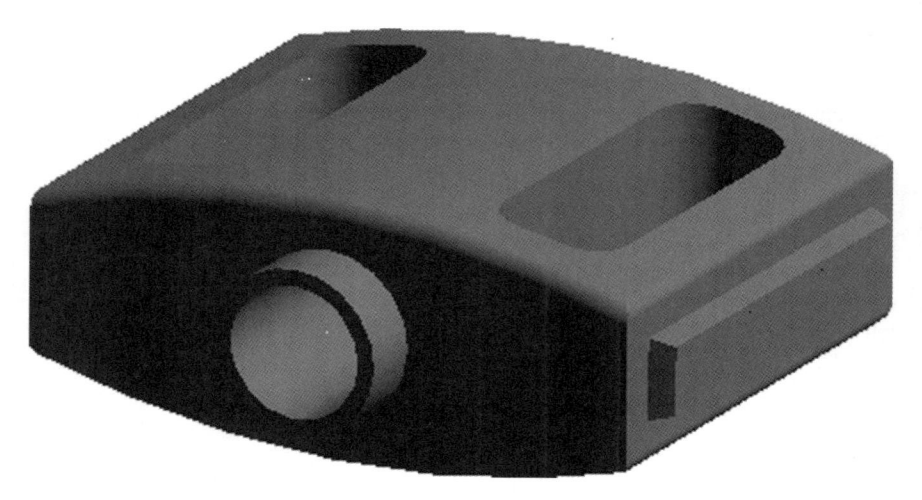

FIGURE 7.1 Example of a part model.

3-D CAD

Assembly model This is a model made up of a collection of part models. The assembly model is often just a list of names of part models and where they are located with respect to some global or master origin (where X, Y, and Z are zero). Although the assembly model will appear to contain many parts, each of those parts would still be a model in their own right, and the assembly model is merely pointing to or referencing those part models. Figure 7.2 shows an assembly model. It has 2 part models of a pedal and a crank part model connects them.

Solid model This is a type of part model that is assumed to have volume. There are no open faces or free edges on a solid model; all the faces of the model are connected to other faces. Figure 7.1 is an example of a solid model. For a long time, what is now called 3-D CAD technology was actually referred to as solid modeling; this was to differentiate it from 3-D CAD systems that used a technique known as wireframe. However, now it is assumed that a 3-D CAD system will almost always be creating solid models.

Surface model A type of part model that does not have volume. This is a model that is made up of surfaces that do not totally connect at all the edges. It is open, and it appears to be made up of paper thin pieces. Figure 7.3 shows an example of a surface model.

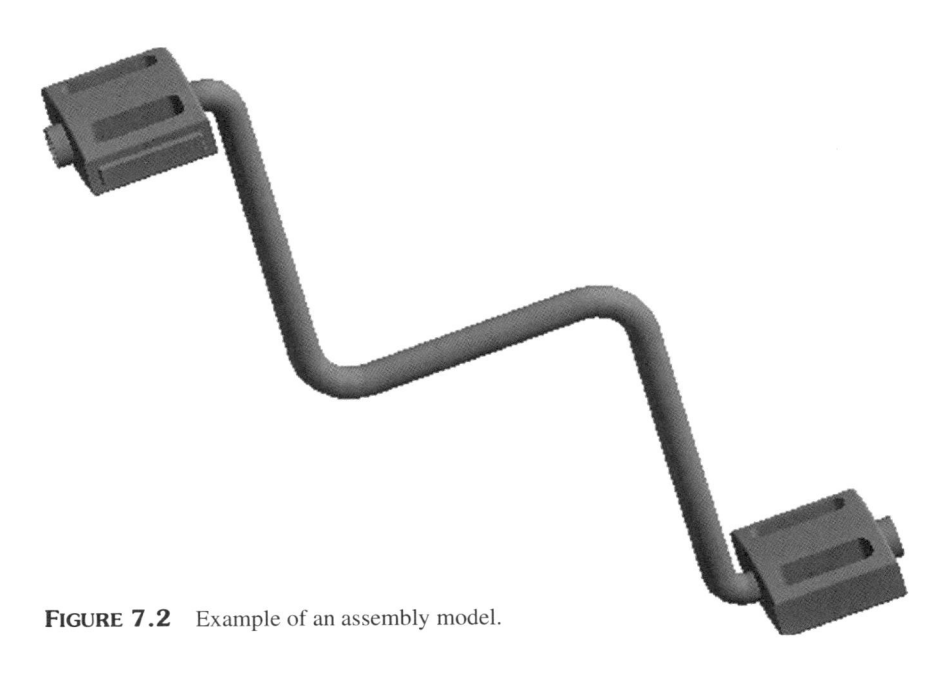

FIGURE 7.2 Example of an assembly model.

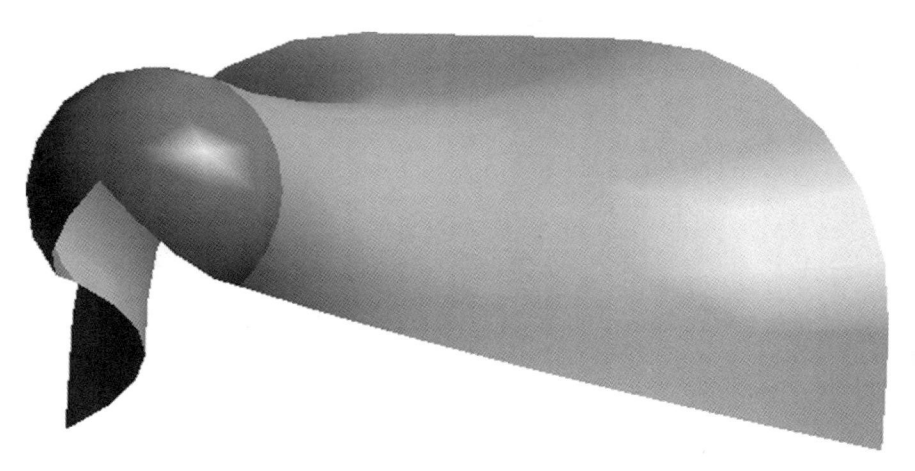

FIGURE 7.3 Example of a surface model.

7.2.2 Geometric Terminology

Section A section can be thought of as a cross section of a part model or a cross section of a feature of a part model. A section that is closed encloses a 2-D area in much the same way that a solid model encloses a volume. Often, CAD systems refer to section properties. These properties include area, the centroid (basically the geometric center or CG of the section), and inertias (a calculation that weights the distribution of the area with respect to an axis). Figure 7.4 shows an example of a section. The section is the bold line that looks like the basic cross-sectional shape of the part shown in Figure 7.1.

Vector A vector is a mathematical device for indicating that something has a direction as well as a value. It can be thought of as an arrow. The longer the arrow, the larger the value that the vector signifies. The direction, then, is indicated by the direction the arrow points. A commonly known vector quantity is a force. A force being exerted on something has a definite direction, and it has a value (i.e. how much force in Newtons or pounds-force). Figure 7.5 shows an example of a vector. In this example, the length or magnitude of the vector is 163. A common use of the vector in 3-D CAD is to indicate a direction for a construction (such as, protrude in "this direction"). It can also be used to show a distance with a direction. In Figure 7.5, this vector could indicate the distance to an arbitrary location on a part as 163 mm. This 163 mm could also be shown as having components in the X, Y, and Z direction. These components would indicate what the change in distance is from one point in space to another (these changes may also be referred to as delta X, delta Y, and delta Z).

3-D CAD

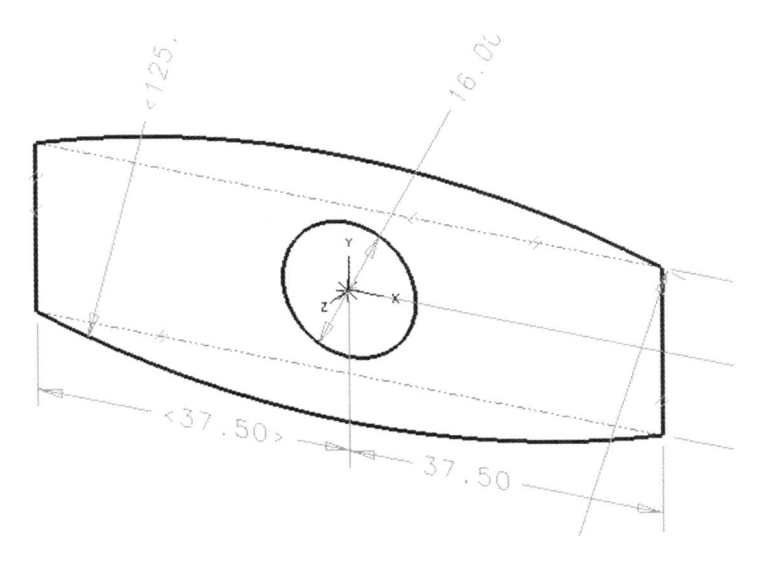

FIGURE 7.4 Example of a section.

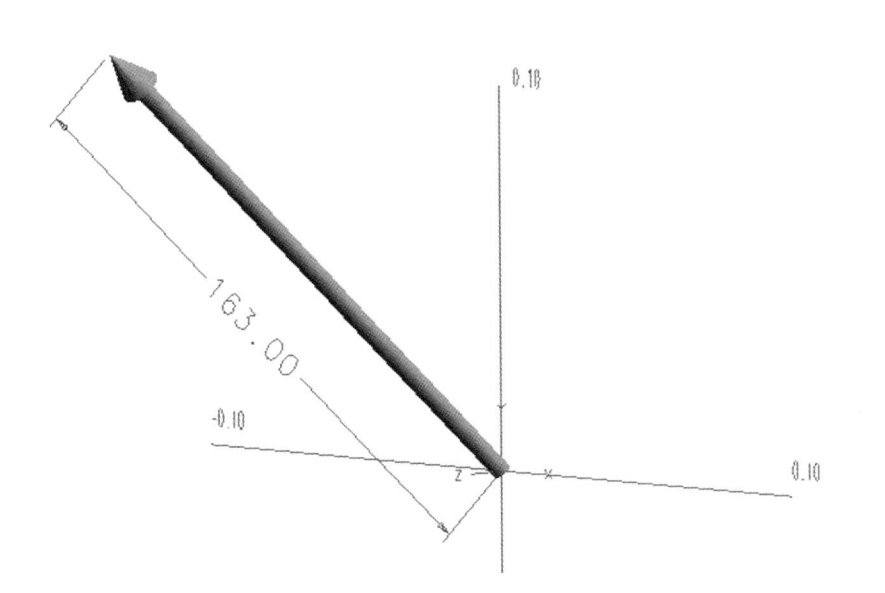

FIGURE 7.5 Example of a vector.

Normal vector   A normal vector is a special case of a vector. In this case, the vector's direction is dictated by a surface. At some point on this surface, the normal vector points perfectly perpendicular from the surface. No matter what plane is chosen that the vector lies in, the vector still points perpendicularly from the surface at that plane's view of the model. Figure 7.6 shows a normal vector; note how it is perpendicular to the surface at the point where the vector touches the surface.

Unit vector   A unit vector is another special case of a vector. In this case, the magnitude or numerical value associated with the vector is 1. In this case, the vector only indicates direction, not magnitude. This is a common operation in 3-D CAD systems when the system needs the user to specify an arbitrary direction. The CAD system may prompt the user for a unit vector.

Surface   A surface (in 3-D CAD anyway) is any arbitrary, infinitely thin boundary. A surface can be totally flat (a planar surface). A surface can be very curved or warped. It can have straight edges. It can have circular or curved edges. It is

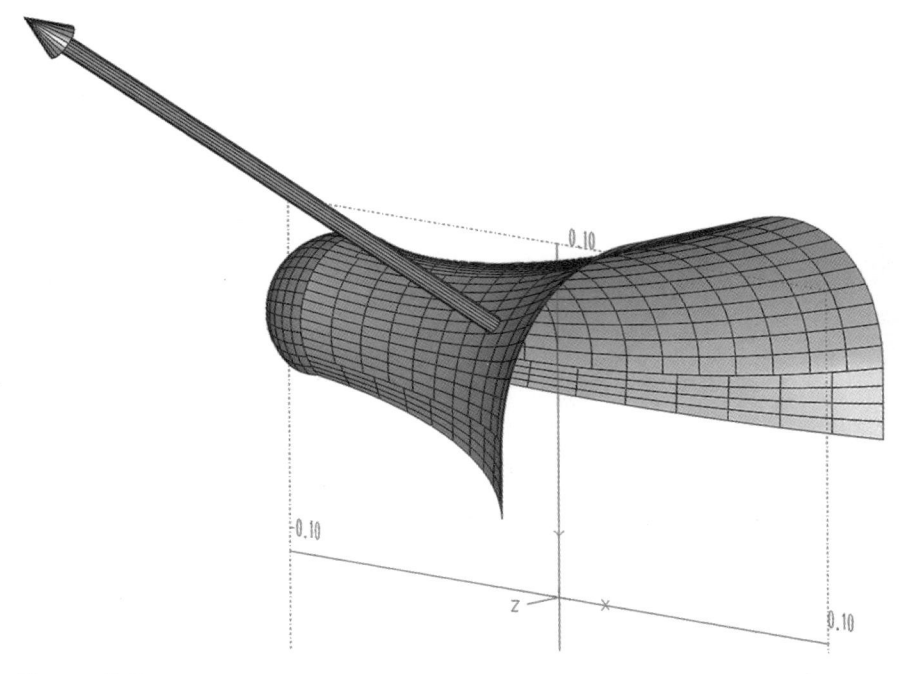

FIGURE 7.6   Example of a normal vector.

3-D CAD

often based on something called a NURBS (which is explained in a little detail in Chapter 9); so some systems or users will use the term NURBS interchangeably with a surface. Suffice it to say that a surface is defined by some specific mathematical equations. Figure 7.3 (an example of a surface model) shows 2 surfaces that are connected together (the rounded front and the saddle-like body are each surfaces).

7.2.3 Graphics Terminology

In addition to the general terminology of the 3-D CAD system, there is a certain amount of terminology that has arisen from computer graphics technology. These terms may have slightly different uses in different CAD systems, but the basic meaning is usually the same as presented here.

Wireframe or wireframe view The most obvious computer graphics term is wireframe. This refers to a 3-D model that is not shown with its surfaces. Instead, it just shows the edges of the surfaces. As mentioned earlier, this was the only approach to 3-D CAD at one time, so for some users, wireframe will mean doing 3-D CAD without solid models. However, this work assumes that this is only a concern for companies that have this older technology in their "legacy" systems (meaning they are not used in the latest product development activities). Figure 7.7 shows the wireframe graphics on the right.

Solid or shaded view This might be called the opposite of the wireframe display. Referring to displaying the 3-D graphics with the surfaces instead of the edges, it shows the state of the design much more clearly than wireframe. However, when a designer needs to work on a part model's interior, it is often very helpful to switch back to the wireframe display. Figure 7.7 shows the solid view or shaded graphics on the left. The important point is that this is just a graphics issue; it does not change the nature of the part model, only how it appears to the user.

Pan The option of changing the view of the 3-D model on the screen by moving left and right, or up and down. This is usually done with the mouse and/or keyboard.

Zoom The option of changing the view of the 3-D model on the screen by appearing to get closer to or farther from the model.

Rotate The option of changing the view of the 3-D model by rotating the model with respect to various axes. Often the model can be rotated with respect to the observer; or the observer's point of view can be changed.

Clipping Cutting away portions of the 3-D model to reveal internal surfaces. This is different than actually using a cutout feature described earlier (which changes the model). Clipping does not change the 3-D model; it only changes the appearance of the model on the computer monitor. Figure 7.8 shows an example of a clipping display.

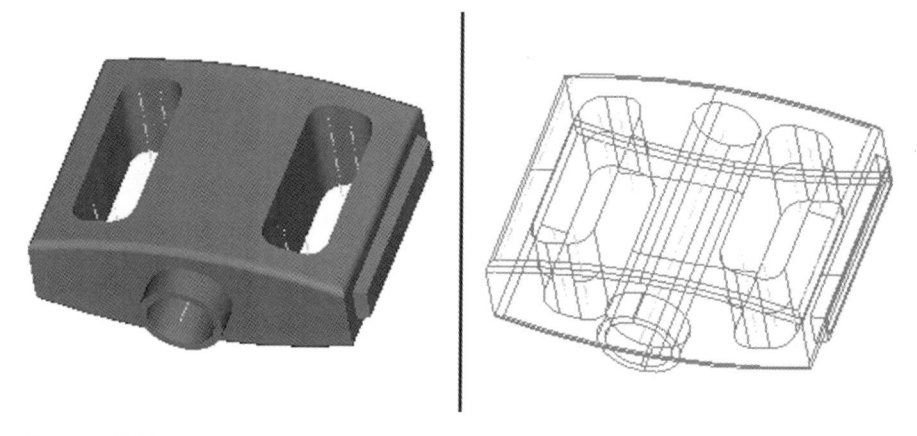

FIGURE 7.7 Comparison of shaded and wireframe display.

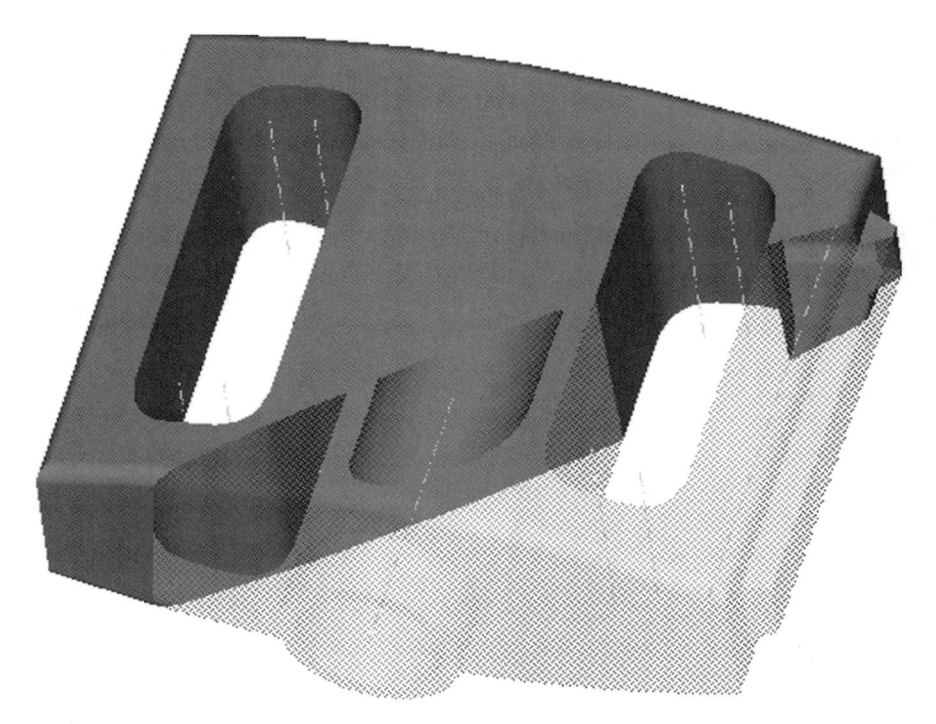

FIGURE 7.8 An example of a clipped display.

3-D CAD

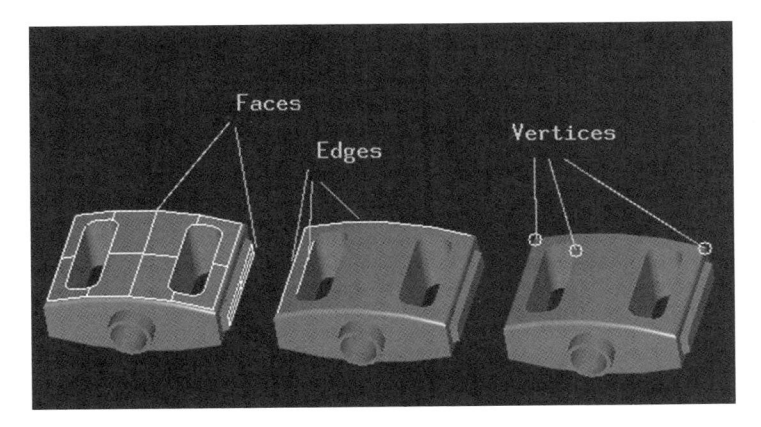

FIGURE 7.9 Examples of faces, edges, and vertices.

Triad A coordinate system that indicates the overall orientation of the 3-D model being displayed. As the model is rotated or manipulated, a triad may be shown on the screen to give the proper bearings of which directions are controlling the display. It appears as an X, Y, Z with some direction lines.

Face A somewhat interchangeable term with surface. Usually, however, a face implies that the surface is used in the construction of a solid model. Some users may consider a face to be planar or flat, but often it can also be any curved or warped surface as well. Figure 7.9 shows some highlighted faces.

Edge Usually refers to the boundary of a surface. Although an edge is clearly found at a sharp corner where faces meet, an edge would also be indicated by the boundary of a surface model that does not meet any other surface. Figure 7.9 shows some highlighted edges.

Vertex Refers to a point at the end of an edge. Again, a vertex certainly is found at a corner where 3 faces meet (as in the corner of a box), but a vertex could also be found where 2 separate open surfaces meet, and vertices would even be found on a single open surface (at each of the edges of the surface).

7.3 DISPELLING COMMON MISCONCEPTIONS

For better or worse, many CAD system vendors have tried to make 3-D CAD systems as fundamental and simplistic as possible. This has certainly expanded the reach of the software, but this unfortunately insulates the user from grappling with a number of issues that should not be ignored. These issues often relate to making a good mathematical model that unleashes the underlying power of the 3-D CAD system (instead of just making a model that looks good at the moment).

What designers really need to do is make models that can automatically store what they meant or their design intentions. They also need to know how the 3-D CAD modeling works so that they can debug the model when it can't comply with the desired design intent. In other words, it is important that users gain a basic understanding of how the modeling software really works (as opposed to just mimicking the models from an on-line tutorial). In order to gain this deeper understanding of the software, there are three misconceptions that most users need to overcome. The remainder of this section attempts to address these misconceptions.

7.3.1 "You Can't Do 3-D Without 2-D First"

It may come as a surprise to some users that good 3-D models almost always require a good 2-D foundation. So, time spent understanding 2-D drawings or planar mathematics is not a waste of time. And, although some 3-D CAD systems have the option to create arbitrary 3-D surfaces without any 2-D geometry (using points, lines, splines, etc. in space), building part models in this fashion is often not productive. Instead, the 3-D CAD user needs to figure out what 2-D geometry can be used to build up the 3-D model. This often relates to finding various planes within the imagined part that can be used as a place to sketch 2-D geometry.

In Figure 7.10, one can see the underlying 2-D entities (lines, dimensions, etc.) that make up a particular section or feature of a 3-D model. The plane on which these 2-D entities are created should not be thought of as a drawing, however. The 3-D CAD system will refer to it as a sketch or sketch plane or sketch pad or sketching datum. The sketch is more informal than a drawing since it is not expected to be seen by anyone besides the designer (unlike a drawing that is sent to manufacturing, vendors, customers, etc.). However, the sketch is more formal than a drawing in a mathematical sense. The geometric entities on the sketch plane are "intelligent." The 4 lines that form a rectangle on a 2-D CAD system are really just 4 separate entities with no formal relationship to each other. In the 3-D CAD system, however, 4 lines that are created for the rectangle shown in Figure 7.10 will have relationships to each other. The 3-D CAD system knows that the end points of the lines are connected. This allows the 3-D CAD system to create 4 faces based on the 4 lines and know whether the 4 faces touch or share an edge.

Beyond the relationships between the various entities on the sketch plane, the sketch plane usually also offers the ability to parameterize dimensions. This means that mathematical formulae can be programmed into the sketch plane so that one dimension drives another dimension. For example, a hole that is supposed to be 25% of the distance between two edges, can be programmed to be

3-D CAD

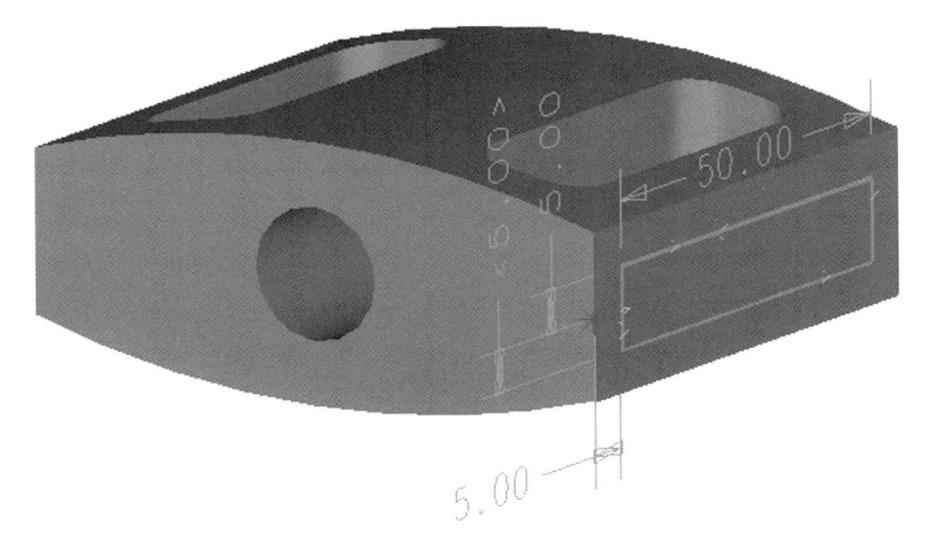

FIGURE 7.10 An example of the 2-D foundation for a 3-D feature.

located at 0.25 times the dimension between the edges. Then, when the edges move, the hole will automatically reorient itself to be at the correct location. This is generally what is meant by making a parametric or intelligent model, and this is an example of capturing design intent. Although this concept can be expanded to include parametric relationships between many of the features of a single part model, or even to the relationships between totally different parts models (such as in an assembly model), one should first master this parametric concept at the 2-D level within the context of the sketch plane, and then build on the possibilities offered in the full 3-D part modeling and/or assembly modeling.

So, as one learns to master 3-D CAD, one should not be mislead into thinking that mastering 2-D geometry is no longer needed. Indeed, understanding 2-D mathematics may be even more important now.

7.3.2 "Solid Models Are Really Just Special Surface Models"

As mentioned earlier, there are different types of 3-D models. For instance, a surface model is a model that is open (refer to Figure 7.3). The faces or surfaces of the model do not completely enclose a volume of space. It is also basically a physical impossibility. No physical part is going to have infinitely thin walls like the surface model.

The more common type of model is the solid model. The solid model is physically realistic since it does have a volume associated with it. However, it is really a misconception to think that they are different kinds of models at all. In fact, the solid model is a surface model. It is merely a special case of surface model where all the surfaces meet at mutually shared edges (within a tolerance or limit). For the solid model, it is said that the surfaces are completely stitched, and the surfaces become faces of the solid model.

There are a number of advantages to understanding the nature of the surface model within the solid model. First, there are solid models that are best made by starting with a surface model. In this case, an open part is created that has the basic shape needed (with no thickness applied). Then, an amount of thickness is added all over the surfaces. This is usually referred to as shelling a part. Figure 7.11 shows an example of this type of part; notice that there is now wall thickness.

Another case of using the surface nature of solid modeling to one's advantage is making a normal solid model, then making it open, and then making the open part solid again (refer to Figure 7.12). In this case, one surface was removed from the solid model. This exposed the internal surfaces. These surfaces can then be used as the basis for a part that can then be shelled to regain a solid model.

Finally, understanding the underlying surfaces in solid models helps one correct or debug part models that are not working as desired. For instance, when one sketches a 2-D shape to be extruded into a solid (such as a rectangle to be

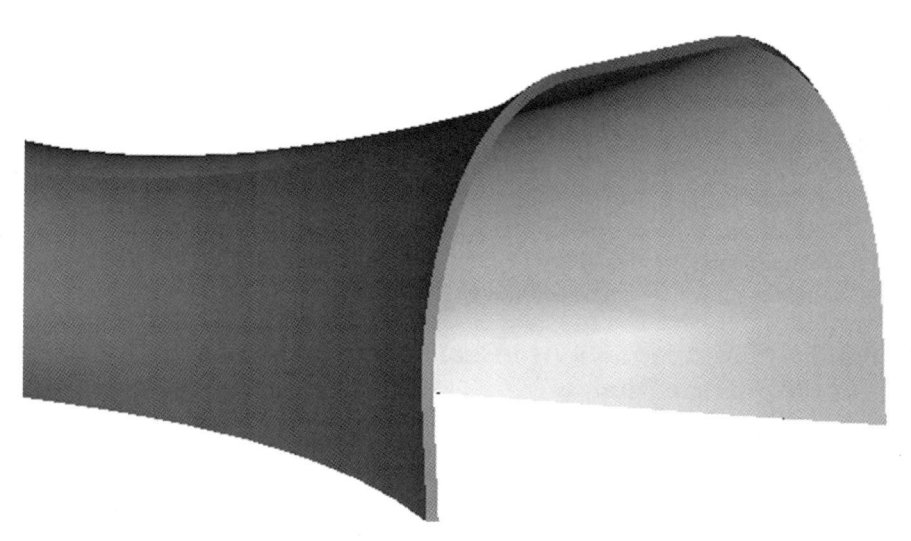

FIGURE 7.11 An example of a shelled surface model.

3-D CAD

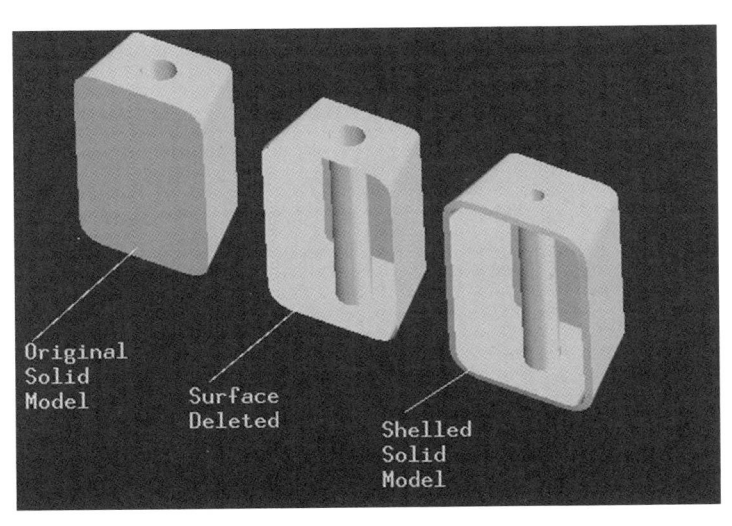

FIGURE 7.12 An example of making a solid part after deleting a surface from the original model.

extruded into a block), but one does not make sure that the lines of the 2-D shape do not connect at their ends, the CAD system may wind up making a surface model instead of a solid model. As shown in Figure 7.13, a small gap between the edges of the surfaces is left where the original 2-D lines did not connect. Sometimes it is not clear that this has occurred (particularly for small features on a part). One way to correct the problem is to have the CAD system look for free or unstitched edges of surfaces.

So, as one learns to master 3-D CAD, one needs to realize that the solid model is really a special case of a surface model.

7.3.3 "You Are a Historian, Not a Sculptor"

The next concept, called part history, is is probably the most difficult concept for new users of 3-D CAD systems to appreciate. This is probably because there is a natural tendency to view the 3-D model creation as an exercise in sculpting. New users often create a basic shape and then cut away segments of that shape until it has the desired final shape. Although this may be an appropriate modeling technique in some cases, there are usually much better approaches. These better techniques are usually superior in their ability to permit fast and easy modifying or tweaking the part over the design iteration process.

Instead of viewing modeling as sculpting, users should consider part modeling as creating history. The 3-D CAD system does not electronically store tril-

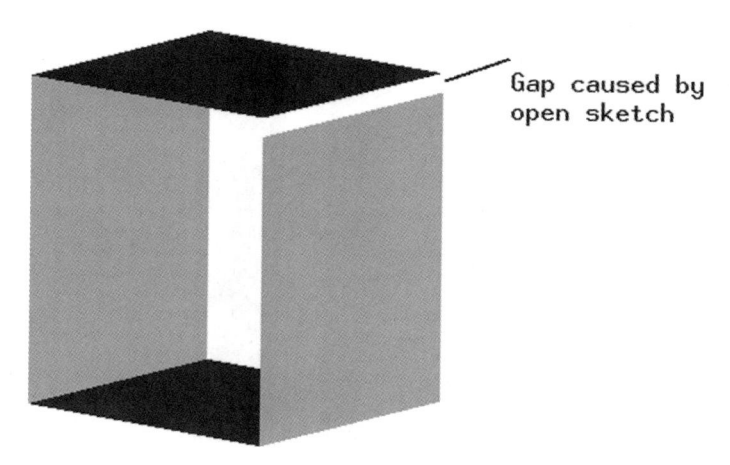

FIGURE 7.13 An example of a surface model created by mistake.

lions of little specks that the user carves away (as if sculpting with 3-D pixels). Instead, the 3-D CAD system records or tracks or captures the steps that a user has followed during the creation of the 3-D part model (whether it is a surface model or a solid model). Recording these steps does not really take up much space on the computer system (as would the 3-D pixels in the sculpting idea). The history of steps is also a concise means of storing the intelligence of the mathematical model of the part model. It also retains a logical approach to how a part has been modeled. The designer starts with the some big first step that gives the basic geometry, and then one adds, removes, or builds upon that base to get the desired level of detail for the final 3-D part model.

Although there are some 3-D CAD systems that do not use this history-based modeling method, the vast majority of systems and users do use this method in one way or another. CAD systems may refer to this recording of steps as part history, model history, history tree, model tree, feature list, etc. They all mean the same thing—the user's sketching and modeling process for the part is stored in some sort of sequential format. Figure 7.14 shows some steps in the formation of a part. The steps proceed from the top of the list to the bottom. These are steps that would be used to create the part in Figure 7.1.

This approach to 3-D modeling might also be considered features-based modeling. This is because each of the major steps in the part history may be called features. A boss protrusion sticking out of a part would be a feature; a particular hole drilled through the side of the boss would be a feature, etc., and each of these features would be found as a step in the part history. However, although

3-D CAD

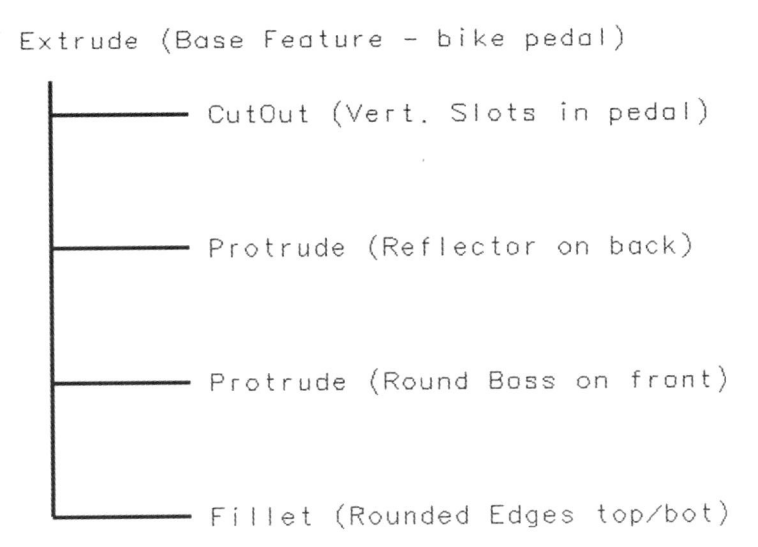

```
Extrude (Base Feature - bike pedal)
    |──────── CutOut (Vert. Slots in pedal)
    |
    |
    |──────── Protrude (Reflector on back)
    |
    |
    |──────── Protrude (Round Boss on front)
    |
    |
    |──────── Fillet (Rounded Edges top/bot)
```

FIGURE 7.14 An example of part history.

all 3-D models should have features, some systems may not track this history as fully as others. So it seems to be more accurate to refer to what most 3-D CAD systems are doing as history-based, not just features-based.

As with the other misconceptions that are trying to be dispelled here, it turns out that there are significant advantages to the user if they understand what the history really means. First of all, the terminology associated with part modeling makes more sense. The 3-D part models can be rolled back, replayed, updated, re-generated, stepped through, etc. All these terms are referring to activities relevant to the part history. Refer to Table 7.2 for a simple explanation of some of these terms.

An extremely important concept that needs to be understood with respect to history-based modeling is that the order of steps is usually very important to making a good 3-D part model. One can not just pick any order to follow for the steps. For instance, if one wants to create a hole in a part using a specific flat planar surface as the sketch plane, then that surface must exist at the moment that one needs to make the hole. If the planar surface turns into a curved surface by some other previous step in the history, then the flat plane is lost. So, the hole is probably going to be "messed up" in some way. By the same token, once the hole is created in the history, then it often does not matter if the plane is destroyed in a later step in the history. What is important is that the plane existed at the appropriate time.

TABLE 7.2 Part History Terminology

Terminology	Basic Explanation
Feature	A step in the history of a part.
Replay or regenerate	Replaying a history-based model refers to having the CAD system recreate the 3-D part starting with the first step in the history.
Rollback or midprocess	Rolling back refers to viewing the start of a part model in the middle of the history. This term may also refer to allowing the creation of features or steps in the middle of the history.
Reorder	Reordering refers to changing the order that features or steps appear in the history.
History-supported	An operation or change to a part that is "tracked" in the history of the part. Some steps are actually shown on the part, but cannot be found in the history, and thus cannot be manipulated as an independent feature of the part.

So, as one learns to master 3-D CAD, one needs to plan an approach or strategy for part modeling. Sometimes, it can help to think about how the part is going to be manufactured, but sometimes even this is not sufficient. Designers need to gain enough experience with modeling their type of parts, and to be able to look at or imagine the part and then mentally plan how to create the 3-D model. This can be a daunting task at times, but it can also be a fun challenge.

7.4 CHAPTER EXERCISES

1. Try to determine when your company or university first started using 3-D CAD. Record the name of the software package that was adopted.
2. How much did the software cost per user (or "per seat")?
3. Compare and contrast three of the currently available 3-D CAD systems. Look at issues such as their cost per user, special capabilities, technical support, supported platforms, customer base, etc.

7.5 CHAPTER REVIEW

In your own words, define the following terms:

1. Model
2. Part model
3. Feature
4. Assembly model
5. Surface model
6. Section
7. Vector
8. Normal vector
9. Unit vector
10. Surface
11. Clipping
12. Triad

8
Part Modeling

8.1 INTRODUCTION

Part modeling uses a 3-D CAD system to create a single model for a single, stand-alone object. These models can be referred to as parts or part files. Also, designers familiar with 2-D design will sometimes refer to these items as details (meaning they were documented on a detail drawing). In this work, however, they will be referred to as parts or part models.

    Recall from the previous chapter the number of misconceptions often encountered by users of 3-D CAD systems. Try to keep these issues in mind during the discussion of this chapter to see how they are relevant to part modeling in general. These issues were: 1) one must often create 2-D geometric entities before 3-D features can be created; 2) the 3-D model is not really solid, but really a set of paper thin surfaces that meet at common edges (the next chapter on surface modeling discusses this in more detail); 3) a more complicated 3-D model is created by following a series of steps called a history, and the system remembers and tracks these steps.

Part Modeling

8.2 THE THREE BASIC STEPS

In many cases for many different 3-D CAD systems, there is a basic three-step process that is employed in creating and then expanding a 3-D part model. Table 8.1 lists the 3 steps in a brief list.

Step 1   Selecting the Sketch Plane

As mentioned before, many 3-D features start with 2-D geometry. Obviously, this 2-D geometry cannot be created unless the CAD system knows what 2-D plane is being employed. In a 2-D CAD system, this is no problem; one draws on the virtual paper. In a 3-D CAD system, however, this is an important issue; one needs to think about where to sketch in a 3-dimensional space.

So, Step 1 is concerned with picking or creating this sketch plane. This sketch plane may be an existing face of a part (meaning that an additional feature is being created on an existing part model), or the sketch plane may be an imaginary flat space or plane shown on the screen, or the sketch plane may be a plane from a blank or prototype part that just has axis planes, datums, or coordinate system planes.

Step 2   Creating 2-D Geometry

Once the sketch plane is specified, the CAD system can now be used to create the base or foundation 2-D geometry. For example, if one wants to make a cylinder, then one needs to sketch a circle first. If one wants to make a block, then one needs to sketch a rectangle. Figure 8.1 shows a sketch example. Keep in mind, though, that a 3-D CAD system will go beyond the simple creation of these 2-D geometric entities; it will also permit intelligent modeling or constraining for the 2-D geometry. Constraining is explained in more detail in a later section of this chapter.

TABLE 8.1   The Three Basic Steps

Step	Description
Step 1—Selecting the sketch plane	Picking a plane to start from
Step 2—Creating 2-D geometry	Sketching some base or foundation 2-D geometry (see Figure 8.1)
Step 3—Making the feature	Turning the base geometry into a 3-D feature (see Figure 8.2)

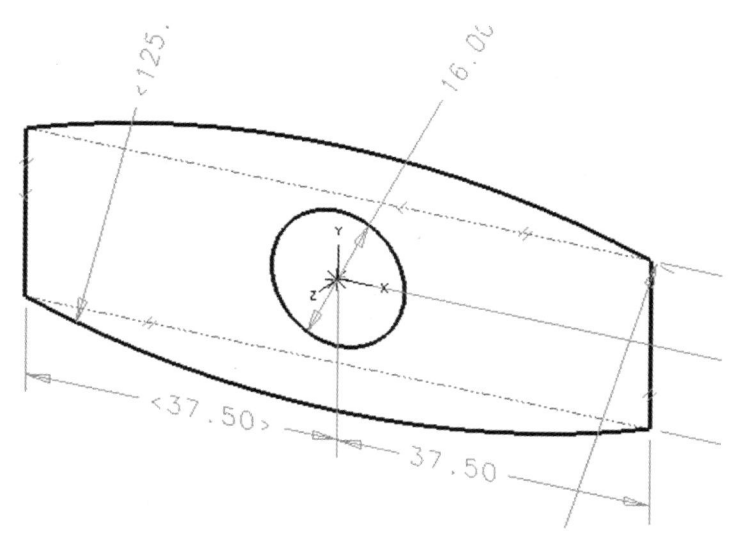

FIGURE 8.1 Creating or sketching 2-D geometry.

Step 3 Making the 3-D Feature

Once the 2-D geometry has been created (on the sketch plane, of course), then this geometry can be used in Step 3 to create or expand the part model with new 3-D geometry. Basically this involves selecting what segments of the 2-D sketch geometry (from Step 2) are to be used in making the new feature and then selecting the method to create the feature (protruding, revolving, sweeping, etc.). Notice that there are some horizontal construction lines in Figure 8.1 that are not part of the bold lines (or section) that are used to create the feature shown in Figure 8.2. It is often best to leave these construction lines in the sketch permanently (sketching is not an exercise of making a drawing).

Table 8.2 lists the basic 3-D feature creation methods that usually require initial 2-D geometry (later on there are some other methods that don't actually require 2-D sketching). Figure 8.3 shows what they create using a circle as the initial 2-D geometry.

Recognizing and understanding the basic three-step process should be helpful in optimizing the potential of 3-D part modeling. It is simple and logical, and it lends itself to the understanding of how history-based modeling works. Some CAD systems will make the three steps recognizable in the process of modeling, while other CAD systems will not make it obvious. Regardless, the user needs to understand that this process that may be occurring behind the scenes.

Part Modeling 191

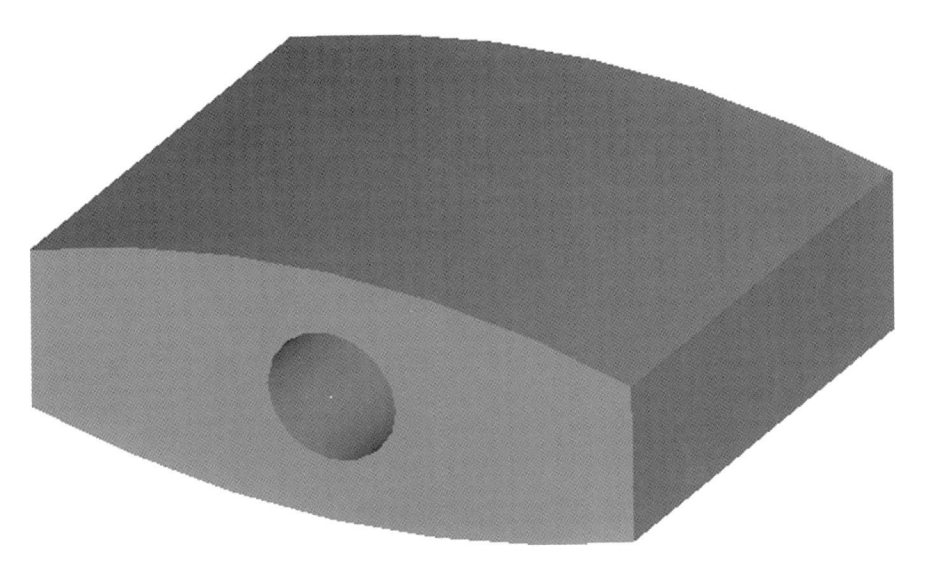

FIGURE 8.2 Making the feature.

TABLE 8.2 The Most Common Methods of Making a Feature

Method	Description
Extruding	Extruding means taking the 2-D geometry and creating surfaces that follow along a single direction or vector. Depending on the user's selection, this can be used to create a protrusion (a new positive volume) or a cut (removing existing volume from a solid). This is probably the most common 3-D feature creation method.
Revolving	Revolving means taking the 2-D geometry and creating surfaces that revolve about an axis. If this is done to add a feature to an existing part, then this operation can be used to add volume (as in a protrusion) or remove volume (as in a cut).
Sweeping	Sweeping means taking the 2-D geometry and creating surfaces by dragging the geometry along some sort of arbitrary, but smooth path. A typical example is a pipe or tube with bends.

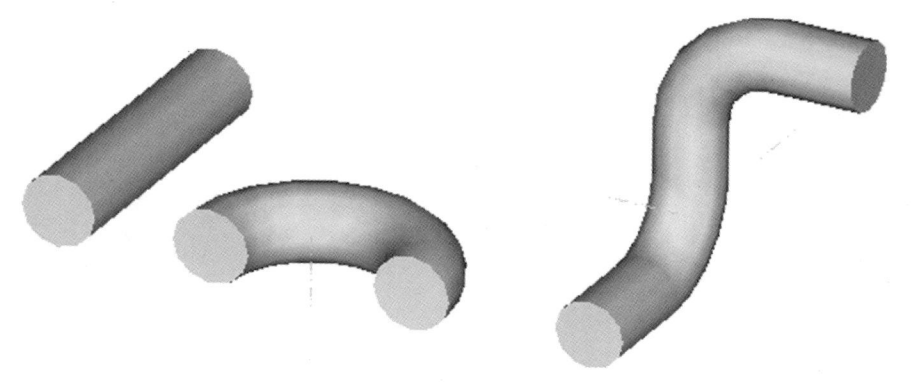

FIGURE 8.3 Examples of the common methods of making a feature: extruding, revolving, sweeping.

8.3 CREATING NEW FEATURES

The three basic steps just explained are going to be useful in not just starting or creating a new part model, but also in expanding upon a part model. Although one generally thinks of the word "feature" meaning something that is added onto an existing part, "feature" frequently refers to the first history step that starts a part model. In this case, though, it can be referred to as the "base feature" or perhaps the "base node" of a history tree. In any case, creating the part model with the base feature is usually just the beginning of the part model. Although some very simple parts (such as flat plates) might only require the base feature, most 3-D parts are going to require that new features be added onto the base feature.

To create these added features, one can often apply the three basic steps once again. However, in Step 1, one must be careful to pick a sketch plane that is related to or associated with the base feature. This may seem obvious, but many new 3-D CAD users miss this point. If the 3-D CAD system allows the user to create a base feature by just sketching out in space, then the user may be tempted to just start sketching out in space a second time to make the next feature for the part. However, this really skips Step 1 (selecting a sketch plane). If the user does not select this plane carefully using geometry from an existing part model, the system will simply assume that the user wants to start a second, separate part model. Although it is possible to have these two part models connected later (using a technique known as a Boolean operation), it is usually preferable to have new features automatically connected with the base feature. So, to do this, the

Part Modeling

user needs to be careful to pick a sketch plane (or to be careful to think of all three steps in the three-step process).

8.3.1 Sketch Plane Based on a Face

The simplest way to be sure that the sketch plane is connected to the base feature is to pick a face of the base feature. This face would be a flat surface (a planar surface) on the existing part model. Figure 8.4 shows a sketch plane selected from an existing face of a part.

8.3.2 Sketch Plane Based on a Datum Plane

Obviously, sometimes there are no visible faces of a part that supply a plane in the orientation or location needed for the new feature. At other times, the use of a face for a sketch plane is considered too risky. For instance, a face that could be used as a sketch plane is not yet fully designed or is still subject to change. In these cases, a different approach is needed for the sketch plane. Often, the solution is to create and/or select a sketch plane based on a datum, reference geometry, or construction geometry.

For drawings, a datum refers to a plane that is considered a foundation or baseline for dimensions on a drawing. This plane is given a letter designation in a notation system called Geometric Dimensioning and Tolerancing (GD&T). This datum concept is easily transferred to the 3-D model for a selecting a face of the

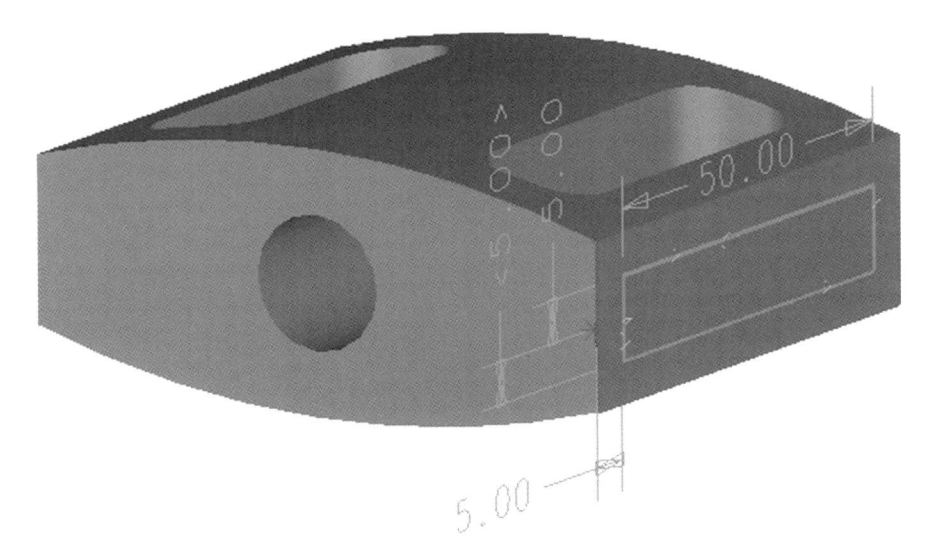

FIGURE 8.4 An existing face as a new feature's sketch plane.

part. But this idea is often expanded upon in the 3-D CAD system to include planes that can be created based on arbitrary X, Y, Z coordinates in space or planes created based on various geometric techniques. For instance, a datum plane can be created by selecting any three points on a part, or a plane can be created by specifying 1 point and a normal vector to the plane. In each of these cases, a plane is being created that can be used as a sketch plane, and it is created based on geometric entities that already exist in the part. This is the essential concept if one wants this new feature to be part of the history of the existing part. Take care when creating these types of sketch planes that they are not just out in space and unconnected to the existing part.

Figure 8.5 shows a datum-type of plane created by using the centerline of the circular feature and then a midpoint of one of the vertical edges. This plane is basically located at the center of the part.

8.3.3 Sketch Plane based on a coordinate system

Another common approach for sketch planes is the use of coordinate systems. A coordinate system usually specifies the X-, Y-, and Z-directions (refer to Figure 8.6). Obviously, the X-, Y-, and Z-directions are all perpendicular to each other (generally referred to as being orthogonal). Also, the Z-direction should always be in a consistent direction based on the X- and Y-directions (if one draws X and Y on a piece of paper in the "normal" orientation of X to the right and Y going up, then the Z direction is out from the paper toward the viewer; this is dictated by something called the "right-hand rule" or "right-handed coordinate systems").

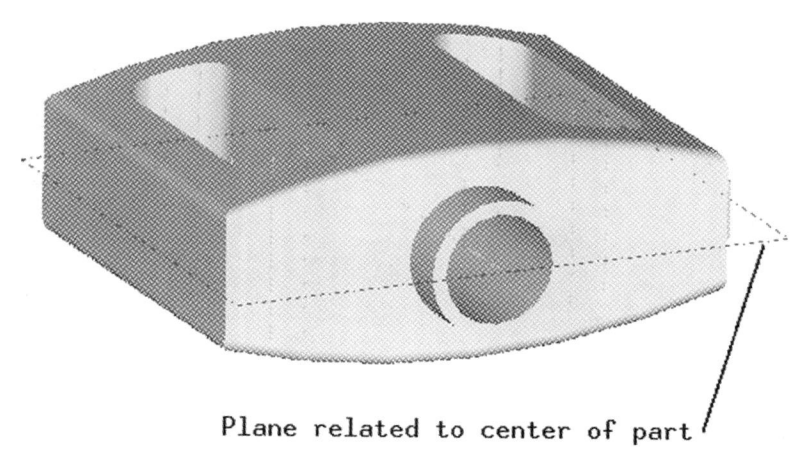

FIGURE 8.5 A datum or reference plane that can be used as a sketch plane.

Part Modeling

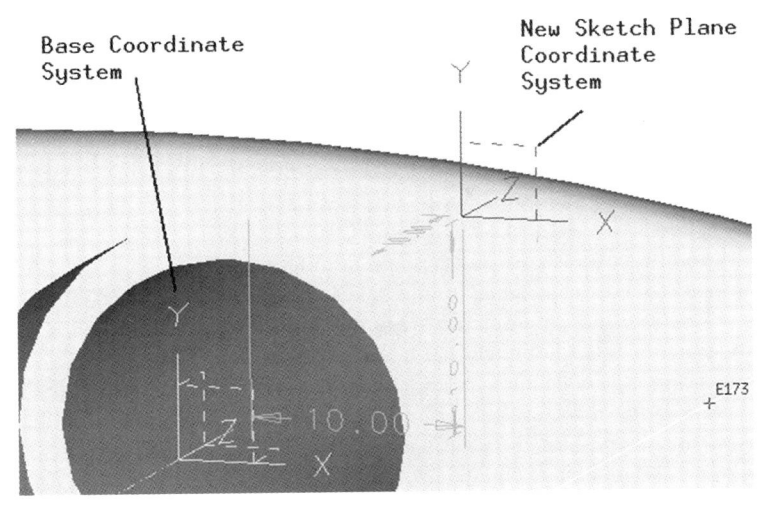

FIGURE 8.6 A coordinate system axis plane as a new feature sketch plane.

Usually, these coordinate systems are shown in the 3-D CAD system with axis planes, as well. These axis planes are planes that are formed by the 2 lines of the X-, Y-, and Z-directions. For instance, the XY plane on the coordinate system would be the plane that contains the X- and Y-directions (extending infinitely in each direction). The XZ plane contains the X- and Z-directions, and the YZ plane contains the Y- and Z-directions. Clearly, these axis planes can be quite useful for providing sketch planes. Often the X-, Y-, and Z-directions are the basis for many features (these parts may be referred to as prismatic parts). Indeed, some 3-D CAD systems (either by default or by standard user practice) always create a coordinate system for parts (whether the user realizes it is there or not).

If a default coordinate system for a part does not provide the needed axis planes for sketching a new feature, then the CAD system may allow new coordinate systems to be created based on other geometry of the part. This is quite similar to the creation of a datum or reference geometry; however, coordinate systems are often better since they supply three potential sketch planes (the axis planes) instead of just one plane (the datum plane). Figure 8.6 shows a coordinate system that has been offset from a base coordinate system. The first one is for a base feature that is already created. The second one (which is being shown as a sketch plane for some new 2-D geometry) is offset from the first one by three linear dimensions in the X-, Y-, and Z-directions. This is a quite useful technique since it gives easily modifiable dimensions to control the new feature.

8.3.4 Sketching for the Added Feature

Once the associative sketch plane has been successfully selected, the next step in adding the feature is doing the sketch. As before, this involves creating 2-D geometry that forms the foundation for the new 3-D feature. Normally, this sketching is no different that the sketching that made the base feature or started the part model. One uses the 2-D sketching tools or commands to create lines, arcs, splines, etc. as necessary.

Normally the process of constraining and adding parametric or design intent information is exactly the same as for the base feature, as well. However, this process becomes a little more interesting since the new feature being created is surrounded by existing 3-D geometry that can be used in interesting ways. For instance, if one wants a new feature to start where another feature left off, or if one wants an edge of the new feature to be perpendicular to the edge of another feature, then this may be possible by relating or constraining the new 2-D geometry to the existing 3-D geometry. One needs to be careful, though, that the existing 3-D geometry is actually from the same part model that is being sketched on. Otherwise, changes to the existing feature will not drive changes in the new feature (ignoring for now how assembly models might accomplish this). Figure 8.7 shows how a 2-D point (shown as an asterisk) has been projected onto the sketch plane by using an existing 3-D vertex. Since the 3-D vertex is from a preceding feature in the part history, changes to that 3-D vertex should automatically

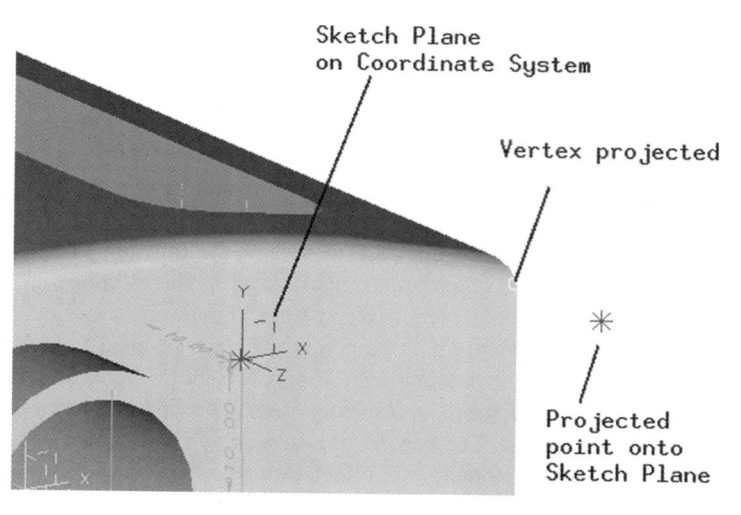

FIGURE 8.7 Using an existing 3-D vertex for use with sketching 2-D geometry.

Part Modeling

update the sketch plane point; then the new feature being created will also adjust automatically.

8.3.5 Making the New 3-D Feature

Following the three-step process once again, the next step after creating the 2-D sketch geometry is actually making the new 3-D feature. As with the base feature, the common methods shown in Table 8.2 are very often used. One could use an extrusion to make the 3-D feature prismatic (it sticks straight out); one could use a revolve to make the 3-D feature revolute; one could use a sweep to make a path-following 3-D feature, etc.

It is very important to notice, however, that indicating a method like "Extrude" is not enough for added features. This is because the new feature is going to operate on the existing part, so the CAD system needs to know how the new feature is going to affect the existing feature. The obvious example is deciding if the extrusion is a protrusion or a cutout. If the user indicates that the extrusion is to protrude, then the CAD system joins the new 3-D surfaces to the existing part, and it sticks out from the part as a protrusion. However, if the user indicates that the extrusion is a cutout, then the CAD system makes the new 3-D surfaces cut into the existing part. Table 8.3 lists the usual operations for new features acting on existing part models.

Sometimes it seems unnecessary to have to indicate whether the new feature is to protrude or cutout since the CAD system often will often prompt the

TABLE 8.3 Some Operations that Added Features Can Perform

Operation	Description
Protruding	This operation means that the new feature is going to add volume to the existing part. After extruding, revolving, etc., the new 2-D geometry forms its own 3-D geometry, and then the system combines the 2 volumes and figures out how they connect together to create a new single volume.
Cutting	This operation means that the new feature is going to remove volume from the existing part. After extruding, revolving, etc., the new 2-D geometry forms its own 3-D geometry, and then the CAD system takes away the new volume from the existing volume to create a new single volume.
Intersecting	This operation means that the new feature is going to be a junction. After extruding, revolving, etc., the new 2-D geometry forms its own 3-D geometry, and then the CAD system will create a new volume that is made up of the space that is shared by both new volume and the existing volume.

198   **Chapter 8**

user to indicate a direction to extrude. If one makes this direction away from the existing part, then one would think it will always be a protrusion. If one makes the direction into the existing part, then it should be a cutout. However, there are some cases, where the new extrusion will interfere with other features, and it will not be clear what the user intended based on the direction of the extrude alone. In Figure 8.8, for example, a new rectangular feature is shown being created from the sketch plane coordinate system shown in Figure 8.7. On the left, the new feature is extruded as a protrusion; on the right, the new feature is extruded as a cut. The direction alone would not be enough to let one know what the designer intended.

8.3.6 Surface Issues for New Features

Although indicating whether a new feature is a protrude, cutout, or intersection is very often all that is needed to create the desired feature, there are some underlying surface issues that need to be considered. This deeper understanding will allow the user to figure out why certain 3-D feature operations are successful, while others fail.

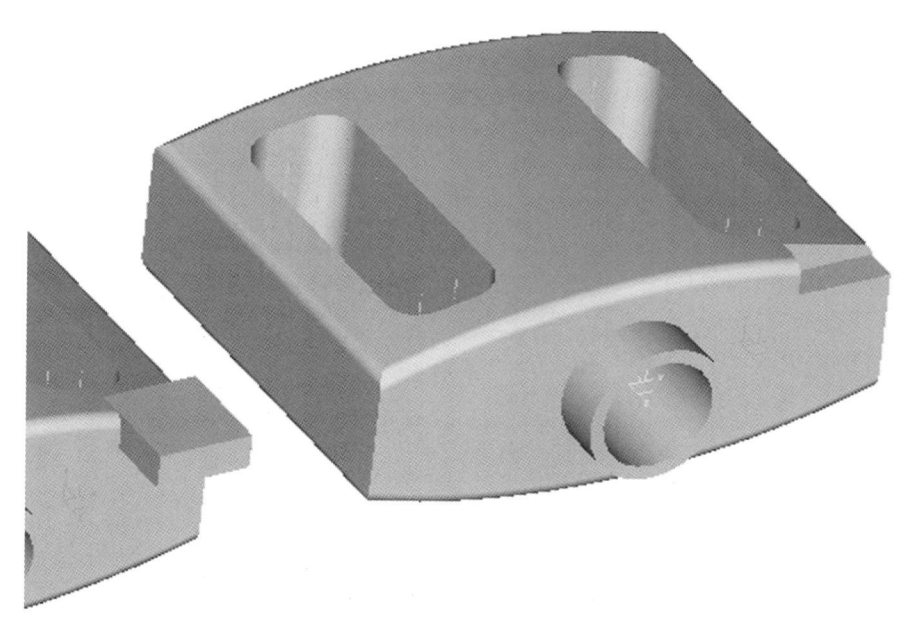

FIGURE 8.8 Example of different features using the same extrude direction (left is protrude; right it cutout).

Part Modeling **199**

Recall that one of the major misconceptions about 3-D modeling (even when it is called solids modeling) is that the 3-D models are not solid at all. Instead, they are a construction of surfaces that meet at their edges. This process of having the edges meet is often called stitching. For instance, a cylindrical 3-D model has a cylindrical surface that is the body of the model, and then there are two circular surfaces that represent the ends. The edges of the circular surfaces (which are circles) are considered stitched or sewn up by the CAD system. The system deems them this way since the edges of the surfaces are found to be coincident within a very small amount (this small distance is usually called the basic part tolerance). Figure 8.9 shows the surfaces that would be stitched to create a rectangular solid.

8.3.6.1 "Stitching" for New Features

This issue of stitching can often be ignored since the CAD system takes care of this automatically. But to really master 3-D modeling, it is important to think about this issue at times. For example, Figure 8.10 shows a cylindrical feature

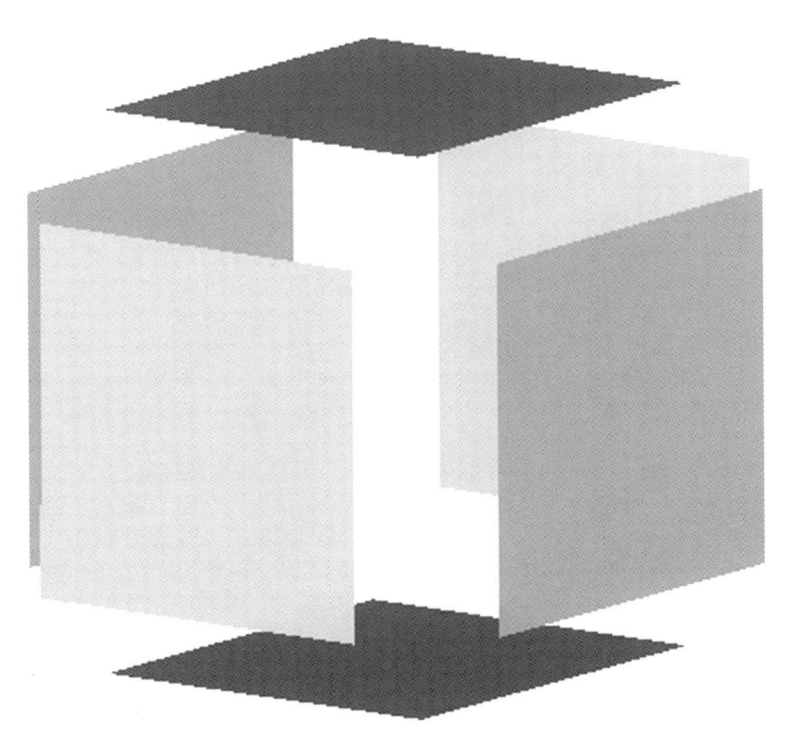

FIGURE 8.9 The surfaces that would stitch to form a cube-like solid.

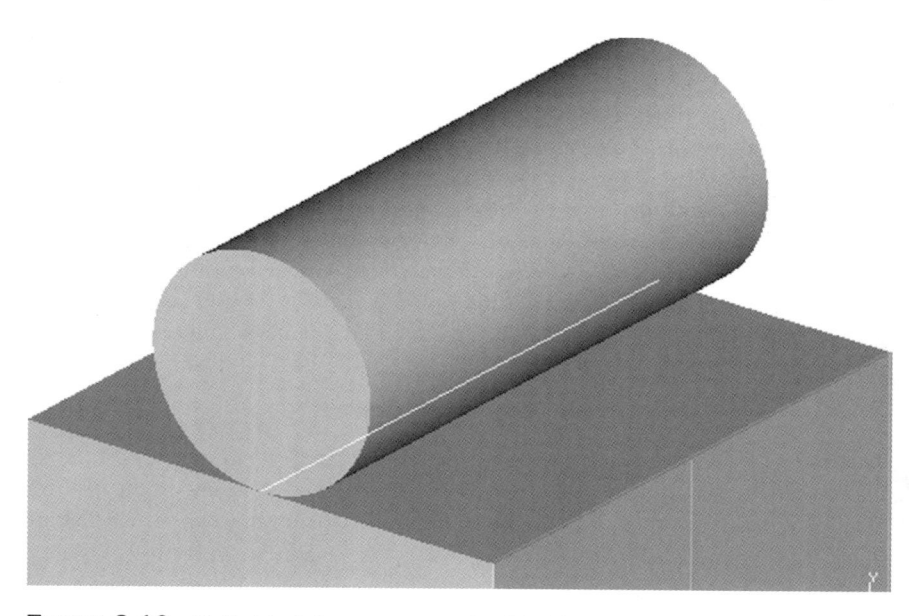

FIGURE 8.10 Cylindrical feature grazes the block's top surface (stitching may be undefined).

that has been added to the top of a block (using the block as the base feature). This was done by sketching a circle that was precisely tangent to the top edge of the block. In this case of the sketch circle's tangency, there is just one point (one local X- and Y-value) on that face that is both part of the straightedge at the top of the block and the circle. Now think about how that cylinder is extruded across the top surface of the block. It means that there is an exact line (infinitely thin) that just splits the top surface of the block (this edge is shown highlighted in Figure 8.10).

Now, if the circle had been sketched down into the block, then there would be more interference between the cylinder and the block, and the CAD system would find that there is not an exact line shared between the block and the cylinder. Instead there would be a finite region of the top of the block that can be used to create a new top surface. The top surface of the block would simply have a rectangular region whose edges would be dictated by the location of the cylinder. This new top surface of the block would contain both the flat top surface of the block and also the cylindrical area. This would be a new surface whose edges can then be stitched to the boundary of the cylinder that intersects into the block. Refer to Figure 8.11.

Part Modeling

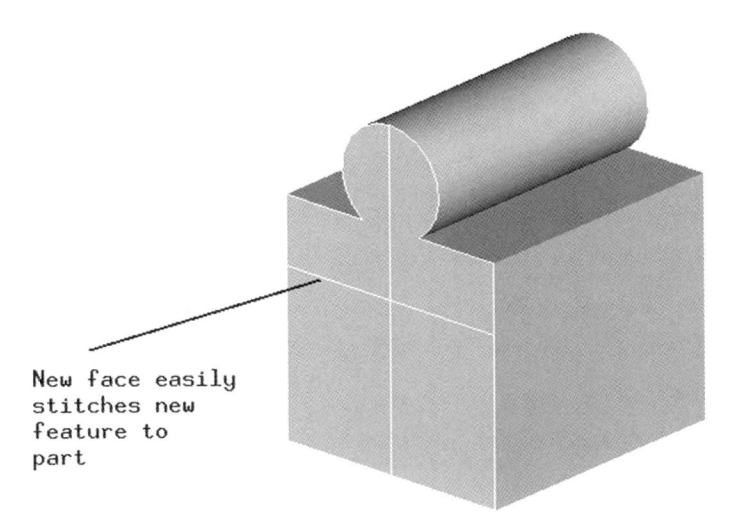

New face easily stitches new feature to part

FIGURE 8.11 Cylindrical feature interferes with block (stitching clearly defined).

Of course, if the cylinder were located above the block, then there would be no interference or shared space between the block and the cylinder. In this case, there would be no effect on the surfaces of either feature. The part model would simply contain two totally separate volumes (which is sometimes the desired result).

Now, returning to the case where the cylinder is exactly tangent or just touching the block, note that there is no shared volume, but yet they still touch. Should the CAD system consider this another case of two separate volumes? Or should it consider the two volumes joined together like Figure 8.11? The answer is not clear. In this sort of case, the CAD system may issue messages, warnings, or errors to the user. Or, the CAD system may simply make the two features closer together or farther apart by a small amount to force a solution to the problem. In either case, it really is not a good idea to have the CAD system making assumptions about what the designer really means.

Understanding how the perfectly tangent situation causes a problem for stitching, the user could think ahead about this issue and deal with it accordingly. Perhaps this part is going to have a shaft welded to the top of the block, so there is really weld material between the two features. If the weld is actually modeled (by adding some new 2-D sketch geometry with the circle), then the tangency issue can be avoided. Refer to Figure 8.12. Understanding stitching is often the key to knowing which method to use.

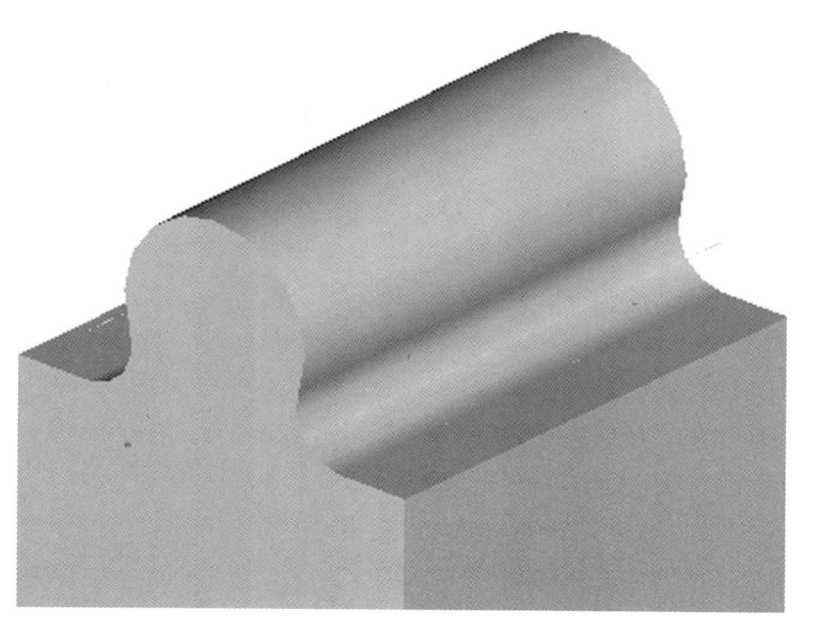

FIGURE 8.12 Modeling as manufactured can avoid the tangent surface problem.

8.3.6.2 Joining Versus Adding

Earlier in this section, protruding was defined as an operation on a part that created new volume for the part. It is the opposite of cutting. Note that it is assumed that the protrude operation is going to force the CAD system to figure out how all the surfaces stitch together (as was just discussed). When the system does successfully calculate the stitched surfaces, the feature can be referred to as joined. Although join seems to mean that it only applies to protruding, the need for stitching surfaces also exists when the feature is a cutout.

However, some CAD systems may allow a new step in the part history, but without the stitching process. This is sometimes referred to as adding (as opposed to joining). In the add operation, the history of the part simply considers the new volume of the new feature to be a step in the history, but the new feature has no effect on any other area of the part. Clearly this is not a good idea in many cases (interference is being ignored). For instance, the CAD system cannot reliably calculate the volume of such a part (the 2 volumes might be interfering significantly). On the other hand, if the user just wanted to show some 3-D geometry as just a "reference" (such as weld material in the earlier example), then the add

Part Modeling

operation might be satisfactory. The part will look right, in this case, and perhaps the weld is not supposed to calculated as part of the volume or weight of the part anyway.

8.3.7 Summary: Creating New Features

At this point, the critical issue that the user needs to understand is that each time a new sketch plane has been selected, one has started the process for creating a new feature. And, if the process is completed, this feature is then going to be shown in the part history data in some fashion. Also, it is essential to realize that one has, at some point in the process, specified an operation (such as join, add, cut, intersect, etc.) for how each new feature affects the state of the part model.

Realizing this, it is essential that the user somehow formulate a plan or strategy for part modeling. A good plan of attack can make a part that replays more quickly, and changes to features can be made more easily. With no planning, one can count on eventually having a rather frustrating experience trying to rearrange or completely redrawing the part model.

8.4 SKETCH CONSTRAINTS

Constraints or constraining is somewhat of a point of contention between various 3-D CAD systems and their users. It is controversial since some CAD system vendors insist this process is absolutely essential and any part not fully constrained is useless. On the other hand, some 3-D CAD systems do not have robust constraint solving algorithms and their 3-D part models wind up being incapable of being intelligently revised and re-used as designs evolve. This book assumes that constraining is a very important feature of 3-D CAD systems, and they should be used whenever it makes sense to do so.

8.4.1 What is Constraining?

As the name implies, constraining is a process of limiting or restricting the geometry that is being created. The simplest example is showing a dimension in a drawing. Imagine a drawing with a part that has a hole, and the drawing has a dimension that indicates the diameter of the hole. This dimension fixes or restricts the hole size. If a hole is shown without a dimension, one knows that it is a circular feature, but the diameter could be anything. So, the diameter dimension constrains the hole size, and this dimension can be referred to as a constraint. Other types of constraints are more purely geometric. They would be constraints such as parallelism (forcing 2 lines to be parallel), or coincidence (forcing one line segment's endpoint to be connected to another line segment's endpoint).

8.4.2 Dimensional Constraints

Referring to Figure 8.13, one can see that the hole has a dimension for its diameter (15.00). Although the size of the hole can now be considered constrained, there needs to be more information to fully define the sketch. In particular, there needs to be an indication of where the hole would be located with respect to the rest of the part. This can be accomplished by more dimensions such as the horizontal dimension (45.00) and the vertical dimension (35.00) shown. These dimensions constrain the location of the hole, and again, these dimensions can be considered constraints.

8.4.3 Geometric Constraints

Referring to Figure 8.14, the circular hole has now been replaced with a rectangular-type of slot. Once again, dimensions can be shown on the sketch to constrain the size of the slot. However, notice that the horizontal and vertical dimensions that position the slot (45.0 and 35.0) are to the center of one of the arcs for the slot. It does appear, at first, that this would completely describe the feature, and that anyone reading this information as a drawing would know how to manufacture the part. But the problem is that there is no way to be sure that the slot can not rotate about the center of that arc. The dimensions were to just a central point. The slot can be imagined to be free to rotate about that point with an axis of rotation sticking out from the page. There is nothing in the dimensions that explicitly demands that the slot be parallel to the bottom edge of the plate.

Understanding a problem like this rotating slot sketch is crucial to benefiting from the constraining process. Almost anyone would assume from a drawing that looked like Figure 8.14 that the slot is supposed to be parallel to the bottom

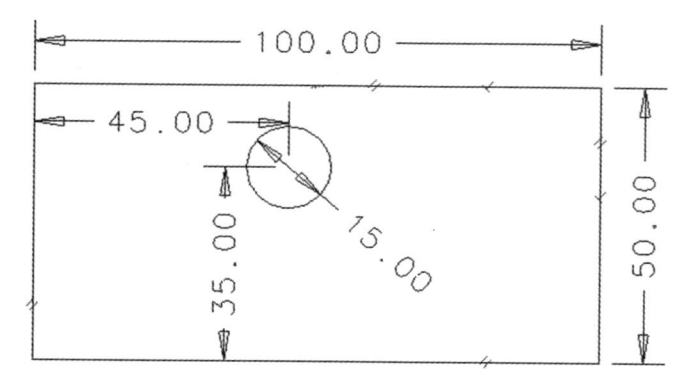

FIGURE 8.13 Example feature sketch with dimensions as constraints.

Part Modeling 205

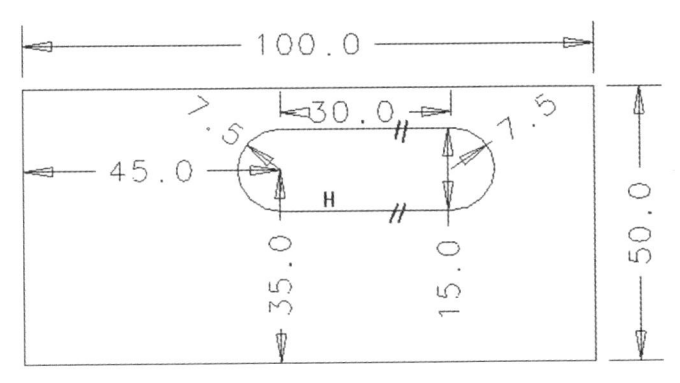

FIGURE 8.14 Example feature sketch with some geometry-based constraints.

of the part. But, this is just an assumption or an impression. People who read drawings make assumptions like this all the time. But 3-D CAD does not always allow these assumptions. With the 3-D CAD system, the central idea is to not just create a 3-D part, but rather to create a part MODEL that contains a mathematical system that can change and evolve intelligently and predictably. Geometric constraints are an essential ingredient in this process.

So, to completely constrain the slot with the dimensions shown, new constraint information needs to be added. One possibility would be to indicate that the slot is horizontal. Another possibility would be to indicate that the slot is parallel to the bottom edge of the part. Looking at Figure 8.14, one can see that there is an H symbol near the bottom line of the slot. This symbol indicates to the designer that the line is forced to be horizontal. The CAD system usually automatically detects when lines are drawn in a horizontal position and adds the constraint for a line automatically to be horizontal (or vertical, parallel, etc.), but users still need to be aware of the meaning of these constraints. The H is shown on the bottom line only, so that would only properly constrain the bottom line. Looking at the top line of the slot, there is a parallelism symbol (2 short lines together). It has a partner symbol on the bottom line. These 2 symbols form a pair and would be identified as a single constraint. It indicates that the top line is parallel to the bottom. The exact symbols and their meaning differ for the different CAD systems; but these symbols are typical. These added geometric constraints should "fix" or control the problem of rotating the feature about the dimensions to the point for the arc.

With the horizontal and parallel constraints in place, the sketch is said to be better constrained. One would now think that the sketch is fully defined and that the geometry is under control. Unfortunately, there is still something missing. The arcs at the ends of the slot are shown as "meeting" or "connecting with" the straight lines, but there is no certainty that the arcs are perfectly tangent to the lines at the exact point where they connect. The tangency condition becomes a necessary constraint to make certain the sketch shape is under full control.

8.4.4 Degrees of Freedom

It should be clear from the previous example that there can be a fair amount of thought and planning required in creating a good system of constraints for even a simple 2-D sketch. One way to analyze this situation is referred to as removing degrees of freedom (DOFs). In any 2-D sketch, there are 3 degrees of freedom available for each entity drawn. These degrees of freedom are translation in the X-direction (sliding horizontally), translation in the Y-direction (sliding vertically), and rotation about the Z-axis (as in the slot rotating example).

For each entity, then, the designer can try to imagine if that entity is free to move in each of the 3 Degrees of Freedom. In other words asking, "can it move in X?," "can it move in Y?," "can it rotate about Z?." If it can move in one of these ways, then constraints can be added to remove this degree of freedom.

8.4.5 "Inheriting" Constraints

Many times, when constraints are added to a sketch, the CAD system will automatically include other constraints. In a sense, the sketch is inheriting the added constraints since they are necessary to have a consistent network of constraints. The simplest example is adding a linear dimension between two lines as shown in Figure 8.15. Obviously, the dimension is between the two lines, but consider the fact that this dimension is meaningless if the lines are not forced to be perfectly parallel. If they are not parallel, then the lines will eventually cross (and the dimension becomes wrong). Therefore, when the linear dimension is added between the lines, if there isn't already a parallel constraint present, the CAD system will most likely add the parallelism automatically.

Unfortunately, it can take quite a while to become accustomed to how the CAD system implements these inherited constraints. They usually are necessary, but sometimes the reason they are needed can be rather subtle. Also, sometimes the user applies constraints in a slightly different manner, and the system may respond quite differently. For instance, in Figure 8.15, the linear dimension was created by picking the two lines. The system applies the inherited parallel constraint. However, what if the dimension was created by picking one of the lines and then an endpoint of the other line? Now, the system may figure that the user wanted to allow the one line to be able to rotate about that endpoint. So, the dimension is applied (but it is only between a point and a line), and the system may not create the inherited parallel constraint. This is a subtle difference when the dimension is created, and the dimension may look the same to the user, but the sketch may easily become scrambled since the line is free to rotate.

Part Modeling

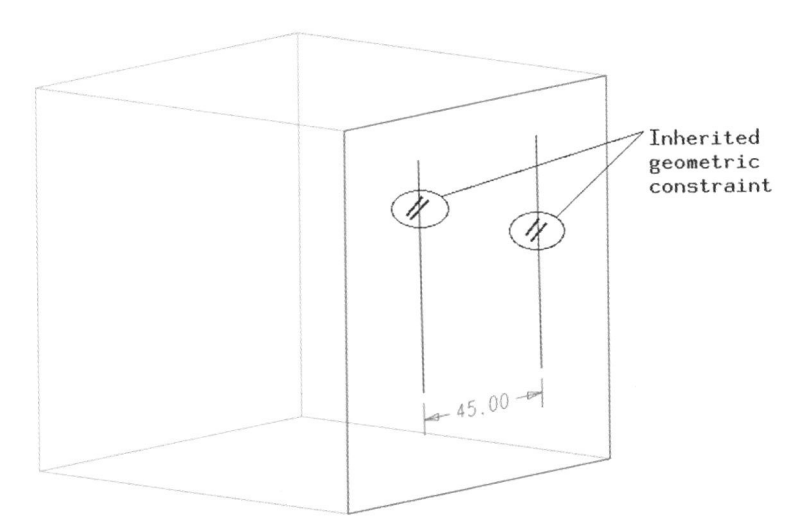

FIGURE 8.15 Example of inheriting a geometric constraint.

It is not really possible to consider all the ways that the constraints are inherited or all the ways that simple mistakes in creating constraints can cause a sketch to become unstable. It is usually simply a matter of practice and learning from mistakes.

8.4.6 Types of Constraints

Table 8.4 lists a variety of constraints that may be explicitly created and/or inherited. It is not possible to list all the degrees of freedom that can be removed by the various constraints, but the documentation included with the CAD system may provide this. The CAD system documentation should also indicate the various symbols or icons that are shown for each of them.

8.4.7 Sketch Regeneration or Updating

When constraints are created or modified, the CAD system may or may not automatically "solve" the constraint network. Whether it is done automatically and/or immediately by the system, eventually all the constraints that are applied to a sketch will have to be checked for consistency, and any pending changes implemented. It is done this way so that many changes can be made individually, but the sketch does not change until all the desired changes are made. Then, when all

TABLE 8.4 Various Types of Constraints

Constraint	Application	Description
Parallel	Applies to two straight lines in a sketch.	It forces the two lines to be parallel.
Perpendicular	Applies to two straight lines in a sketch.	If forces the two lines to be perpendicular.
Coincident	Applies to two points of various kinds. For instance, between endpoints of lines, arcs, splines, etc. It can also be applied to centerpoints of circles and arcs.	If forces the two points (from two separate geometric entities) to share the same X, Y location. This X, Y location may move as the entities move, but the two points stay together.
Collinear	Applies to two straight lines.	If forces two lines (which are really line segments) to lie on the same mathematical line (which has infinite length).
Grounded, Fixed, or Rigidified	Applies to lines and points of various kinds.	For lines, it forces them to not move or rotate. For points, it just forces the points to not move at all.
Tangent	Applies at least to lines and arcs, arcs to other arcs, or perhaps to lines or arcs and splines.	If forces the 2 entities to just touch at a point; they each share exactly one point, and they seem to just graze each other.
Dimensions	Applies to points, lines, arcs, etc.	It forces the various entities to be separated by the amount shown in the dimension value.

the dimensions or other constraints are as desired by the user, the user can then regenerate or update the sketch.

In the regenerate or update process, some CAD systems may remove constraints in much the same way that they were inherited earlier. The system may determine that some constraints are redundant or unnecessary based on other constraints. It is important that the process complete successfully before the sketch is used to create a new 3-D feature or the part. Indeed, some CAD systems will force the user to resolve any problems with the sketch immediately. However, other systems will allow the user to continue to create the feature even though there are unresolved, minor problems with the sketch.

Part Modeling

Although it is beyond the scope of this book, the regeneration or update process does involve solving a system of equations. Each constraint (of all the various types) represents an equation with a number of unknowns which usually cannot be solved by itself. However, the process of regeneration or updating solves all these equations together or simultaneously so that a solution for each equation (representing where each line, arc, etc. belongs in the sketch) can be found. This solving process can be complicated, and sometimes the system of constraints can not be solved so that one single stable sketch is created. The next few paragraphs explain some of these failure situations.

8.4.8 Overconstraining

One possible result from the attempt to regenerate the sketch is the identification of redundant constraints. This problem can arise when geometric entities are overconstrained. Referring back to Figure 8.13, it is clear that the hole is positioned by the dimensions that locate the center of the hole with respect to the edges of the plate. Since a dimension is shown from the center of the hole and the left edge of the plate, then there is no need to show a dimension from the right edge of the plate. If this second horizontal dimension is placed on the sketch, then the position of the hole is overconstrained. There is more constraint data than necessary.

Different CAD systems may treat this situation differently. Some may simply fail to solve. Others may simply not allow the dimension to be created in the first place. Yet other CAD systems may allow the dimension to be shown, but it will automatically consider the extra dimension as a reference dimension. This means that it is shown for reference only, and it is not really controlling the geometric entities. A reference dimensional value of the constraint may be shown in parentheses, or a "REF" may be shown at the end of the dimension value.

The case of the hole with a dimension to the left- and right-edge of the plate is somewhat oversimplified. It is clear that the second dimension is redundant. However, as a sketch becomes more complicated, it may be rather difficult to actually identify such redundancies and overconstraints. Also, some CAD systems will simply not allow the new 3-D feature to be created for the part until the redundancy and/or overconstraint it resolved. Sometimes, though, it is an advantage to allow the extra dimensions to be shown on the sketch (as long as the extra dimensions are clearly identified with REF or parentheses as not controlling the sketch geometry). Perhaps the distance from that other edge was of interest to the designer, so after a regeneration successfully completes, then this distance will be automatically recalculated after the original, controlling dimension is modified. In this case, the redundant constraint is just supplying interesting information, and it really causes no problems.

8.4.9 Underconstraining

Another result from the attempt to regenerate the sketch is that it may be found to be underconstrained. In this case, there are not enough constraints. This would be the case for the hole in Figure 8.13 where one of the dimensions for the hole is missing, for instance, if the horizontal dimension was present but there was no vertical dimension for the center of the hole. In this case, there would be no way to be certain that the CAD system could always locate the hole vertically.

The problem that arises in this case is that the CAD system may or may not demand that the vertical dimension be present. As the system solves the simultaneous equations, the solution will not yield a single reliable answer for the vertical location of this hole. Therefore, the hole may actually move up and down unexpectedly during regeneration. Of course, the CAD system may simply examine where the hole has been placed on the sketch and assume that there is a dimensional constraint there, even though the user did not explicitly create one. In this case, the CAD system will be able to work with the hole as already shown and allow the creation of the new feature. However, other CAD systems are written so that the user is forced to indicate the dimensional constraint; this type of system forces the designer to fully constrain.

In reality, allowing underconstrained sketches entails some risk, but sometimes it can be valuable. One example is that when a company manufactures a specific product, they often don't design everything within that product. Instead, a fair amount of the design would be components purchased from suppliers. If the company uses 3-D design methods for the product, then the company will need 3-D models for parts that they do not actually design and manufacture. It could be quite a waste of resources to use a CAD system that forces them to use the fully constrained approach for these parts. Creating a model based on just sketching something reasonably close and not bothering with constraints could save valuable design time.

Another argument for not fully constraining would be creating legacy 3-D models for parts that are not new enough to warrant a fully constrained approach. These parts may be quite unlikely to be revised any more. And, again, it could be quite a waste of time to be forced to do the fully constrained approach for these types of parts.

8.4.10 "Multiple Solutions"

Another issue from attempting to solve the constraint equations may be the discovery of multiple, yet potentially equally valid solutions. In this case, there are enough degrees of freedom removed, but there is some ambiguity in how a constraint is interpreted.

Figure 8.16 shows an example of this. The two lines through the center of the sketch shown are fully constrained; they can not move. They are drawn, how-

Part Modeling   **211**

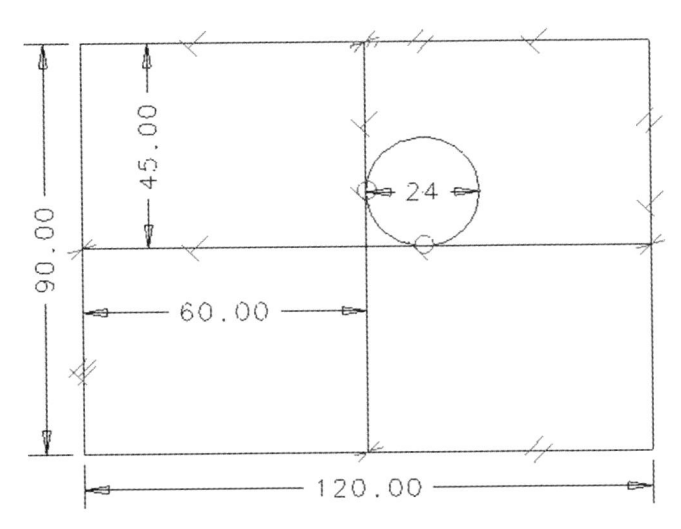

FIGURE 8.16 A potentially ambiguous constraint system.

ever, just to help set the position of the circle. The circle, in turn, is constrained by being tangent with the two lines, and it has been located in the upper right quadrant formed by the lines. One can imagine that the circle cannot move; it seems fully constrained. Although the circle can not actually move into any arbitrary location, it actually would be equally valid to show that circle in the same position in any of the other three quadrants (lower right, for instance). It would still be tangent to those two lines in the same way. Therefore, in this case, there are four equally valid solutions to the constraints (even though the degrees of freedom were seemingly removed). Different CAD systems may handle this situation differently, but in the end, the user simply needs to think through the problem and deal with it accordingly.

8.4.11 Inconsistent Constraints

Finally, a result from attempting to solve the constraint system or network may be an inconsistency. In this case, there are constraints that conflict in such a way that no stable solution is possible. Unlike the underconstraining, overconstraining, and multiple solutions scenarios, inconsistencies must be corrected to have a totally valid model. The inconsistency situation can lead to erroneous designs.

As an example, Figure 8.17 shows a dimension which tries to fix the location of the circle example from Figure 8.16. This can be a valid dimension since

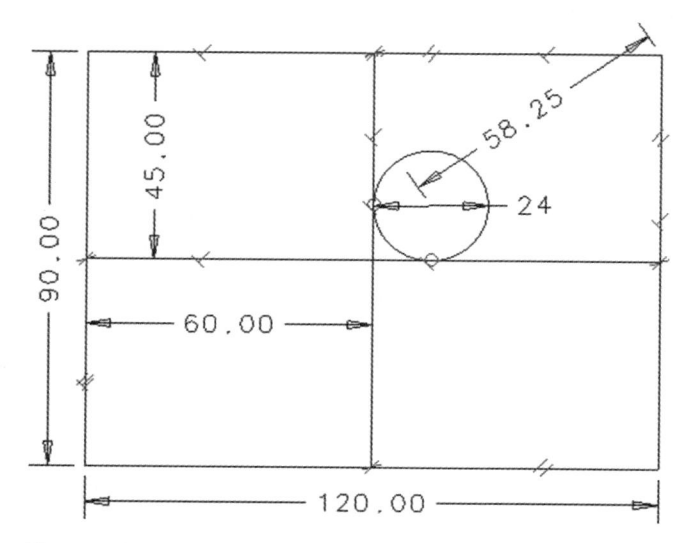

FIGURE 8.17 A dimension that may cause inconsistency.

it removes the ambiguity of which quadrant to place the circle. Therefore, the CAD system should allow this constraint system to regenerate or update.

But, what if the designer changes this dimension to some other value besides the four valid values (for the four quadrants)? This constraint system could not be solved. The dimension cannot be too low and still maintain the tangency to both lines. If a "too low" dimension value is placed on the sketch, the system of constraints is considered inconsistent. And, whatever the CAD system shows after an attempt to solve the constraints may be totally invalid. The CAD system may keep trying to find a solution and then give up, at that point leaving the geometry in the last position it tried. Situations like this need to be corrected by the designer.

8.4.12 Parametrics

A natural extension of the constraining capability in 3-D CAD systems is applying mathematical relationships (such as equations or formulae) between dimensional constraints. This can be referred to as creating parametric models. It can also be referred to as storing design intent, using varying geometry, using family tables, or using driving dimensions. In any case, this is a very powerful capability. It allows the designer to create a complicated sketch that can be changed in significant but predictable ways by only changing one or a few vital dimensions. This can save a large amount of time in a design project.

Part Modeling 213

Referring to Figure 8.18, there is a hole in the part that is located parametrically with respect to the height and width of the part. This is indicated by the equations shown where the dimension value would typically be for the location of the hole. For instance, the horizontal location of the hole is shown as the W*2/3 (or two-thirds of the rectangular width). This means that whatever the width of the rectangular section of the part, the hole's horizontal dimension will be ⅔ that amount and thus the hole will keeps its proper location based on the width parameter. There is a similar relationship shown for the height of the part and the vertical location of the hole. This same capability can be used to "match" dimensions (where 2 dimensions forced to be the same); in this case, the equation would be something like Dimension1 = Dimension2.

Using these relationships or equations is obviously going to be helpful as the design progresses or as new parts similar to this need to be created. Most likely, the most important design issue for this part is the overall width and height; the hole is simply placed in an appropriate location based on those overall dimensions. This is a very common occurrence in the design process. There are often important global characteristics that need to be considered as variable, but then the smaller details are driven by these global characteristics.

As usual, it can take a fair amount of effort and planning to implement these relationships for a complicated part, but it can save a great deal of time and effort later in the design process. This technique should be used whenever possible.

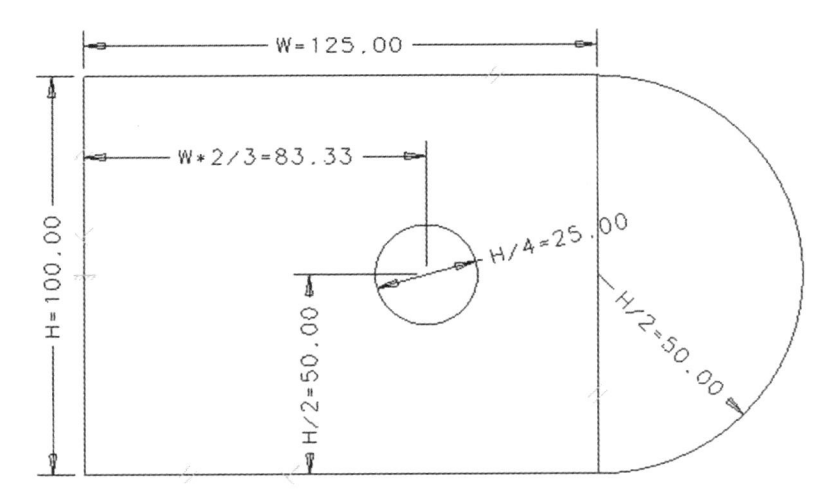

FIGURE 8.18 A simple example of parametric dimensions.

8.5 SECTION PROPERTIES AND BEHAVIOR

Recall from the previous chapter that part modeling may involve creating sections. When using a 3-D CAD system, understanding how sections behave can be valuable in creating robust part models as well as understanding why some parts are not created as expected. Also, sections provide some useful geometric properties. As usual, some CAD systems will make using sections an explicit activity, while others may hide this mathematical foundation from the user.

The first issue to consider in the behavior of sections is whether they are open or closed. A closed section means there are no breaks in the geometry that forms the section. Since the section is like a cross section for a 3-D feature, one can imagine that feature can't be solid if there is a break in the section (see Figure 7.13). Of course, it really isn't a section if it isn't closed, but the CAD system may refer to an 'open section' just the same.

Another important behavior to consider with creating sections is that not all the geometry created in the sketch plane must be used in the section and/or 3-D feature. It may be desirable to create some basic geometry (particularly to get good constraint behavior) that is not going to be included in the section, and thus not in the 3-D feature. Figure 8.19 shows a sketch plane with some geometry created for a new feature, and although the new feature will have rounded corners (since the bold geometry indicating the section excludes the sharp corners), the sketch still retains its "sharp" corners. This is important since the dimensional constraints shown are to the ends of these corners. So the width and height of the

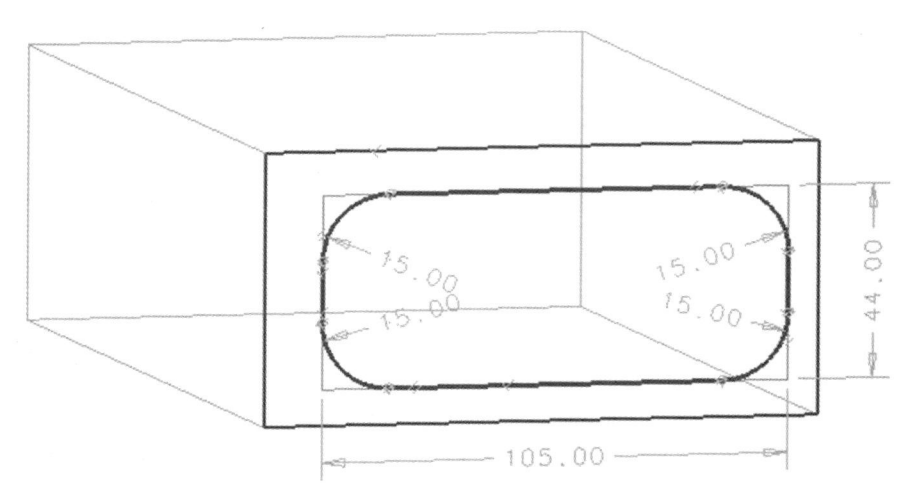

FIGURE 8.19 Construction geometry (light lines) used for constraints, but not section building (heavy lines).

Part Modeling 215

rectangle can be the driving dimensions, and the fillets can adjust accordingly. But again, it is the section that is going to be used to create the new 3-D feature. The other parts of the lines that are not part of the section simply remain as construction geometry that is controlled by the part (assuming the CAD system handles this approach).

If one considers a closed section as containing or bounding an area, and this area can be used to create a solid feature (since there are no breaks in the section), there are some situations that should not be allowed for sections. One such situation is "self-intersecting." Referring to Figure 8.20, it is seen that the section is crossing over itself. This is a problem since if one imagines this section being extruded to form a 3-D feature, there is going to be a singularity problem where an edge is going to be part of two separate, but just touching volumes. Or, the system may consider a part to have negative volume, or it may not be able to resolve what side of the faces are inside the part and which are outside. In any case, the system needs to correct this situation by rearranging the entities, or it should force the user to correct the self-intersecting situation.

There are a number of properties associated with sections (at least proper sections that are supposed to be closed). Not surprisingly, these are known as section properties. Table 8.5 lists the various values that can be calculated by the CAD system based on a section. Keep in mind that these are all 2-D properties that are derived from the section that has been created on the sketch plane. Later

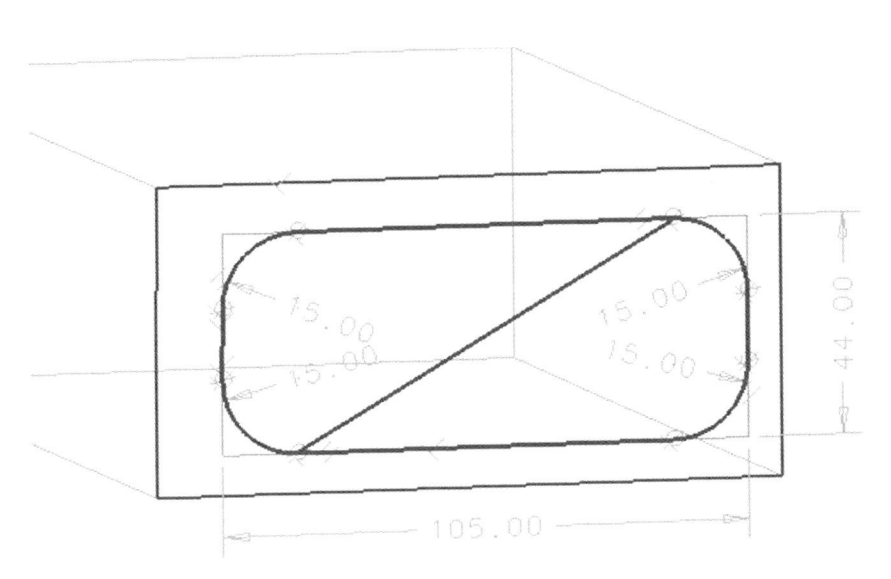

FIGURE 8.20   A self-intersecting section.

TABLE 8.5 Geometric Properties Derived from a Section

Property	Description
Area	A closed section will define an area. This may include composite section behavior where holes or other geometry subtract from the overall area.
Centroid or "CG"	The centroid is like the center of the area. It may also be referred to as the Center of Gravity or the CG. Once a section is created, the CAD system can do calculations to identify this point.
Intertia	The inertia or I_{xx}, I_{yy}, I_{xy} is a calculation that weights the distribution of the area with respect to some axis. This is often used to assess a part's ability to resist deformation about that axis. Keep in mind that the axis needs to be clearly defined. Sometimes, it is assumed to be through the centroid and then with respect to the local horizontal and vertical axes. Hopefully this is clearly indicated by the CAD system.

in this chapter, there is a list of 3-D geometric properties that are derived from the complete solid part model.

8.6 THE BOOLEAN OPERATIONS

Up to this point, the creation of 3-D features has all been based on the idea that a new feature is sketched on a plane that is attached in some way to an existing part (whether that plane is a face of the 3-D part, a datum plane, a coordinate system, etc.). This is the most important aspect of the features-based modeling approach. However, there are some other techniques that may be included in 3-D CAD systems that do not require this approach at all. They do not necessarily require a plane for sketching, or perhaps no plane at all. These techniques are generally known as Boolean operations.

The term Boolean generally refers to the mathematics and logic of combinations (unions, intersections, ANDs, and ORs, etc.). In this case, it is applied to the combining of two separate 3-D part models. Boolean operations for 3-D models are listed in Table 8.6. Figure 8.21 shows three different parts that result from the *join, cut, and intersect* of two separate original parts shown.

Just like the normal features-based approach, the Boolean techniques should be thought of as operating on a base feature of the part model. As such, these Boolean operations create steps in the part history in some fashion. However, instead of one long progression of a single line of operations (as with the pure features-based approach), a Boolean operation can bring together two separate histories into a single part with its own history. Each of the parts (with their histories) become subhis-

Part Modeling

TABLE 8.6 Boolean Operations

Boolean operation	Description
Join	This operation brings two separate 3-D models into a single new model. This operation will usually determine where the two models meet and create new edges and surfaces so that only one single volume is bounded by the new model. If the two models do not actually touch or interfere, then the new model will look like two parts but is treated as a single part by the system.
Cut	This operation cuts one 3-D model with another 3-D model and results in another new 3-D model. This operation will be forced to determine all the junctions to create new edges or surfaces. If the two parts do not interfere, then a sort of nothing part is the result (or the system does nothing).
Intersect	This operation creates a new 3-D model using only the volume that results from where two separate 3-D models are interfering or intersecting.

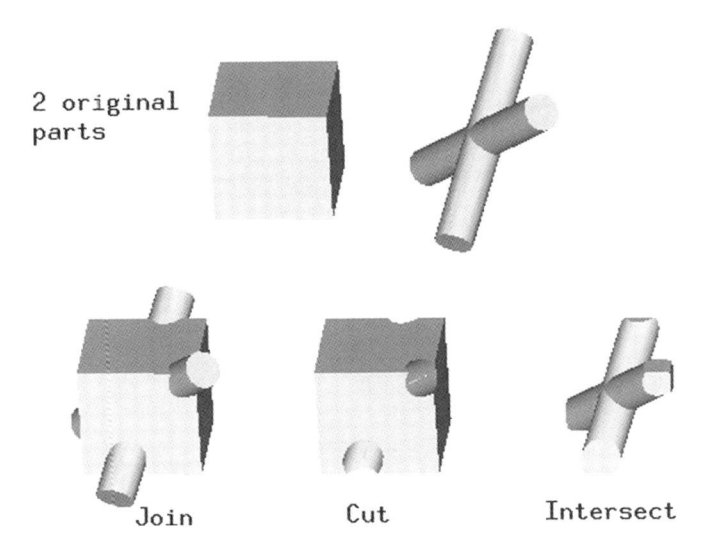

FIGURE 8.21 Example of the *join*, *cut*, and *intersect* Boolean operations.

tories to the overall new part history (or they could be thought of as subparts connected into a master part). This can very helpful with complicated parts since it can be broken down into simpler pieces that can be created first. It can also help the part to regenerate more quickly. Figure 8.22 shows a part history that used a Boolean technique to connect two parts together into one new part.

Keep in mind that the use of these Boolean techniques breaks the idea of the basic three step process. There is no picking of a sketch plane to signify the beginning of the new feature. Also, one needs to consider how the two separate parts are positioned prior to the Boolean operation. Some CAD systems will allow the two parts to be positioned with respect to faces or planar surfaces that are going to be shared by the new, larger part (such as a face where they can be thought of as mated). In this case, sketch constraints (such as dimensions, parallel, perpendicular, etc.) can be used since the separate parts are sharing a plane (where 2-D sketch constraints can be applied). Other CAD systems may allow the 2 parts to be just arbitrarily moved into position relative to each other and have no constraints at all for their connection with each other (although this is rather risky since there is no clear design intent stored with the part).

Since the use of the Boolean techniques imply that the user has a good grasp of how the 3-D parts are made, it is a good idea to use the Boolean techniques only after becoming pretty comfortable with the features-based approach (and the basic three-step process).

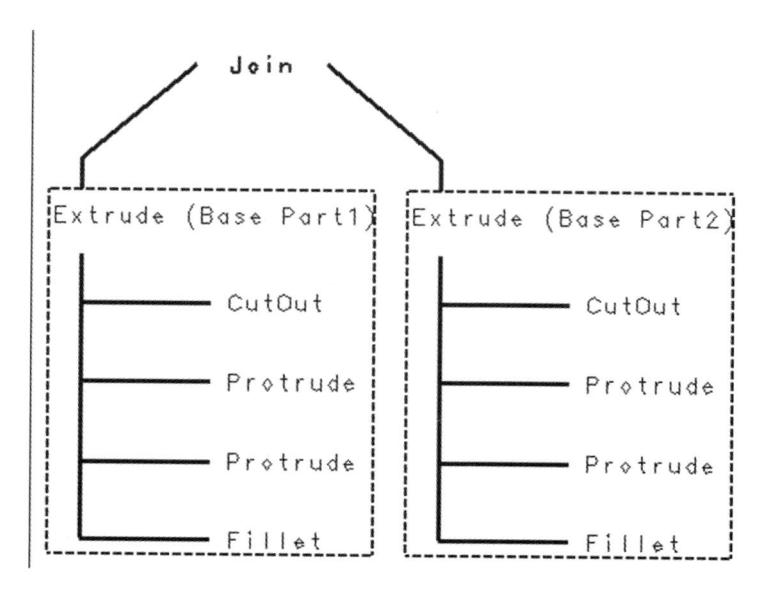

FIGURE 8.22 A part history that used a Boolean operation.

Part Modeling

8.7 CREATING STANDARDIZED FEATURES

Although the Boolean operations can be important techniques for 3-D part modeling, there are also other techniques that may not follow the basic three step process. For instance, most CAD systems will provide packaged or simplified methods to create very common features on 3-D part models. For example, holes, fillets, and patterns may be created by just picking from some standard menus. Table 8.7 lists a some of these standardized features.

TABLE 8.7 Some Often Automated Standardized Features

Feature	Description
Holes/slots	Beyond holes that cut through the entire volume of a part, the CAD system may automate holes that are given countersink, counterbore, and drilled and tapped. Countersink indicates a conical cut-out along an edge of the hole. Counterbore indicates a cylindrical cutout along an edge of the hole. Drilled and tapped means that after the hole is cut through the part, threads are cut into the cylindrical cutout's walls. Holes and slots remove volume from a part.
Fillets	Fillets are rounded edges where two surfaces meet at an angle of some sort. Fillets are generally a circular blend between the surfaces with a radius specified, but other shapes are possible such as elliptical or other conical sections.
Chamfers	A chamfer is similar to the fillet, but instead of a rounded blending, there is a new surface at a sharp angle to each of the two original surfaces. This is often specified by giving an angle for the new surface with a distance along an existing surface, or with two distances along the two existing surfaces.
Ribs	Ribs are protrusions that connect two surfaces with a new segment of material. A rib would specify a distance along each of the existing surfaces and then a triangular web connects the endpoints of these distances at the vertex where the surfaces meet. Ribs add volume to a part.
Bosses	Bosses are protrusions that will have a circular or slotted cross section. They may be totally solid, or they may have a hole within the cylindrical feature. Bosses add volume to a part.
Patterns	Patterns may be used on individual features or entire part models. As the name implies, they are used to create a repeating sequence of the selected geometry. The advantage of patterns is that one master set of geometry can drive the many copies that create the feature. This allows many features to be changed by modifying the master. Patterns could be applied to cutouts or protrusions, so they could remove or add volume to a part.

Even though the creation of these features is automated, they should still be included in the part history. This means, that the user needs to take care to apply them at the right stage of the part construction to get the desired results.

8.8 3-D GEOMETRIC PROPERTIES FOR PART MODELS

Earlier in this chapter, section properties were discussed as geometric information that could be derived from sections or from the 2-D foundation for 3-D features. However, even more valuable information can be derived from the complete 3-D model. Table 8.8 lists some of these properties.

TABLE 8.8 Geometric Properties for "Complete" 3-D Models

Geometric property	Description
Volume	Clearly the most common geometric property for the part model is volume. If a solid model has been properly created, then the 3-D CAD system should be able to immediately calculate the volume of the part. This is immensely helpful in comparison to 2-D drawings. The 3-D CAD system should be able to calculate the volume even when the part is very complicated.
Mass/weight	Assuming the 3-D CAD system permits the application of material properties (such as density) to part models, then the mass and/or weight of the part can also be immediately calculated. This is another immensely valuable capability.
Surface area	Even if a part is not a solid model, the total of all the surface's areas can be readily computed and shown to the user. The 3-D CAD system may also indicate if any surface area is open (not completely stitched). This means that there are free edges (possibly shown graphically). This open surface area calculation can help verify whether a part is really solid or not.
"CG" and inertial properties	As with the section property's inertia, there is a similar calculation for 3-D models. It is a more involved calculation, though, based on how volume is distributed in all 3 dimensions. There should be at least 3 values shown (I_{xx}, I_{yy}, and I_{zz}), and the meaning of the x, y, z directions should be clearly defined. The inertias will change based on their directions, and cross-product inertias (such as I_{xy}, I_{yz}, I_{xz}) may appear. At the rotation of x,y,z directions where the cross-product inertias are zero, inertial principal axes are indicated. The location of the point

Part Modeling 221

TABLE 8.8 Continued

Geometric property	Description
"CG" and inertial properties (cont'd)	where there is equal volume in every direction or regardless of cross section is the center of the volume. If the part is made of a consistently dense material (usually an appropriate assumption for a part model), then this center point becomes the center of mass, or depending on the terminology implied, it may be referred to as the CG or Center of Gravity.

8.9 PART MODELING SUMMARY

Hopefully this chapter has given some valuable insight into how a 3-D CAD system creates and manages part model geometry. Obviously, the particulars of how to create these models is quite dependent on the particular CAD system being used, but the issues presented here will appear likely regardless of the system. The issues are pretty much inherent in the mathematics and logic that the systems are based upon.

The remaining chapters will get into even more issues than build on the basic ideas of this chapter. Reread sections of this chapter as needed to become comfortable with the concepts within the context of one's particular CAD system (before proceeding any further with this work).

The following highlights the most important ideas:

- 2-D geometry techniques are still needed to help make 3-D part models.
- 3-D solid models are really a collection of paper-thin surfaces that are connected at the edges (at least if the part is supposed to be solid).
- The creation of a complete 3-D part model involves first making a base feature that starts the part, and then a sequence of operations or features are performed on the base feature in a history or part history or feature list.
- The most common means of creating a feature is a basic three-step process. The first step is selecting a plane for sketching; the second step is sketching the needed 2-D foundation for the 3-D feature; the third step is turning the 2-D foundation into the 3-D feature by selecting a process or operation (e.g. extruding, revolving, protruding vs. cutting, etc.).
- CAD systems usually offer a means of constraining the 2-D sketch geometry. This will at least include making dimensions. This may also include the ability to have intelligence or parametric data applied to the geome-

try so that they automatically update and change based on varying part needs.

CAD systems also usually offer means of creating 3-D features that are not based on the 2-D sketch plane operation at all. Some of these techniques are Boolean operations that allow 2 separate part models to be connected to each other to form a new larger part model. Another means for creating 3-D features without the sketch plane are to make standard features such as holes, fillets, etc.

8.10 CHAPTER EXERCISES

1. Begin learning and using your 3-D CAD system. Refer to the documentation and tutorials supplied by the CAD system vendor, your company, or your university.

2. Record whether the CAD system can follow the basic three-step process discussed in this chapter.

3. Record whether coordinate systems can be connected with a part and used for sketch planes.

4. Record whether new features (sketched on a sketch plane) can be created that *join* to the existing geometry of the part, *cut* the existing geometry of the part, or *intersect* with the existing geometry of the part.

5. Record whether features can be *Added* (intersections of the 3-D geometry are not determined).

6. Try the problem shown in Figure 8.10 where a cylinder (sketched as its own features) grazes or is just tangent to the top of a block. Record how the CAD system deals with the singularity.

7. Try the problem of sketching and constraining shown in Figure 8.16 where multiple distinct (but valid) solutions can be found.

8. Record whether the CAD system can take two separate part models and use the Boolean operations of *join, cut,* or *intersect* to make them become one part model.

8.11 CHAPTER REVIEW

1. What are the 3 basic steps that often are used in the features-based approach to part modeling?
2. What are some potential sources of planes that can be used for sketching? What are some of the advantages/disadvantages of these different sources?
3. What are some situations where it may not be necessary to fully constrain the geometry in a part model?

Part Modeling

4. What is meant by stitching for 3-D part models?
5. Explain why an *intersection* operation might be considered the result of a Boolean "AND"? What Boolean would apply to *join*? What Boolean would apply to *cut*?
6. Explain some advantages to using the Boolean operations that make larger part models based on smaller part models?

9

Surface Modeling

9.1 INTRODUCTION

Surface modeling or *surfacing* is often considered a separate activity within the 3-D CAD system. As mentioned before, normal solid modeling can really be thought of as surface modeling as well. It is just that in solid modeling, the 3-D CAD system is automatically doing all the surfacing functions and creating a closed volume part where all the surfaces connect or stitch or sew up. This chapter will look at this issue in more detail.

In surface modeling, the user takes on the burden of understanding how the surfaces are created. The user has the choice of figuring out if, and when, the surfaces are to be stitched together. The user has the choice of figuring out how properties such as thickness or perhaps material side are going to be applied to the part model. Clearly, this is going to make it more difficult to make basic 3-D part models. On the other hand, surface modeling can succeed in making part models where all other approaches fail.

Typical applications of surface modeling of parts include sculpted parts, molded parts, and swept parts. Sculpted parts would be parts that are somewhat free form or wavy. A car body panel would be a kind of sculpted part. These kinds of parts have surfaces that are sometimes referred to as Class A surfaces (surfaces the consumer first sees; they are not hidden within the product). It is

Surface Modeling

certainly going to be difficult, if not impossible, to find a single flat surface on a car body panel that could be used for sketching and extruding. So, clearly the basic three-step process presented in the previous chapter is not going to be useful.

Molded parts could also probably be considered sculpted, but they have the added issue of being created in a mold (such as in injection molding). Their surfaces need to be modeled in such a way that the mold can be split into two halves, and the surfaces of the part must be drafted (having an angular slope so that the part can be removed from the mold). Also, this is an interesting part modeling application because the part can just be surfaces (it does not need to be solid) since the mold cavity really only needs to have the outer skin of the part model. Often, 3-D CAD is used to make the mold as well as the parts, and 3-D CAD may even be used to interface with a package that creates the NC instructions to actually machine out the cavities for the mold. In this case, there might be no real human intervention with the creation of the physical part at all, and the mathematical definition of the surfaces carries right through to the final product.

Swept parts are created by modeling an arbitrary path for sweeping. This path is used to guide some geometry along the path. The geometry forms a sort of cross section for the part. A tube or wire that makes various turns and bends in arbitrary directions would be an application of a swept part. For simple parts such as a pipe that has a few elbows, surface modeling may not be needed since standard extrudes and revolves could be used instead.

Another class of parts that might be considered an application of surface modeling is called sheet metal. These parts are meant to be thin sheet metal parts (such as used to form a metal box or enclosure). The physical parts are made by cutting out a shape on a flat sheet of metal (this geometry is usually called a flat pattern). Then, at specific locations, the metal is bent to form the final part. Sheet metal part models in the 3-D CAD system need to have this ability to fold and unfold in order to develop the flat pattern (taking into account the proper bend allowances where material is deformed). At the moment, however, there does not seem to be any standardized approach to developing these kinds of part models among CAD systems. The reader must consult the documentation available for their specific CAD system for more information. Of course, these part models are still subject to the issues of creating solid models based on properly shaped and/or stitched surfaces.

Finally, surface modeling can be quite complicated, and this chapter is not intended to be a complete work on the subject. If the reader is not interested in these kinds of parts at all, then this chapter could be skipped altogether. It is not necessary to understand everything in this chapter before proceeding with the remaining chapters.

9.2 SURFACES

Since surface modeling is concerned with creating 3-D part models by working with individual surfaces, it is clear that users should have some idea of what is meant by a surface. Although different 3-D CAD systems may define them somewhat differently, they have some basically common characteristics.

9.2.1 Common Characteristics

First of all, surfaces are very general shapes. A flat plane (such as a face of a cube) is a surface; it can be referred to as a planar surface. The curved shape of a cylinder is a surface; it can be referred to as a cylindrical surface. The curved shape of a cone is a surface; it can be referred to as a conical surface. And, very arbitrary or free-form shapes such as the body panel of a car are surfaces; these can be referred to as free-form surfaces. The planar, cylindrical, and conical surfaces are typical of the extruding and revolving procedures presented in the previous chapter (these types of parts are often referred to as prismatic parts). Surface modeling, on the other hand, often implies the use of the free form surfaces. These surfaces are very 3-D in nature (since they can warp and twist). Figure 9.1 shows an example of a free-form surface.

Individual surfaces also have no thickness. They are "paper thin" and have no volume. However, the 3-D CAD system will often keep track of the side of the surface that is supposed to be on the inside of a solid model (versus facing the outside). If surfaces are used to construct a solid model, then eventually there is

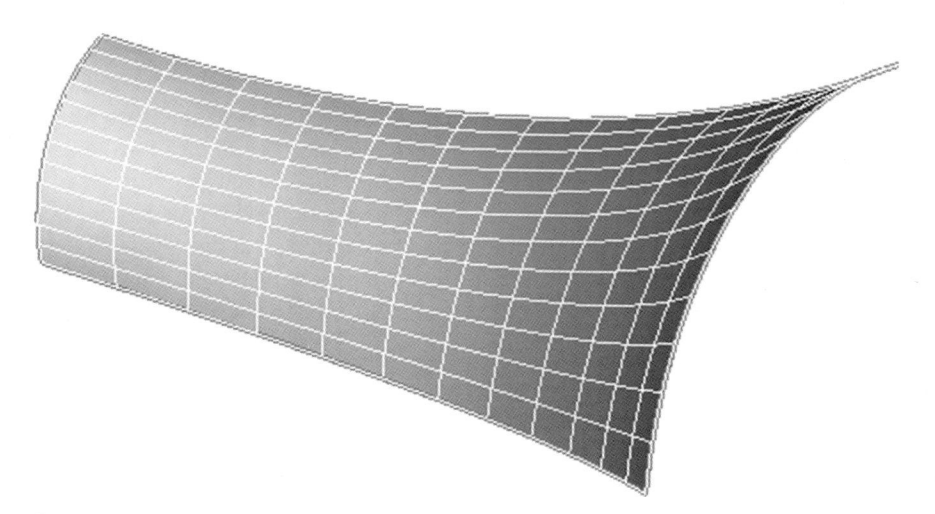

FIGURE 9.1 A free form surface.

Surface Modeling

an inside and an outside. But, when working with the surfaces alone, one may notice that one side of a surface is very dark, but the other side is bright. This is because the CAD system or the graphics adapter is assuming that the dark side is the inside of the part, even though there is no real inside or outside yet. Usually, the system can be forced to make both sides bright by enabling an option called backlighting. In this case, the surface will be easier to see, but there has probably been no change in the way the CAD system recognizes the surface in an analytical sense. Another issue to consider is material side. This affects how the surface operates on a model by indicating which side or sides of the surface can be considered to have material for cutting.

Next, surfaces have clearly defined boundaries or edges. This allows them to be used in the creation of the solid models. The CAD system knows where every edge is in 3-D space (these edges being defined by some sort of mathematical relationship or formula with appropriate constants or coefficients). Then, as surfaces are brought together through various modeling methods, the CAD system can then attempt to figure out how the edges of those surfaces can go together to form a solid (a stitching process). Another important use for edges is that they form the geometrical data needed for creating drawings from 3-D models. It turns out that what is shown in drawings for parts is really driven by the edges of surfaces.

Surfaces should also generally be considered as analytical objects. The surfaces are stored by the 3-D CAD system as numerical data and mathematical relationships that are totally refined or accurate. Therefore, in a mathematical sense a curved surface can be "perfectly curved" and accurate within the CAD system itself. However surfaces are often displayed to some given minimum accuracy on the monitor (indeed one might see small flat faces on curved surfaces). But, as the user "zooms in" closer to the model, the surface's appearance can be regenerated by the CAD system to a greater and greater accuracy (with less and less approximation shown on the monitor) because the surfaces are analytical.

9.2.2 NURBS

It turns out that most 3-D CAD systems use a mathematical definition for the surfaces based on something called NURBS. A full explanation of NURBS is beyond the scope of this book, but since it is a common term with respect to 3-D CAD systems, a little information is presented here. NURBS stands for Non-Uniform Rational B-Splines (and B-Splines refers to Basis-Splines). Recall from an earlier 2-D chapter that splines are special geometric entities that can be found in drawings and 2-D CAD. Splines are special because they can define very arbitrary shapes. They can loop back on themselves, for instance. They are also special because the user can pick a sequence of points and then the CAD system would create a smooth curve automatically to follow those points. Splines also have control points that can guide the shape from a distance. These control points can have different weights to indicate how much the curve should be influenced by them. These special characteristics of splines are applicable to the NURBS, as well.

In the case of the 3-D surfaces, the NURBS representation is used to contain the mathematics of a whole surface (instead of just a 2-D spline). NURBS are very flexible; they can form all sorts of shapes (from planar surfaces to very warped free form surfaces). The NURBS mathematics can model the different shapes by using the mathematical concept of orders or degrees. Lower orders can represent simple surfaces (such as a conical surface), and then higher order relationships can be used for free-form surfaces (such as a car body panel). NURBS can even be used to create a surface from a large number of points in 3-D space (called a *point cloud*). All this makes NURBS very advantageous for CAD system programming.

9.2.3 Trimming

Another important issue for the surfaces in a 3-D CAD system is that they are usually trimmed. The mathematical definition of a surface (such as the NURBS) really defines a rectangular-type of surface. It is like a rubber sheet that starts out flat and rectangular, and then no matter how it is deformed or stretched, the 4 rectangular edges remain. Since many surfaces for part models are not rectangular in shape, the CAD system allows curves to be overlaid onto the rectangular surface. Then these curves trim the full surface down to the shape needed. Figure 9.2 shows an example. This trimming issue is usually taken care of by the CAD system automatically, but occasionally trimming can become a problem for surfaces, too. It often arises from importing 3-D data from one CAD system to another and the two systems do not have the same algorithms or tolerance for the

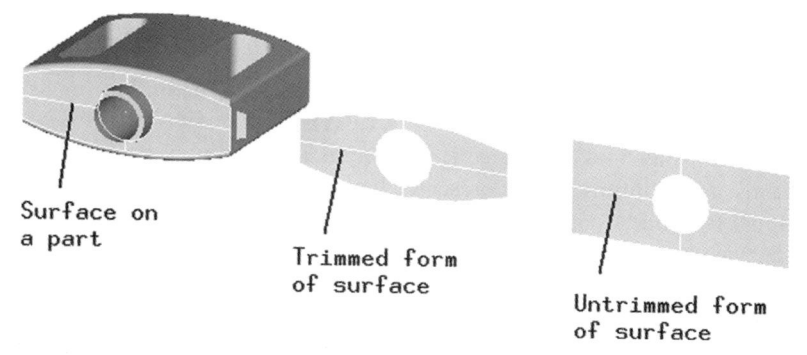

FIGURE 9.2 Example of a trimmed and untrimmed form of a surface.

Surface Modeling

curves and/or surfaces. In this case, undesirable untrimmed surfaces may appear in models.

9.2.4 Surface Qualities

Imagine the untrimmed rectangular surface with an X- and Y-value to each point on the surface. However, instead of the X-value being a horizontal distance and the Y-value being a vertical distance on a flat drawing, these values are usually a percentage of the travel along each of the edges of the rectangular surface. These values are often shown as the variables s and t, and the location of any point on the surface would be identified by an (s,t) pair. These values of s and t can vary from 0 to 1 on the untrimmed surface, where 0 is at the beginning of the edge direction, and 1 is at the end of the edge direction. However, the trimmed surface's edges will often not reach the 0 or 1 values since they will be inside the bounds of the untrimmed surface.

At each (s,t) location, 3-D CAD systems allow the properties or the quality of the surface to be evaluated. An important property of a point on a surface is curvature. This is a parameter that indicates how sharply the shape of the surface is changing at that point. Of course, a free-form surface's topography is changing in many directions, so one may select a plane or planes at this point in which to evaluate the curvature. A good surface is one that would show a smooth change in curvature in many directions. This becomes a particularly important issue at points where surfaces are stitched and a degree of continuity is desired between separate but mating surfaces; there may need to be a smooth transition from one surface to the next by having the curvature property the same for both surfaces at the location of the stitching.

9.3 SURFACE MODELING OPERATIONS

There are a number of operations that are used in surface modeling. Recall that in part modeling, a typical technique is extruding. This creates what is generally called a prismatic part. The operations for surface modeling, though, generally create the free-form surfaces and parts. Extruding can still be used in the context of surface modeling, but surface modeling uses more advanced techniques such as open part modeling, lofting, sweeping, and surfaces by edges or points.

9.3.1 Open Part Modeling

Open part modeling can be done in a number of ways. In this technique, the part model is simply not solid at some point in the part history; there is no bounded volume in some way. Once an open part is created, then an operation such as shelling can apply a material thickness to all the exposed surfaces (in a direction that is normal to the surfaces). If multiple open parts are created, then they can also be brought together to form a solid. Of course, if only the outer face of an object is of interest (like the surface of a bottle in a mold), then the open part model may be all that is really needed.

Figure 7.12 shows an example of open part modeling followed by shelling. The first part of the figure shows a normal solid part with a hole. The second part of the figure shows a face deleted from the part which then creates the open part. Note that the cylindrical surface for the hole is exposed from within the part. The third part of the figure shows the shell operation. Now the part is solid again; it has volume. Note that the hole is now a hollow shaft in the middle of the part.

The deletion of a surface is just one technique of creating an open part. Table 9.1 lists some other techniques for creating open parts.

Open part modeling can also involve the use of the Boolean operations mentioned in the chapter on part modeling. In this case, separate part models can be created for different areas of a part. Then they can be brought together to be a new single part. Figure 9.3 shows an example of taking two open parts, using an *Add* operation to bring the surfaces together, and then using a shell operation to make the part solid. Note that the operation to put the surfaces together was *Add*. If surfaces are just brought together like this, and the CAD system is not going to figure out where edges are shared, then it can be referred to as just an add operation. In this case, the user must then do the *stitch* operation manually to make the connection between surfaces where they share edge geometry (if that is what is desired). Instead of the add operation, the *join* operation probably could have

TABLE 9.1 Open Part Modeling Techniques

Open part technique	Description
Deleting a surface	In this technique, a normal solid part is created and then a face or faces are deleted from the part.
Extruding an open section	In this technique, an open section is extruded. Recall from Chapter 8 on part modeling that when 2-D geometry is sketched in preparation for making a 3-D feature, the 2-D geometry usually needs to form a section. If the section is totally closed (no disconnected segments), then a 3-D feature results. If this section is not totally closed, then an open feature results. This open feature can be used for open part modeling.
Revolving an open section	This is the same operation as extruding an open section; however, in this case the open section is revolved.
Building panels or surfaces	In this technique, a closed boundary for surfaces are created in 3-D space or on a sketch plan, and then this boundary is used to create a trimmed surface.

Surface Modeling

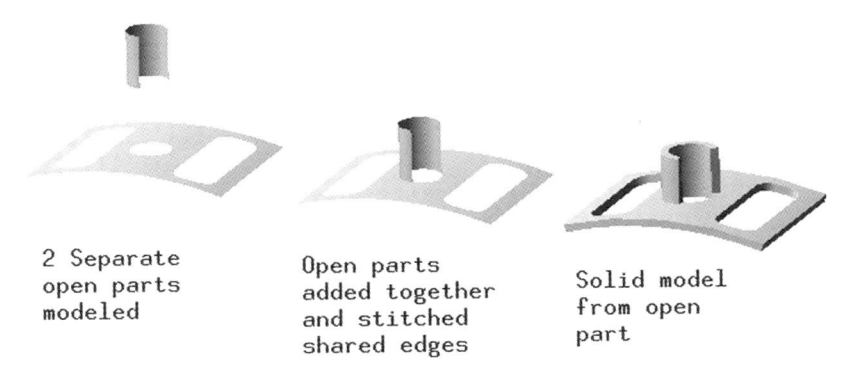

FIGURE 9.3 A process of using Boolean-type operations with open part models.

been used. In this case, the *join* would do the stitching automatically (at least if the surface edges are within the default tolerance).

9.3.2 Lofting

Another common technique used in surface modeling is called *lofting*. Lofting involves creating a set of cross-sectional geometry. Then the 3-D CAD system creates surfaces that loft or stretch across to connect them. Figure 9.4 shows an example of some cross sections. Note that the cross sections could be dimensioned and/or constrained, and note that the cross sections are somewhat dissimilar shapes. Figure 9.5 shows a lofted open part that results from these cross sections. A different part can be created from these same cross sections depending on the tangency dictated at the cross sections. Figure 9.6 shows a part model that forces the surfaces to be tangent along the Z direction at the two larger cross sections.

9.3.3 Sweeping

The next technique discussed for surface modeling is called *sweeping*. Sweeping could also be considered solid modeling depending on whether the beginning and/or ending of the sweep has surfaces to close off the volume (i.e. the part has endcap-like surfaces). Sweeping involves creating geometry for a path (either 2-D or 3-D geometry), and it involves creating the cross sectional geometry that follows the path (some CAD systems will support 2-D or 3-D cross-sectional geometry). Figure 8.3 shows a simple example of a solid swept part.

There are a number of special considerations to keep in mind with swept surface models. First of all, the nature of the path is important since it affects how the part may or may not be constrained. Recall that basic features-based model-

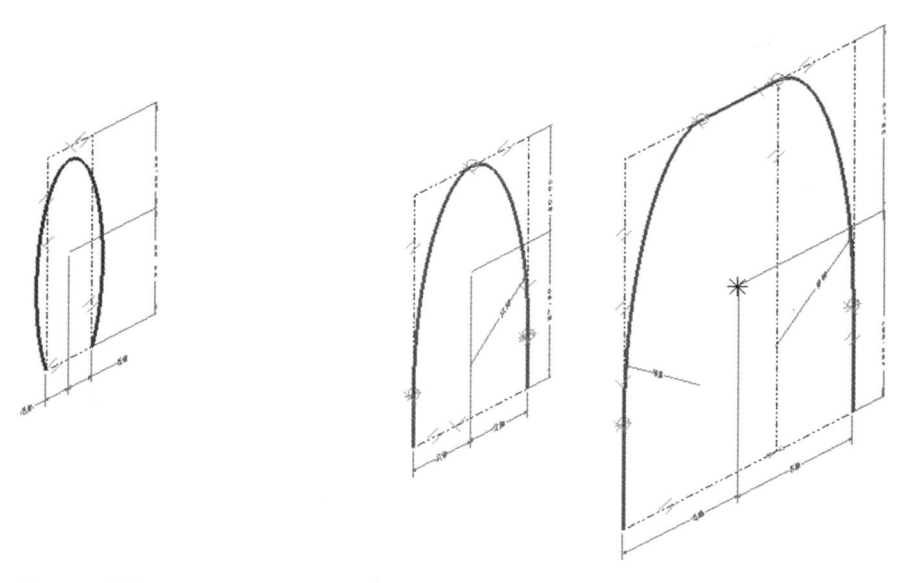

FIGURE 9.4 An example of cross sections for a lofted part.

ing was dependent on the use of sketch planes, and these sketch planes were used to create and manage constraints (dimensions, perpendicular or parallel relationships, etc.). So, making the path for a swept-part using 2-D planes may allow the swept part to be better constrained in some cases because it could be controlled by the standard sketch plane constraints.

Secondly, the path can not have too high of a curvature (or the rate of change of the direction vector along the path). The curvature should not be so high such that the cross sections begin to interfere along the path. If the cross sections are small, then very sharp bends (high curvature) can be tolerated. If the cross sections are large, then the curvature must be reduced (at least to make a physically realistic part that does not have surfaces interfering). Of course, the worst case of this is to have a sharp corner in the path (or an infinite curvature). This is generally called a discontinuity, and discontinuities can prevent the CAD system from creating the swept surfaces at all. Figure 9.7 shows a case where the curvature is too high, and surfaces are interfering.

If an arbitrary 3-D path is needed for the sweep, then often 3-D splines are used to create the path. The 3-D splines, in turn, need to have 3-D points to follow (and perhaps tangency vectors). In this case, 3-D dimensions from one point in space (or relative 3-D dimensions between points) could be used for locating

Surface Modeling

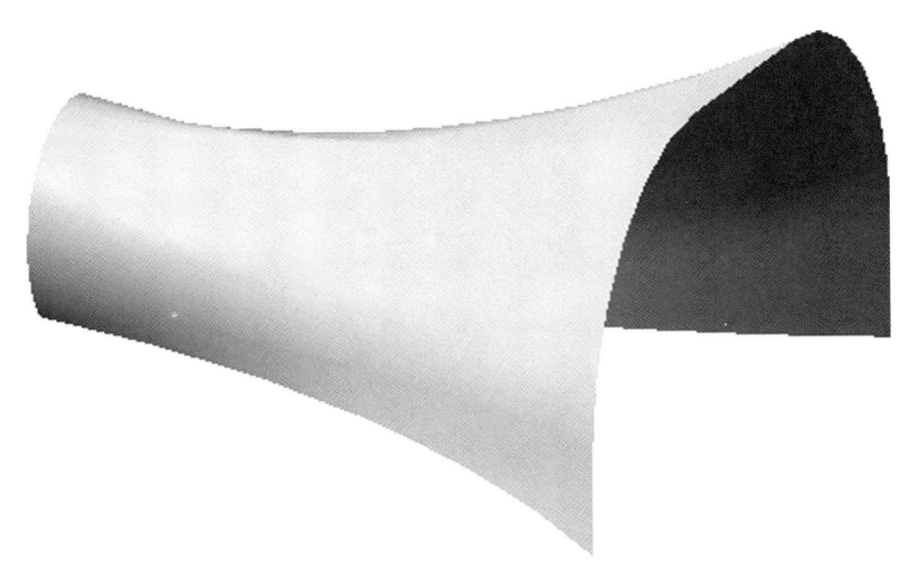

233

FIGURE 9.5 One open part model that can result from the cross sections.

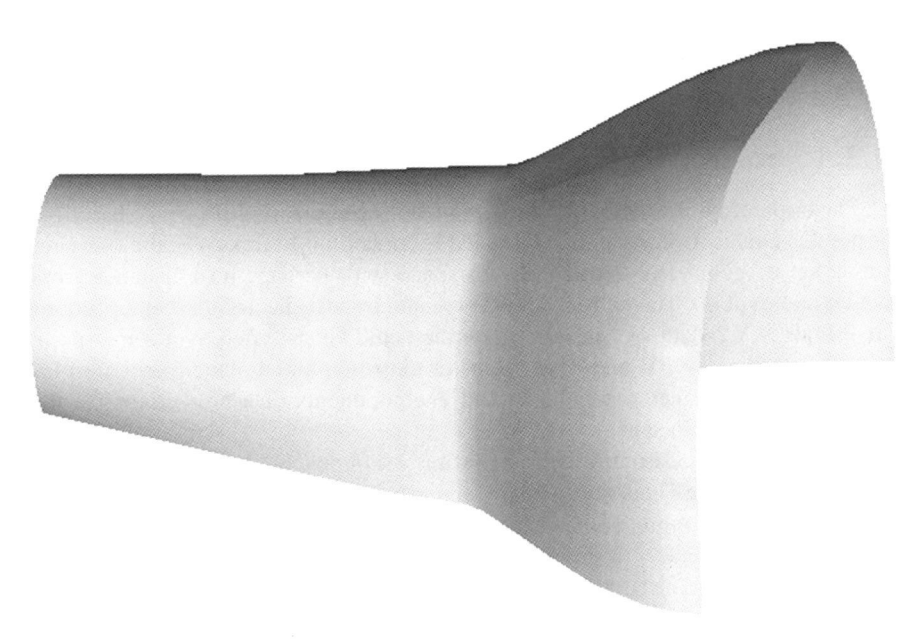

FIGURE 9.6 Another open part model resulting from a different tangency at 2 of the cross sections.

Surfaces interfering
due to curvature of
path too high

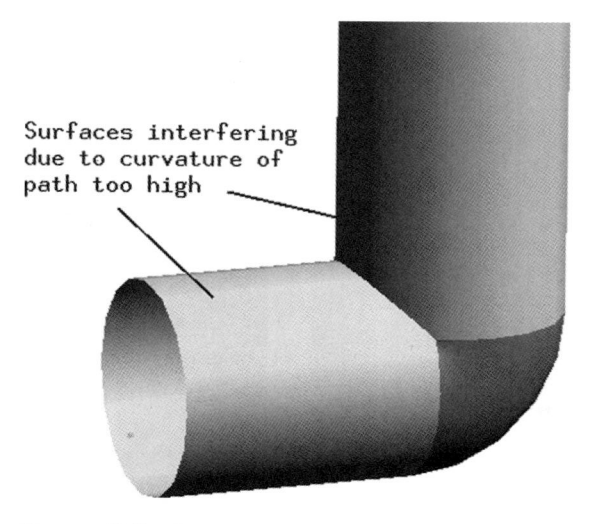

FIGURE 9.7 Example of curvature too high on a swept surface.

these 3-D points. Then these dimensions would control the spline to some degree. However, the weight or tension of the spline will also need to be specified and/or constrained to fully define the path. Figure 9.8 shows how multiple 3-D splines can be made between the same constrained 3-D point locations.

9.3.4 Surface by Edges or Points

The last operation presented for surface modeling is creating a surface by edges or points. This is a very manual approach to modeling. In this case, the user creates a set of geometric entities in 3-D space with perhaps no assistance from sketch planes at all. The entities may be created by just specifying their appropriate 3-D or X,Y,Z data. For instance, a circle could be specified by a center point with three values (X, Y, and Z), a radius or diameter, and a plane orientation for the circle (for instance using X, Y, and Z components of a unit vector for the plane's orientation in space).

The left half of Figure 9.9 shows some 3-D lines that have been created in the modeling space. This sort of data can be referred to as a wireframe model. Assuming the end points of these lines are connected, then a very general surface could then be created to fit these edges (assuming the 3-D CAD system provides this functionality). The right half of Figure 9.9 shows a surface that could be fit to the edges shown. After more such surfaces are created, they could be stitched into a solid part. Obviously, creating these wireframe-type entities and figuring out how to connect them as surfaces could become quite cumbersome for even

Surface Modeling

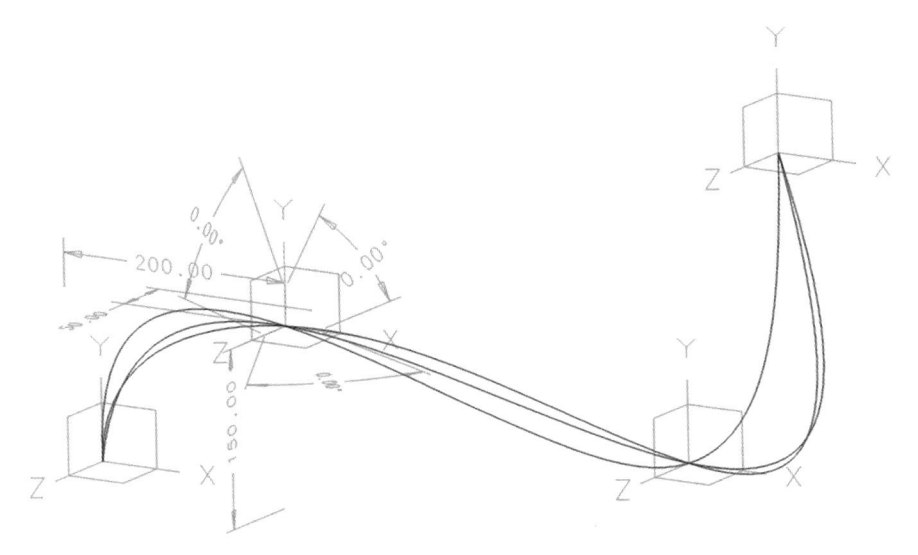

FIGURE 9.8 Multiple 3-D splines through the same 3-D points.

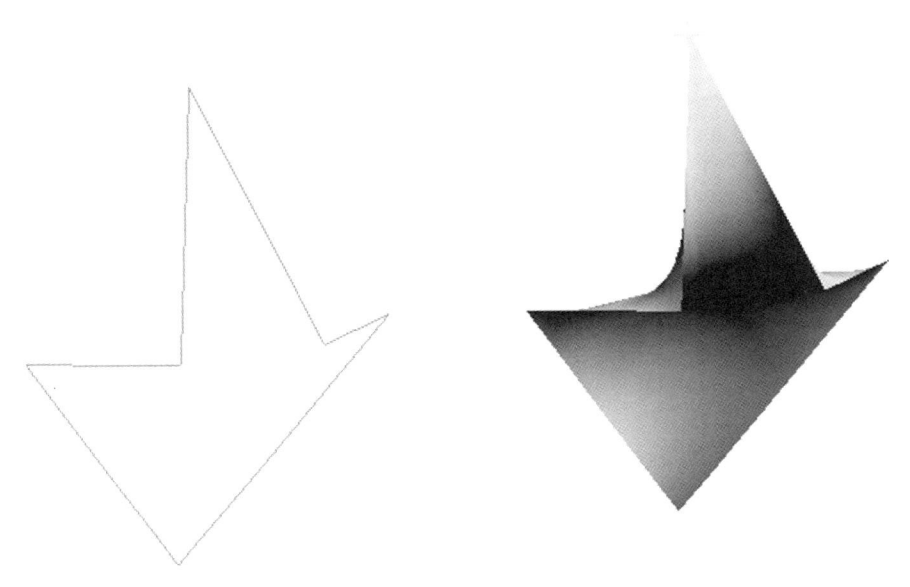

FIGURE 9.9 Example of 3-D entities in a "wireframe" model and a surface that can be created from them.

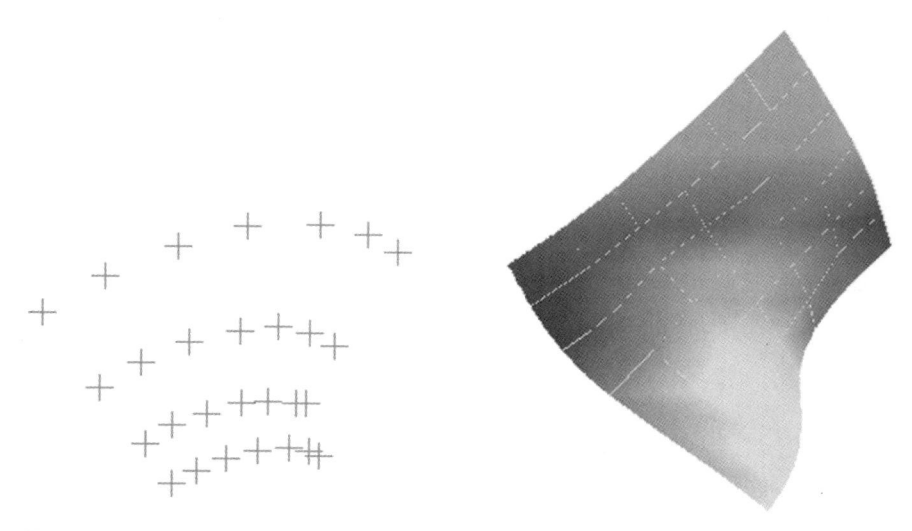

FIGURE 9.10 Example of a surface from a "point cloud."

the simplest sort of parts. Therefore, this approach would generally only be used in cases where all the other methods have been exhausted. Constraining or controlling these parts would present even more difficulties (since there were never any sketch planes).

Another method of creating surfaces with basic 3-D geometry is with points. In this case, there is a relatively large number of points that are positioned in 3-D space. This grouping of points can be referred to as a point cloud. As with the edge method, the 3-D CAD system creates a rather general purpose surface by using these points. This approach can be used when parts are reverse engineered. There are also 3-D laser scanning devices that can be placed in a physical environment. The environment is scanned and recorded as a very large set of 3-D points. These points can be used as the basis for creating 3-D surfaces. Figure 9.10 shows an example of a surface created from a set of points. Notice that the surface is untrimmed. This surface does pass through the points, but it is obviously covering more area than the original points.

9.4 STITCHING AND BASIC PART TOLERANCE

Once surfaces are created by surface modeling techniques, they can be used to stitch together a solid model. Stitching is an essential step in turning an open part

Surface Modeling

into a closed part (assuming this is what the designer desires). Stitching may also be needed to merge surfaces together so that a new surface with new and different surface characteristics can be created.

A fundamental issue for the stitching operation is that 3-D CAD systems are designed to determine successful stitching using a basic part tolerance or part modeling tolerance. This is some small value of distance that is used to assess the distance between the edges (a typical value is 0.01 mm). If the edges are within the basic part tolerance, then the stitching would be considered successful. If there is some amount of gap where the edges of the surfaces spread apart more than the basic part tolerance, then the stitching would be unsuccessful. If it is unsuccessful, then either the edges need to be modified, or the basic part tolerance needs to be increased (i.e., a looser tolerance). It can be quite dangerous to change the basic part tolerance. There may be cases where the system determines that surfaces stitch due to a loosened tolerance, but then the underlying mathematical model does really produce a solid. If other designers work with the part, they may not know its tolerance became nonstandard.

Of course, if very small parts are modeled, then the basic part tolerance may need to be modified just so that the parts can be stitched at all. In this case, the tolerance would probably be tightened to an even smaller value, so in some sense, it is safer.

Basic tolerance is not standardized between 3-D CAD systems. This often causes problems when transferring data between systems. The system with the tighter tolerance may not stitch together surfaces since their edges are not found close enough together.

9.5 ABSTRACTIONS

Surface modeling can also create unrealistic parts. These parts are not physically realistic, but they may be useful for various analysis methods, visualizations, or calculations. They can be referred to as *abstractions*.

One example of abstraction is a midsurface model. In this case, a normal solid part may have been created by surfacing or other methods. For analysis or visualization reasons, a surface may be needed that is just between the pairs of faces (surfaces) of the part. Figure 9.11 shows a midsurface sandwiched between two real faces of the part model.

Clearly this midsurface can not be physically manufactured into the part. However, it would allow calculations to be made for the overall characteristics of the design in this region without being concerned with the thickness of the part in this region. This can be a very useful calculation. Also, notice that the midsurface edges meet at the middle of the outer faces (not at edges); there is a sort of T

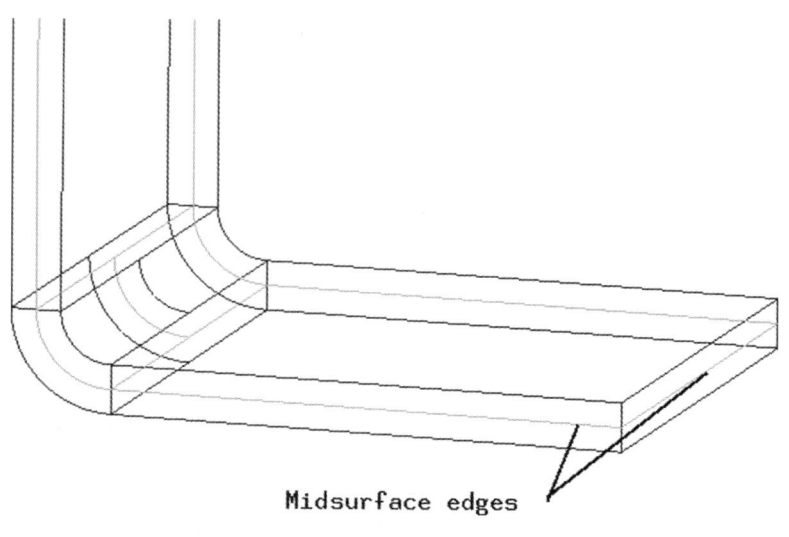

FIGURE 9.11 Example of a midsurface "abstraction."

junction between the surfaces. This is a special type of violation of standard solid models.

For proper stitching of a solid model, the edges of one surface must meet the edges of other surfaces. And, all the edges must meet just one other edge. This proper stitching creates what can be called a manifold part; the violation of this principle such as at the midsurface can be called nonmanifold. Some CAD systems will not allow the creation of nonmanifold parts (since they may not support the advanced analysis methods associated with something like a midsurface abstraction).

Another example of an abstraction is using a surface as a sort of place holder for a real feature. For example, a hole that is drilled into a plate may be tapped (meaning that the screw threads are cut into the sides of the hole). These threads are regularly not shown in the 3-D part model (since it can consume too much time and computational resources). Instead, an extra (and nonmanifold) surface may be created that is concentric with the hole. It merely indicates to designers that a tapped hole is located there. It would also create an associative circle in the drawing to help indicate a drilled and tapped hole condition. Figure 9.12 shows an example of such an offset surface for a hole.

Surface Modeling 239

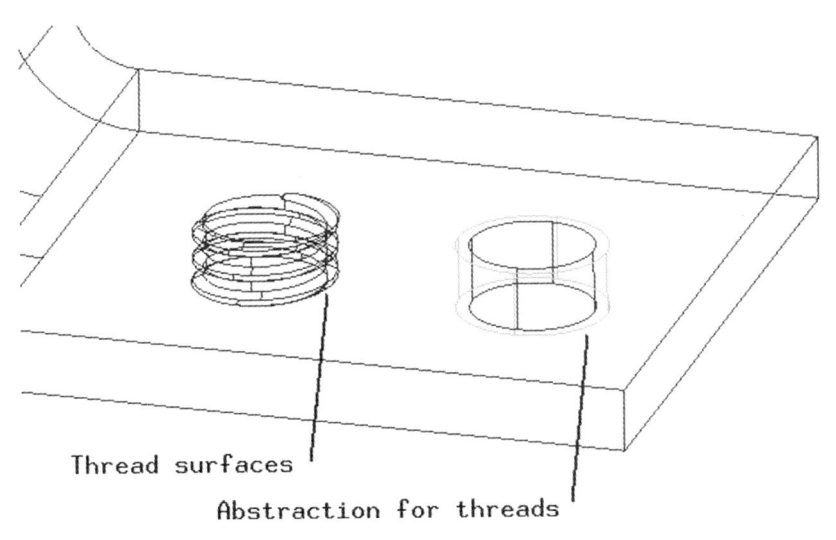

FIGURE 9.12 Example of an abstraction to simplify screw thread representation.

9.6 PART HISTORY ISSUES

The final issue presented with respect to surface modeling is how part history is supported for the surface modeling process. Recall from normal part modeling that the part history or part structure or feature list is simply the sequential list of operations performed to create the part model and the relationships between the various steps and features. Part history was fairly simple for the basic three-step process where sketch plane selection is used to set up the connection between the features (each feature basically being a single step in the history). But for surface modeling, part history can present difficulties.

For instance, one needs to consider whether surface modeling operations are tracked at all by the part history. In some CAD systems, each surface modeling operation merely creates or alters existing surfaces and there is no relationship or tracking among the surfaces. In this case, the part model is really just a list of surfaces. Each surface is identified in some way, but there is no attempt to figure out how the various surfaces are related to each other.

This situation often arises when 3-D part models are translated between CAD systems. The translated model may have no part history; no list of steps; no constraints; no stitching. The part model may "look" correct since the surfaces are trimmed and in the right location; but there is no way to modify the size of a feature, for instance, by modifying a dimension somewhere in the part model. Some CAD systems will refer to this history-less part model as an orphan or orphan part. All the parent/child relationships between the features are lost.

In general, however, techniques such as lofting and sweeping are more likely to be supported with part history than techniques such as surfaces by edges or points. Note from Figure 9.4 that sketch planes and dimensions can be used in a technique such as lofting. To change the lofted part, the designer would find the lofted feature in the part history and modify the needed dimensions. However, looking at Figure 9.10, it would be more difficult to constrain this surface to be connected to the points in space that created the surface. If this were done, one could then move some of the points in space and the surface would change accordingly.

In Figure 9.9, the surface was created from lines in 3-D space. Imagine that these lines were really edges of other surfaces already created for a part model. In this case, the new surface could be related to that existing part model. However, the part history for this operation may or may not be supported. If it is supported, then when the original edges change, the new surface constructed from the edges would change automatically. However, if history is not supported for this operation, then when the original edges change, the surface would simply stay the same and appear to tear away from the rest of the part model.

9.7 CHAPTER EXERCISES

1. Record whether your 3-D CAD system can show untrimmed surfaces.
2. Record whether the CAD system can calculate and display surface properties such as curvature.
3. Create an open part model by removing a surface from a solid part model. Record whether the surface removal appears in the part history.
4. Create an open section and attempt to extrude it to an open part.
5. Attempt to create a part model for a bicycle seat.
6. Attempt to create a hollow tube with three bends (the tube being straight between the bends). Try to control the shape of the part with linear and angular dimensions such that the tube can be changed by only modifying these dimensions.
7. Record whether the CAD system can create a nonmanifold part (e.g. in Figure 9.11 and 9.12).
8. Record the default or standard part tolerance for the CAD system.

Surface Modeling

9.8 CHAPTER REVIEW

1. Explain how a surface model can become a solid model.
2. What are some reasons for using a surface model instead of a solid model?
3. What geometric entities are analyzed to determine if two surfaces can be stitched?
4. List some uses for abstractions of parts. What risks could arise from the use of the listed abstractions?

10

Assembly Modeling

10.1 INTRODUCTION

Assemblies are simply groups of parts that are assembled or brought together in some fashion. They may be welded together (such as flat plates joined into a beam), or they may be fastened together (such as bolts holding a wheel to an axle), or they may just be parts that form a working unit (such as a barrel and rod in a hydraulic cylinder). Assembly models, then, are simply a 3-D CAD–based representation of these groups of parts. Figure 10.1 shows a example of an assembly model.

10.2 CHARACTERISTICS OF AN ASSEMBLY MODEL

There are a number of essential characteristics of assembly models contained in a 3-D CAD system. If these elements are not present, then one needs to question whether the system is really doing assembly modeling at all. It may just be allowing many 3-D part models to be shown simultaneously, but with no real intelligence about the grouping of parts to the extent that it could be considered a real assembly 3-D model. The remainder of this section presents three essential char-

Assembly Modeling

FIGURE 10.1 An example of a simplistic assembly model.

acteristics for the assembly model—assembly structure, positional information, and instancing.

Naturally, assembly models are going to be dependent on 3-D modeling methods. Although a 2-D assembly model would be theoretically possible in a 2-D CAD system, assembly models are really only found in the context of 3-D CAD systems. Realistic assembly design is simply a very three-dimensional activity.

10.2.1 Assembly Structure

The first essential characteristic of the assembly model is that it must have a structure. This is really just a list or perhaps a table (to use a database term). This list or table simply indicates the collection of 3-D part models (or details) within the assembly model. The assembly structure indicates what is actually in the assembly model. Figure 10.2 shows a typical assembly structure. This list could also be called the assembly hierarchy, skeleton, product structure or even a Bill of Material (BOM) (although a BOM can have implications beyond assembly modeling and is explained in greater depth later).

Although the assembly model has this structure, it does not does not really contain any visible geometry of its own. It looks like it is a 3-D model in its own right, but it is really getting this geometry via the part models. The assembly model can be thought of as using its assembly structure as just a list of pointers to the part models. The assembly model points to the part models, and the part models, in turn, are what is seen on the computer monitor. This is important since

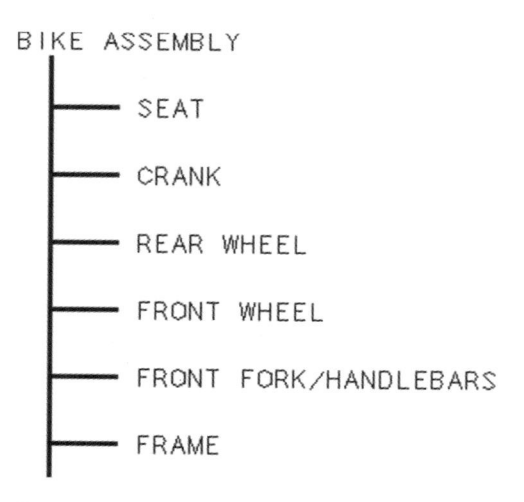

```
BIKE ASSEMBLY
    ├── SEAT
    ├── CRANK
    ├── REAR WHEEL
    ├── FRONT WHEEL
    ├── FRONT FORK/HANDLEBARS
    └── FRAME
```

FIGURE 10.2 Example of an assembly structure.

Assembly Modeling

changes or revisions to the parts need to be reflected in the assembly model (as dictated by the overall design processes), and by using the pointer approach the assembly update or regeneration can happen automatically. As the parts change, the assembly shows these changes appropriately. The idea of pointers becomes even more important with respect to the characteristic of instancing explained below.

10.2.2 Assembly Positional Information

The second essential characteristic of the assembly model is that it uses some sort of mechanism for intelligently tracking and storing positional information about the 3-D part models. It is not enough just to know what is in the assembly. The assembly model needs to know where each of the part models are located with respect to the 3-D modeling space.

Of course, creation of just a part model entails little concern for its location with respect to the modeling space; it doesn't matter where it is located, the part model still is the same (same volume, features, etc.). However, for the assembly model where models are located is quite important. For example, having a part incorrectly located in modeling space could mean that two parts collide or interfere. If the design is not corrected, this could result in a costly production error. The rest of this section discusses the basic positioning of the parts; a later section expands on this concept with respect to assembly constraints.

First of all, the 3-D modeling space needs to be defined and quantified. This includes the specification of the master or global origin. This is the location of a point where the X, Y, and Z coordinates are all zero, and where the direction of the X, Y, and Z directions can be defined (this point may then also be referred to as the global or master coordinate system). Given this global origin, then, any other point can be defined relative to that position with a set of X, Y, and Z coordinates. These coordinates may have some unit system defined (such as mm, inches, m, etc.), or the coordinates may just be normalized or unitless (numerical values the user must interpret).

Now, for each part model in the assembly structure there should be an X, Y, Z position stored and tracked. This gives the overall location of the part, or at least one point on the part. However, location of one point is not a complete picture for the position of a particular part model in the model space. Beyond the point location, there needs to be rotational information as well.

The rotational information would be X, Y, and Z rotation angles for the X, Y, and Z axes or coordinate system that is found at the positions tracked for the part. Some CAD systems will dictate and/or display the location of the part coordinate system as the part model is created. Sometimes this coordinate system is somewhat arbitrary for the part model; but at other times, users will use this coordinate system as a source of datums or other aids during the part model construction. Recall that the first feature for a part model is called the base feature, so this part coordinate system may actually be the base feature and it may be referred to as the base coordinate system or the base node in the part history. On the other hand, some CAD systems will hide the point and/or part coordinate system location from the user since the designer may only be interested in the position of the part in the assembly relative to others parts in the assembly (not relative to the global origin at all). In any case, whether the rotation angles or these part coordinate systems are shown by the 3-D CAD system, these three X, Y, Z coordinates and the three X, Y, Z rotation angles are what the 3-D CAD system is going to use to keep track of the location of the part models.

Knowing the part's coordinate system X, Y, Z position (translational location) and rotation angle values (rotational location) can be sufficient for a simple assembly model where there is one assembly model with its list of constituent part models (such as is listed in Figure 10.2). In this case, each part is oriented with respect to the global origin. However, assembly models are often more complicated than this. They often have smaller assemblies brought together into larger assemblies. The smaller assemblies are referred to as "subassemblies"

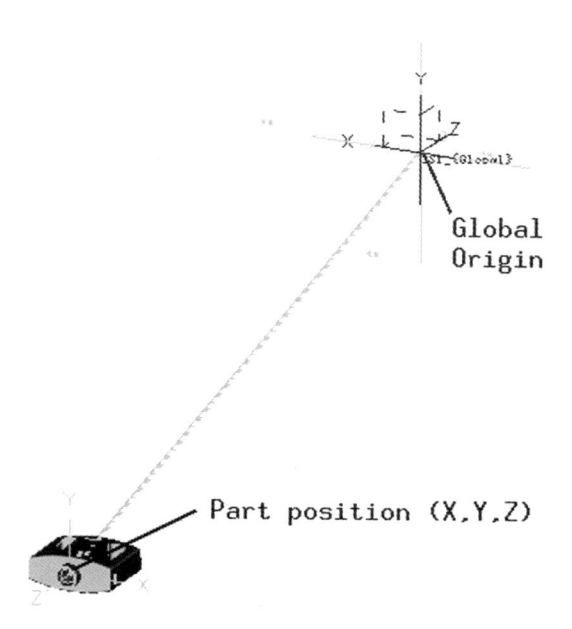

FIGURE 10.3 Example of positional data for a part in an assembly.

Assembly Modeling

since they are under the larger or higher level assemblies. Naturally, the assembly structure must reflect this hierarchy-like behavior as well. Figure 10.4 shows an assembly structure with different levels of assembly.

For these more realistic assembly models, it is best to have a "hierarchy" of the positional information, as well. In this case, the top level assembly (BIKE ASSEMBLY in Figure 10.4) tracks the location of the positions of its parts (SEAT, FRONT FORK/HANDLEBARS, and FRAME). And the top level assembly tracks the location of just an origin or coordinate system for its subassemblies (CRANK, REAR WHEEL, and FRONT WHEEL), but not directly the parts within the subassemblies (such as PEDAL, HUB, TIRE, etc.).

Using this hierarchical approach makes sure that the model is properly broken down into pieces that can properly update or regenerate. The top level assembly relies on each subassembly to track the data for its own part models. Then the top level can use a calculation for a change in frame of reference to show all the parts (regardless of assembly level) in the correct position. This sort of change in frame of reference is handled with a process of summing vectors. Each of the locations of parts within a subassembly can be considered to have a 3-D vector pointing from the subassembly's origin to the part. Then there is a vector that points from the global origin to the subassembly's origin. By adding these vec-

```
BIKE ASSEMBLY
   ├── SEAT
   ├── CRANK SUB-ASSY
   │        ├── CRANK
   │        ├── PEDAL
   │        └── PEDAL
   ├── REAR WHEEL SUB-ASSY
   │        ├── HUB
   │        ├── TIRE
   │        └── RIM/SPOKES
   ├── FRONT WHEEL SUB-ASSY
   │        ├── HUB
   │        ├── TIRE
   │        └── RIM/SPOKES
   ├── FRONT FORK/HANDLEBARS
   └── FRAME
```

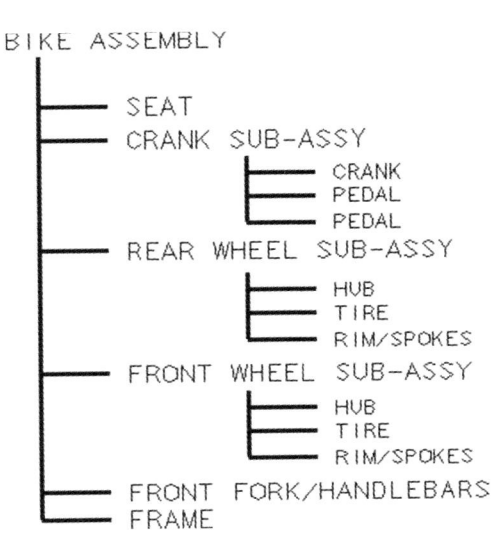

FIGURE 10.4 An example of assembly structure with subassemblies.

tors together using vector sum calculations, the top level can easily locate all the models in the correct location. As parts are moved within the context of a subassembly, the top-level assembly merely points to the subassembly data and can then show the change at the top level appropriately. Refer to Figure 10.5.

10.2.3 Instancing

The third essential characteristic for the assembly model is the use of instancing. An instance is like an intelligent duplicate of an object. This is a common concept in computer programming. If one starts a spreadsheet program twice in a windows-type of operating system, each copy of the spreadsheet window is called an instance of the program. There really aren't two different spreadsheet programs running, it is just like two copies of the same spreadsheet program. Also, in 2-D CAD systems, instancing can be applied to symbols, blocks, or clumps of 2-D geometry.

It follows, then, that in assembly modeling each part model that is included in the assembly structure should really be considered an instance of the part model. This may seem like an unnecessary layer of abstraction if part models are only included in the assembly structure one time (such as the FRAME in Figure 10.4). However, it is quite common to have the same part model used many times. For example, in Figure 10.4, there are two instances of the PEDAL in the CRANK SUB-ASSY. If instancing is not used, then one would be forced to have two separate part models for the pedal. Worse yet, when considering the spokes in a wheel subassembly, one would be forced to create dozens of spoke part models. Using instancing, the pedal or spoke can be modeled just once, and then reused as needed in the assembly model.

An advantage to instancing is that when the real part model (such as the pedal) is changed, all the instances can change or regenerate automatically. Indeed, this is essential to preserving the data management aspects of the assembly model. If a part is revised, then it needs to be shown as revised everywhere it is used. Since the instancing should be based on a pointer or relationship system, the revised part will be able to be shown as revised in all its parent or higher level assemblies. This solves the "where used" problem for drawings where designers have to do research to try to figure out what effect changing a particular drawing will have on the company's products. Without instancing, the designer would have to locate each of the uses of the part and change the part in the context of all those other occurrences.

Although the assembly model can rely on a fair amount of information about an instance of a part coming from the master part model itself, there is obviously some data that must be stored in the assembly model for each instance.

Assembly Modeling

For example, the spokes in the wheel subassembly shown in Figure 10.5 are not all located in the same position and rotation. Instead, each instance of the spoke has been placed into its appropriate location. This positional data, then, cannot be stored with the part model, but rather it must be tracked and stored by the assembly model.

A variety of information in the assembly model needs to be sorted out as to whether it should be tracked based on the master part models or based on the instances of the part models (or even instances of subassemblies). Table 10.1 lists some of these issues. This is not a complete list, but it hopefully shows the somewhat complex nature of how an assembly model needs to keep track of many types of data. It is important to understand how a particular CAD system deals with these issues so that the user knows when and where to work on the assembly model so that the desired results are achieved. This should not be difficult once the implications of instancing are clear.

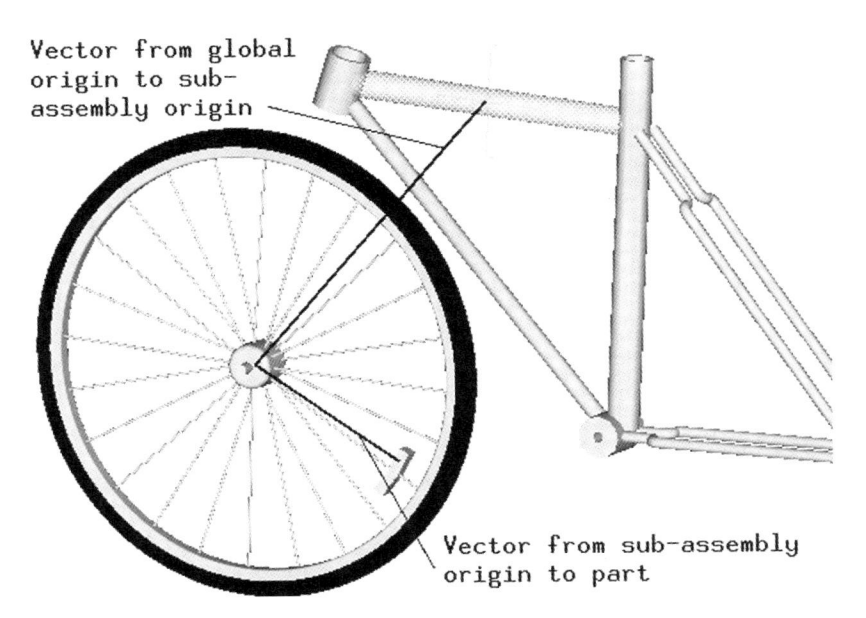

FIGURE 10.5 Locating a position using a transformation or vector summation.

TABLE 10.1 List of Types and Locations of Assembly Model Information

Type of information	Instance or part data?	Description
Part geometry	Primarily part data	The 3-D part model's geometry (surfaces, volume, features, etc.) will certainly be based primarily on the 3-D part model. However, there are cases where part models are altered only in the context of the assembly model. For example, a hole may need to be drilled through two parts only after the two parts are aligned and bolted together. This hole feature would only be seen in the assembly context, so this geometry would need to be tracked by the assembly based on the instances involved.
Positional data	Instance data	The location (translation and rotation) of the various parts in the assembly model will need to be based on each instance (since the same part model may be used in more than one location). Furthermore, positions of part instances within a subassembly are really a case of an instance within an instance (the subassembly itself is an instance; since an entire subassembly can be used more than once as well).
Cosmetic data	Part or instance data	The appearance type of information for part models (such as color, translucency, texture mapping) can be part or instance based. Obviously, the part model information would be acceptable in most cases. But, if the designer wants to see through one instance of a specific part, but not all other instances of that part, then the assembly model would need to allow the translucency to be tracked based on just the instance.
Attributes	Part data	Information such as the material used to make a part would need to be kept with the part model. If a different material were used for the part, then it would typically be considered a totally different part (even if the geometry were the same). Other part attributes might be manufacturing process data (i.e. how the part is to be made).

10.3 ASSEMBLY CONSTRAINTS

Recall from the earlier chapter on part modeling that constraints refer to geometric or mathematical rules or restrictions that are applied to the sketching of 2-D part foundation geometry or perhaps to 3-D feature characteristics (such as depth). The same concept can be applied to assembly models. In this case, though, the constraints are used to control or restrict the location of the part instances within the context of the assembly model.

Although a reasonable assembly model can be made without constraints, it is clearly advantageous to use constraints, particularly if the 3-D design process is dynamic or iterative. If the relative positions of all the part instances in an assembly are clearly understood or static (such as plates that are welded to form a beam), then the assembly constraints may be of lesser value. However, even in this case, constraints can used to make the assembly model intelligent or to capture what the designer intended. With the constraints, as the design needs to be changed (by moving one part instance), other parts can automatically move to new appropriate locations. Without constraints, part instances can be moved to totally inappropriate locations. Therefore, whenever practical, the extra effort should be used to apply assembly constraints to assembly models.

As discussed in the context of part modeling, applying constraints involves the removal of degrees of freedom (DOFs) from the model. When doing the 2-D sketch plane operations, there were 3 degrees of freedom to remove. They were translating or moving in the X-direction, translating in the Y-direction, and rotating about the Z-direction (the direction "out of the page" being an axis of rotation). In assembly modeling, though, there are 6 degrees of freedom to be removed. These are translation in X, Y, and Z, and rotation about X, Y, and Z. Not surprisingly, constraining assemblies can be significantly more difficult. There are simply many more ways that the part instances can be positioned, so controlling these positions involves more effort. For each instance the designer needs to consider, "Can it move in X?," "Can it move in Y?," "Can it move in Z?," "Can it rotate about X?," "Can it rotate about Y?," "Can it rotate about Z?"

To further complicate things, all the issues that can arise for part modeling constraints also apply for assembly constraints. First, there is underconstraining. This could be acceptable, or this could be a problem. If an instance is underconstrained, then it is free to move in some fashion. If the assembly model is for a beam of welded plates, then this would be a problem. But, if the assembly model is for the front wheel on the bicycle, then allowing the wheel to rotate about its bearing would certainly be acceptable.

Next, there is overconstraining. In this case, the same degree of freedom is removed more than once. For instance, if the face of a cube is constrained to not move in the X-direction and it is also constrained to not rotate about the Y-axis, then this is overconstraining (rotating about the Y would make the face stick out in the X-direction, and that degree of freedom was already removed). Again, this could be acceptable, or this could be a problem. As was mentioned with part modeling constraints, the difference depends on how the CAD system solves the simultaneous equations that come from the constraints. So, if the solution process can arrive at a single, stable solution, then the overconstraint can be acceptable. For example, the overconstraint just mentioned for the face of the cube should not be a problem. The constraint to not rotate about the Y-axis is consistent or similar to the constraint for not moving in the X-direction. The added constraint does not violate the existing constraint.

The problems for these constraints arise when the constraints can not be resolved. In this case, the constraints conflict; or, they are inconsistent. If a bolt is dimensioned to be aligned with a hole in a plate, but then that bolt is also dimensioned from an edge of the plate and this dimension is not exactly the distance to the hole, then the bolt is being tugged in two different directions. The constraint solver tries to locate the bolt with respect to the edge and with respect to the hole, but it can't be done, so the constraints need to be changed. This is a simple example, and one that is easy to correct. However, when there are dozens or hundreds of instances involved, it can become difficult to figure out these problems, and the solver is sometimes unable to give a clear indication of what instances are actually causing the problem.

Table 10.2 lists a variety of assembly constraint types that may be provided by the 3-D CAD system. This is not an exhaustive list of possibilities; there may be many others available to the user. Usually, a set of standard or basic operations is recommended for users (such as dimensioning between two planes or faces whenever possible). Figure 10.6 shows an example of a standard dimension between two planar or flat faces of two part instances.

Note that the examples of possible assembly constraints listed in Table 10.2 are generally using geometric entities that are on the physical part (faces, edges, vertices, etc.). However, 3-D part modeling often involves the use of extra geometric entities that are contained in the part model, but would not be seen on the physical part. Datum planes, reference geometry, and part coordinate systems are examples of geometric entities that may be available for a part model. The planes, lines, and points that are found in this type of geometry would certainly be other possibilities for use in the assembly constraints (indeed they can be advantageous).

It is important to realize that some assembly constraints imply the presence of other constraints. For example, the sample dimension shown in Figure 10.6 implies that the planar faces are parallel. It would not be possible to create such a

Assembly Modeling

TABLE 10.2 Some Typical Assembly Constraints

Constraint type	Description
Dimension between two planes	This constraint is probably the best constraint, since it is the most similar to what a 2-D dimension implies in a drawing. Keep in mind it is a linear dimension when the planes are parallel; another possibility would be an angular dimension between planes that are not parallel. This constraint can remove one translation DOF and 2 rotational DOFs.
Dimension between one plane and 1 straight edge	This constraint can be misleading since the part instance that owns the edge would still be able to rotate about the edge. This constraint can remove one translational DOF and one rotational DOF.
Dimension between one plane and 1 point	This constraint will only control one point on the part instance relative to another. The part instance with the point is free to rotate in any direction about that point (sort of like a ball and socket joint), and that is probably only used rarely. This constraint can remove one translational DOF.
Parallelism between two planes	This will keep two planar faces aligned, but they could be any distance apart. This constraint can remove 2 rotational DOFs.
Parallelism between two straight edges	This constraint can remove one rotational DOF.
Perpendicularity between 2 planes	This constraint can remove two rotational DOFs.
Coincident and colinear straight edges	This constraint allows the two part instances to slide relative to each other along the shared edge and to rotate about that edge (similar to a hinge). This constraint can remove two translational DOFs and two rotational DOFs.
Tangency between a curved face and a plane	This constraint allows the part instance with the curved surface to roll along the flat plane. This constraint can remove one translational DOF and 1 rotational DOF.
"Grounding" or "Rigidifying" one part relative to another	This constraint makes two part instances behave as a single rigid body, or they are locked together. This constraint can remove all DOFs between the two instances: they cannot rotate relative to each other or translate relative to each other.
"Grounding" one part relative to the global origin	This constraint makes a part instance immovable. Its location in the 3-D modeling space is completely fixed. This removes all DOFs for that instance.

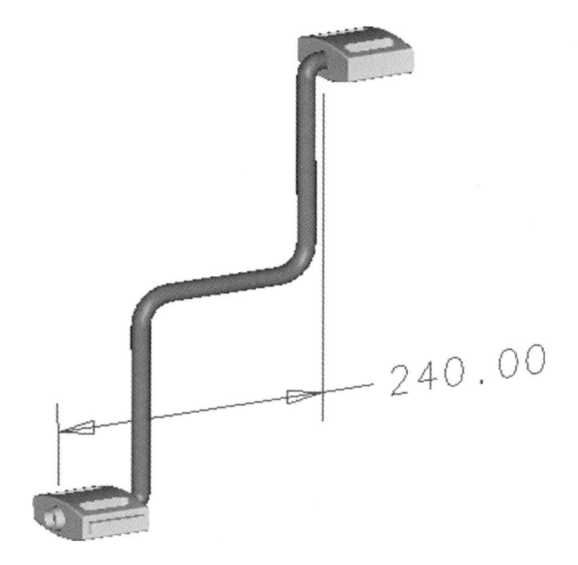

FIGURE 10.6 Example of a typical assembly constraint—a linear dimension between two planar faces.

linear dimension between these faces if they are not parallel (there would be infinitely many distances between two faces that are at an angle relative to each other).

Also, there are problems that can arise from the assembly constraints as the part models are changed. Looking at Figure 10.6, if one of the planar faces participating in the linear dimension were drastically altered to be totally rounded, the dimension would not really be valid anymore. But, if the faces were just given a new feature, such as a hole, then most likely the dimension would continue to be valid. As the design evolves, it would be up to the designer to figure out what ramifications result from the changes to the parts and how to reconstrain the parts to preserve the integrity of the assembly model.

Finally, constraints such as dimensions can be used in the parametric sense. In this case (as with part modeling), equations can be created that relate the values of one dimension to other dimensions. For instance, a dimension to locate one part instance could be set to two times the dimension to another part instance. Thus, when one part instance is moved, the other part instance moves automatically according to the designer's intention. This may also be referred to as having

Assembly Modeling

one assembly dimension drive other assembly dimensions. Keep in mind that these equations (as well as all the constraint data) are another example of information that would be stored in the assembly model based on the instances. That is, the equations may apply to some instances of a part, but not others. The user needs to determine how this data is tracked so that changes to the design can be made in the correct assembly context.

10.4 ASSEMBLY MODEL CALCULATIONS

Once an assembly model is created, there are a number of important calculations that can be performed using the assembly model. This would include volume, weight (if material density is an attribute of the part models), surface area, inertial properties (I_{xx}, I_{yy}, I_{zz}, etc.), and center of mass or center of gravity (CG). The meaning of these properties was explained in the chapter on part modeling.

The only real difference in these calculations versus part models is that the properties need to be summed for the entire assembly model. The volume for an assembly is really just the summation of all the part instance volumes. Assuming all the part models had a density property associated with them, then the weight of the assembly is just the summation of all the weights of the part instances. Of course, the calculation becomes more complicated with the inertial properties and the CG. Now the weight of each instance is not just added together, but consideration is given for the location of the weights, or how the weights are distributed. Even though this is a more complex calculation, it is still a simple matter for the CAD system to do this calculation. This is a very significant advantage over any 2-D CAD systems. Calculating the inertial properties and CG for a large assembly that includes a variety of manufactured parts could take quite a long time without the assembly model.

Another calculation that can be done with the assembly model is interference and minimum clearance. In this case, the 3-D CAD system looks at the various volumes of space occupied by the various part instances. If any part instances are sharing the same volume of space, then an interference has occurred, and the CAD system should be able to indicate this condition to the designer. If part instances do not share volumes of space, then the CAD system may be able to indicate how close they actually are to each other; this is the minimum clearance calculation or just clearance. Figure 10.7 shows some parts that interfere (the solid part interferes "into" the translucent part). Figure 10.8 shows the interference area highlighted for the designer by the CAD system. Obviously, this can be very helpful in working with assemblies and preventing manufacturing problems.

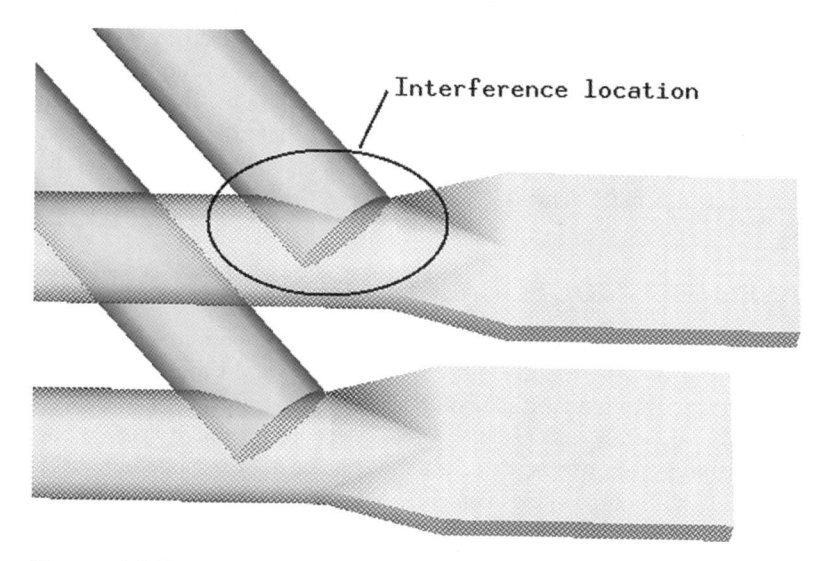

FIGURE 10.7 An example of interference in an assembly model.

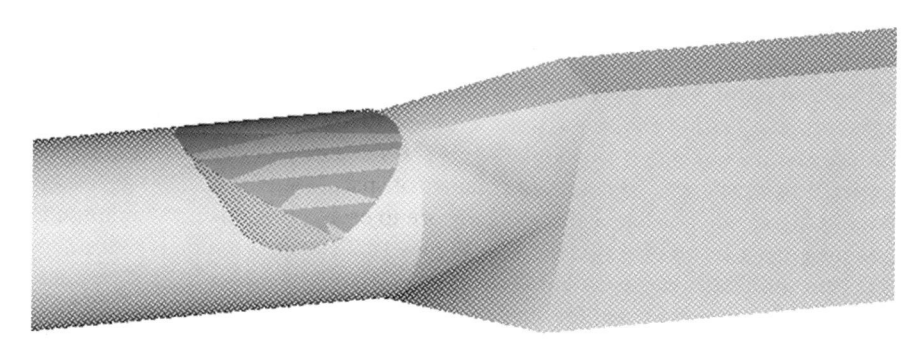

FIGURE 10.8 The interference volume is highlighted by the CAD system.

10.5 ADVANCED ASSEMBLY MODEL CAPABILITIES

Beyond the calculations that can be performed based on the assembly, there are more advanced functions that are possible. This includes animations, part-to-part associativity, and Bill of Material creation.

10.5.1 Animations

Animations are based on sequences of changing part instance positions. For example, when a hydraulic cylinder is extended, the rod of the cylinder moves along the axis of the cylinder. To show this movement, the rod part model needs to be stored at a number of positions along that axis to show a smooth motion.

There are a couple of ways this can be handled by the CAD system. In one case, there may be a dimensional constraint for the stroke of the cylinder. As this dimension's value is changed, the rod moves in and out. An animation of this could be shown in real time as the user changes the stroke from one value to another (the graphics adapter temporarily displaying the rod at intervals of the motion). In this case, the assembly model is showing the change all at once, and the rod is simply left in a new location based on the new dimension value. This animation could be repeated by manually changing the dimension over and over. This sort of animation is quick and easy, but is really temporary from a data management point of view. The animation is not saved as part of the assembly model. Of course, some workstations may allow any motion created by the graphics adapter to be saved in a video file; but this file cannot change automatically if the parts or the assembly were changed or redesigned. It is really just a video clip.

Another way animation could be handled is to have the CAD system calculate and store the location of the rod at the intervals of motion (the graphics adapter not doing the calculations). In this case, the assembly model itself may now store the dimensional values and the graphical information to show the rod at various positions. Also, the user may be able to enter a starting and stopping value for the dimension and the number of intervals desired. Thus the fidelity of the animation can be controlled. Now the assembly model itself can save or store the animation for the sequence, and it can be used again when part models or other parts of the design are changed; it should reflect these changes appropriately.

In both these cases, the user is manually changing the assembly model to show the animation. Another way animations can be created is by building a mechanism or kinematics model. In this case, the assembly works with mechanism-based constraints that mimic a physical system. For example, there may be joint constraints (such as hinges or universal joints), ball-and-socket constraints, cams and/or follower constraints, rigid body constraints, etc.. These constraints would be used to create a physics-based approach to having the part instances move. As the constraints remove degrees of freedom, however, the system will have to be carefully underconstrained (to just allow the motions that are correct). Then, forcing functions must be supplied to fully specify the behavior of the system (to control those remaining degrees of freedom). These functions may be gravitational, or they may be motion functions (such as rotate a gear at 1 RPM for 1 minute). In any case, this mechanism model will need to be solved for a certain number of intervals (such as 1 second), and if it solves successfully at each interval, then the system should be able to show an animation of the system through the simulation. This sort of model can be more difficult to create, but it can handle more sophisticated motions.

10.5.2 Part-to-Part Associativity

Another advanced feature of assembly modeling is part-to-part associativity. Usually part models are self-contained with respect to their geometric information. That is, the surfaces of the parts are totally controlled by the dimensions, geometric constraints, and part history of that individual part model. In the part-to-part associativity capability, one part model in an assembly model actually controls the surfaces in another part model.

In Figure 10.9, a hose is shown in relation to a ramp. But in Figure 10.10, the ramp has raised and the hose has actually changed shape. There is no such thing as flexible surfaces for the part model of the hose; and there is only one hose part model in the assembly. So in this example, the hose part model has simply updated based on geometry that was driven by the assembly model. The control points of the 3-D spline that the hose was swept along are constrained to be driven by the position of the ramp. First, the ramp moved, then the control points moved, then the part model changed based on the new control point positions. This is an example of part-to-part associativity.

This is a rather complicated issue from a data management point of view. The part model is mostly self-contained, but some of the part model data is really driven by the assembly model. This can be a valuable capability. Another common application of this technique is when parts are made by a mold (such as plastic injection molding). The mold cavity really needs to be driven by the part, or vice versa.

10.5.3 Bill of Material Creation

The last advanced capability for assembly models presented here is the creation of Bills of Material (BOMs). There are more advanced capabilities for assembly models that could be discussed, but this is one that is most commonly used. As mentioned at other times in this work, the BOM is a note or document that indicates what items are necessary to assemble or create a particular level of a prod-

Assembly Modeling 259

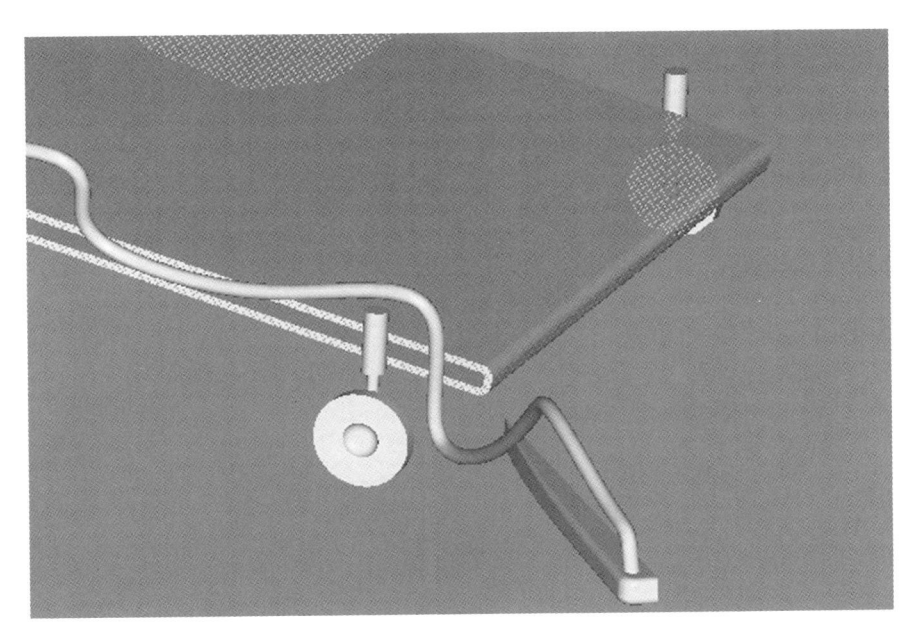

FIGURE 10.9 The hose part model is shown in its normal state.

uct structure. A BOM may be for a low-level assembly such as the pedals and crank for the crank subassembly shown in Figure 10.6. Or, there could be a much larger BOM for the top-level assembly such as the entire bicycle shown in Figure 10.1.

The BOM is a list of items that is usually shown in some sort of logical arrangement. They may be arranged by part number or other identifier; they may be arranged by the assembly and subassemblies; they may be arranged by material type, etc. Often this BOM information is included on 2-D drawings that are created for assemblies to show what items need to come together to form what is shown in the drawing. Figure 10.11 shows a sample BOM. Note that it has a column for QTY or quantity. This makes it easy to understand how many different or unique items go together to form the assembly, and how many of them are required.

The BOM should look pretty familiar. It is basically the same as the assembly structure discussed all along for assembly models. It shows the same hierarchy or parent/child relationships of the part instances and the subassemblies (the levels of the assembly are shown indented in Figure 10.11). The only real difference is that the all the separate instances of the same part are not listed; instead, they are gathered up and their quantity indicated. The BOM in Figure 10.11 is

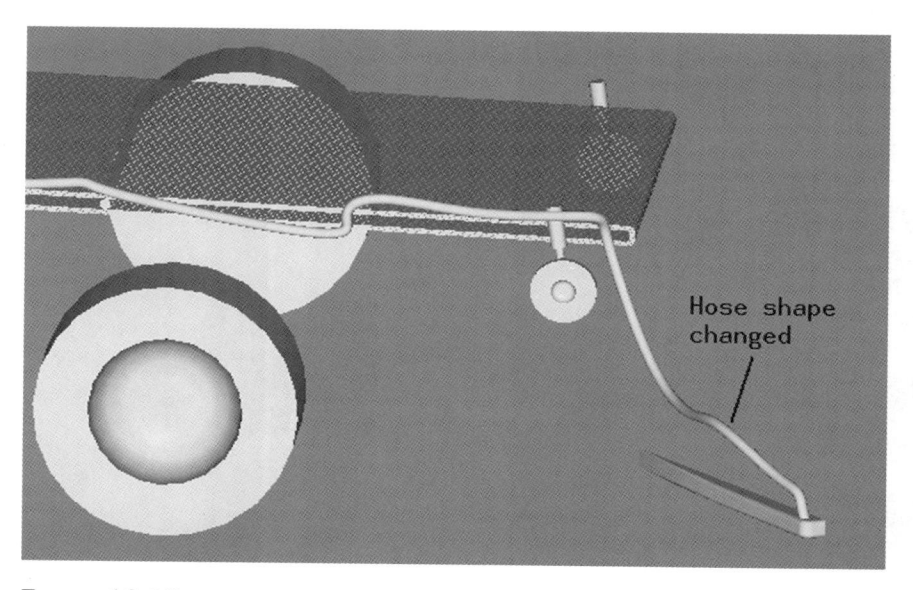

FIGURE 10.10 The hose part model has updated based on the raised ramp.

also an example of an explosion. It shows all the levels with all their constituents from the top level (the BIKE_ASSY) down through each subassembly. In fact, there could be a separate BOM for each of the subassemblies, as well.

The assembly structure information can be used to create a BOM based on the assembly model. This BOM data could then be passed on to the systems that track all the parts for a product or the entire company. Note that there are some extra columns in the listing of Figure 10.11. There is a column for REV (or revision), and one for MATERIAL. These are common attributes that a BOM system will need. It is important to know what revision of the parts or assemblies are being built, and it is important to know what materials are used for these items. Some 3-D CAD systems will allow these attributes to be stored with the part and assembly models, and then they can be included in the BOM data.

One problem with having the CAD system create the BOMs is that there can be data management problems between the CAD system and the BOM system. Usually, the manufacturing or shop of the company will not use the 3-D CAD system, so the BOM directly contained in the CAD system would not be available to them. Instead, there may be a separate system that needs to integrate with the CAD system. There is often little difficulty in doing this in the direction from the CAD system to the main BOM system. Unfortunately, the manufacturing system may also need to drive changes back to the CAD system BOM (such

Assembly Modeling

QTY	IDENTIFIER	REV	TYPE	MATERIAL
1	BIKE_ASSY		ASSEMBLY	
1	CRANK_SUBASSY		ASSEMBLY	
1	CRANK		PART	
2	PEDAL		PART	
1	FRONT_FORK/HANDLEB		PART	
1	FRONT_WHEEL_SUBASS		ASSEMBLY	
1	HUB		PART	
24	SPOKE		PART	
1	TIRE		PART	
1	REAR_WHEEL_SUBASSY		ASSEMBLY	
1	HUB		PART	
24	SPOKE		PART	
1	TIRE		PART	
1	SEAT		PART	
61				

FIGURE 10.11 An example of a BOM.

as a changing a component from manufactured in-house to out-sourced). This process can be much more difficult to handle in an automated fashion.

Another problem is that there may be part models in the assembly model (and thus in the assembly structure) that are needed from the point of view of design, but they are not really part of the final product. For instance, a simple surface model may be created for the seat of the bicycle, but there may actually be many parts that are in the seat. In this case, the BOM from the assembly model will only indicate a single seat, when, in fact, it should indicate all the parts within the seat. Either the seat is going to have to be completely modeled, or a means of editing the BOM from the CAD assembly model will be needed. For another example, a purchased component (such as a chain for the bicycle) may only be shown as a single item in the manufacturing BOM system (to indicate "just buy one chain"). But, to show the location and movement of the chain, a designer may model it as a number of small links put together (as a real chain would be). In this case, the BOM from the CAD system would have too many items, and again, the model needs to be changed or the BOM data from the CAD system needs to be edited. One option in the CAD system may be to exclude items from the BOM data created.

More information on data management issues and BOMs are presented in the next chapter.

10.6 CHAPTER EXERCISES

1. Create an assembly model that includes part models of different densities. Have the 3-D CAD system calculate the total weight and center of mass for the assembly model. Record whether changes to part models instanced into that assembly model are immediately reflected in the weight and/or center of mass for the assembly model.
2. List 5 different assembly constraints and the number of degrees of freedom they remove.
3. Record whether your CAD system has the capability for animations.
4. Record whether your CAD system has the capability for part-to-part associativity.
5. Record whether your CAD system has the capability for Bill of Material generation.

10.7 CHAPTER REVIEW

1. What are the 3 essential characteristics of a proper assembly model?
2. Explain some of the advantages of the use of vectors in the control of parts in an assembly model.
3. Why are some degrees of freedom not removed from an assembly model that is going to analyzed as a mechanism (using a kinematic analysis)?

11

Managing Three-Dimensional CAD

11.1 INTRODUCTION

This chapter presents information on managing or administering a 3-D CAD system. As with the 2-D CAD system, it is best if the person performing the administration tasks has some knowledge about CAD and the tasks designers and engineers are trying to accomplish with the system. However, since 3-D CAD is more demanding than 2-D CAD, it will be more difficult for administrators to become completely fluent in the use of the 3-D CAD system. Hopefully though, the previous chapters on various types of 3-D modeling will at least provide some level of understanding. Administrators should do their best to familiarize themselves with the information in these chapters.

In addition to being more demanding on designers or users, 3-D CAD also tends to be more demanding on administrators. The 3-D CAD system is able to solve many more kinds of problems and organize more data for various parts of an organization. As such, there are then more administration and management tasks. It turns out that the 3-D CAD system can contain and manage more than just design documentation (such as drawings). It can contain types of information such as product structures, standard parts libraries, analysis data, Bills of Material, tooling models, and manufacturing or processing data.

It is critical that an appropriate level of resources be allocated to carefully organize and maintain a 3-D CAD system. This may mean that all users perform some basic administration functions. Or, this may mean that more personnel are dedicated to full-time administration. These needs are probably greatest during the transition from 2-D to 3-D systems. After this transition is complete and all the data management procedures are in place, the demands for administration should be diminished. Regardless, it is essential that the 3-D CAD system be carefully managed; otherwise, a large amount of data will be unorganized and/or lost; this situation should be avoided. Consultants could be retained to assist with creating the proper level of data organization.

11.2 DATABASE MANAGEMENT (THE "ADMIN" PERSPECTIVE)

Probably the most important issue for administration of a 3-D CAD system is the potential for expanded database management activities. 3-D CAD systems can produce a very large amount of data in a short amount of time. 3-D CAD systems also often bring together data from a wide range of activities that may have been previously segregated (so-called islands of automation). The 3-D CAD system database manager could be responsible for tracking 3-D models, drawings, Bills of Material, manufacturing or NC data, and analysis data. These various items may have relationships between them so that relational database concepts are needed, as well. These various items may also require revision control where various obsoleted records need to be retained for future reference by engineering and long-term product support functions (perhaps up to 30 years into the future depending on the longevity of the company's product).

This section discusses some of the background for understanding why the 3-D CAD database needs to operate in certain ways. This is intended to assist the administrator of the system. However, it is important for administrators to regularly consult with the users of the system to make sure that their needs are being met by the database system. A later section of this chapter presents data management (the user's view of the data).

11.2.1 Files Management Versus Metadata

In 3-D CAD systems, the individual entities being tracked by the 3-D CAD database manager software are mostly held in many individual operating system files. For instance, 3-D part models are often PRT files, and 3-D assembly files are often ASM files (these files would be specific to a particular CAD system—there is currently no industry standard PRT or ASM file). The database manager software, then, is going to at least need to control or "track" the location and disposi-

Managing 3-D CAD

tion of these files. This sort of information can be referred to as metadata (meaning data about data).

Although it may be possible to just use an operating system's file manager as a control mechanism, this is not going to provide a good database management environment. In this case, there is no separate database manager program or engine; this can cause difficulties in managing the CAD data. For instance, if an ASM file simply records the directories and file names of a list of PRT files that are contained in an assembly model, but then these PRT files are renamed or moved to another directory, the ASM file is going to be invalidated or corrupted.

It is far better to have a separate database (or table) recording the information about the operating system files and locations. Now, a database manager program can offer an option to move the PRT files to new directories, and the tables can be updated accordingly to keep track of their locations. The ASM file would simply refer to the tables to find out where the PRT files are located (using an appropriate key or item ID scheme). The remainder of this section assumes that this metadata database management approach is being used by the CAD system. Keep in mind, though, that the database management scheme for the CAD system may be based on a special database manager program developed by the CAD system vendor, or it may be based on third-party database programming.

The location of operating system files is just one aspect of the kind of information that can be tracked, controlled, secured, queried, etc. by the 3-D CAD database manager software. Other kinds of information would be similar to the information discussed for 2-D CAD management, such as drawing number or part number, revision level, ECO number, etc. The 3-D model database would also have part numbers, revision levels, and ECO numbers. But, databases for 3-D CAD could also contain information on material properties (i.e. indicating if parts are made of steel, aluminum, plastic, etc.), geometric properties (such as the part being based on a family of parts or from a standard library of parts), and feature properties (such as what manufacturing process is used for a particular feature). The amount and type of this data that is available to be managed will vary depending on the CAD system, but the more information that is controlled by the database manager software, the more processes in the organization that can be made to work together effectively.

11.2.2 Parent/Child Relationships

The hierarchical approach to database management where some items own other items, or where one item is dependent on another item's existence, is heavily used by 3-D CAD system database management. There are two important sources of this hierarchy data.

The first common source of these parent/child relationships is assembly structure or BOM. The BOM is a listing of the items that are needed to create an assembly (refer to the chapter on assembly modeling). For example, a bicycle assembly would need to have a number of parts brought together such as the frame, tires, handlebars, nuts and bolts, etc. It is the BOM that lists each of these items and how many of each item (or quantity) is needed. There is also usually a drawing or 3-D model that shows exactly how the pieces are put together. Figure 10.11 shows how this product structure could be listed. The owning or parent assembly is shown as indented to the left, and all the items within it are shown indented to the right (to indicate that they are under or children of the parent assembly).

Clearly the BIKE_ASSY at the top of Figure 10.11 requires the existence of the parts listed within its BOM or assembly structure. In other words, if the SEAT is a PRT file that is called out by or used by the BIKE_ASSY ASM file, then this PRT file should not be deleted (unless the BIKE_ASSY is altered to no longer need the seat). Also note that the bicycle assembly contains or uses other assemblies (i.e. more ASM files); these are referred to as subassemblies. In Figure 10.11, CRANK_SUBASSY, FRONT_WHEEL_SUBASSY, and REAR_WHEEL_SUBASSY would be called the subassemblies of the BIKE_ASSY, and each of these subassemblies would typically be yet another ASM file using their own PRT files having more parent/child relationships.

Virtually all manufacturing involves these parent/child relationships based on BOMs. The network of relationships between the parts (or details), subassemblies, and top or final assemblies can become quite complex, and there may be many levels of assembly structure. The 3-D CAD database manager software needs to keep track of all these files that may be present and manage the relationships between them (knowing which parts are used by which assemblies; knowing what parts are contained in each assembly).

The second common source of parent/child relationships in the 3-D CAD database is associative 3-D models and 2-D drawings. With a 3-D CAD system, drawings (or at least the geometric entities within drawings) can be automatically generated by viewing the 3-D model. Subsequently, the 2-D geometric entities (such as lines) are associated to 3-D geometric entities in the 3-D model (such as surface edges). Now this means that the drawing (kept in a DWG file of some sort) becomes a child of the 3-D model (a PRT or ASM file). The drawing is dependent upon the existence of that particular 3-D model.

In order to manage these various parent/child relationship, the CAD database manager software uses such devices as pointers, indexes, and relational databases to keep track of the data. The assembly model is going to have pointers to the various part models that go into that BOM. Then, as the part model evolves, the assembly model can easily show the changes since it is only pointing to the part model (or some version of it). Of course, the same part model may be used in more than one assembly model (for various products being designed), so the part model may use its own system of pointers for its parent assemblies.

11.2.3 "Where Used" Functionality

One of the most powerful uses of the parent/child relationship capability is the "where used" functionality. This involves determining how many times and in how many ways a particular item has been referenced or used by various parent or owning items. For instance, the TIRE indicated by the BOM in Figure 10.11 may be used in a variety of the products being manufactured by the company. In fact, a "where used" query for that tire would indicate that it is at least found in the front and rear wheel assemblies of the BIKE_ASSY. The 3-D CAD database manager software should be able to readily provide this sort of information. In some systems, this functionality may reside with PDM software (see Section 11.3.4 below on Product Data Management), but the concept is the same.

The "where used" functionality is important for designers to research how changes to an item (such as a part) would affect the company's various products. Perhaps a part has been used as a standard part in many products, but now the first product that used the part needs to be changed. All the reuses of the part, as well as its original use, need to be checked so that the new part will fit in each application.

11.2.4 Database Integrity

As the complexity of the network of relationships proliferates, the relationship data may not be stored in just the ASM and PRT files, but rather in separate relational database files, or in some combination of files. For instance, it may be better to record the BOM information in the system-wide databases instead of the individual ASM files. Now the BOM data can be generated by simply querying the database (and not having to scan the information in all the relevant ASM files).

Clearly an administrative issue that arises from this approach is database integrity and/or synchronization. Any BOM-type of data in the ASM files will now have to agree with the BOM data in the system-wide databases. Unfortunately, the databases can become "corrupt" in some circumstances. For example, when one computer on a network is attempting to record new database information on the server, but it fails in the middle of the operation (network failure, hardware failure, etc.), then the separate databases may not agree. The CAD system or the database manager software should provide means to recover from these situations. Of course, proper backup procedures are essential to database recovery in some cases. All the CAD data files (PRT, ASM, DWG, DBF, etc.) should be considered mission critical and backed up accordingly.

11.2.5 Version Management

Another issue to consider for how database management is somewhat unique with 3-D CAD systems is version management. Many non-CAD databases or tables are intended to contain only the latest and greatest information about a subject. As new data is created, then the database merely gets the new information and removes old information.

However, design and engineering work has its own life cycle. The data is typically tracked through phases such as conceptual design, detailed design, approval cycles, release to manufacturing, retention and long-term maintenance, and destruction. Sometimes the current disposition of the engineering information is dictated by the new product development process; sometimes it is dictated by the manufacturing process; and, sometimes it is dictated by legal, liability, or even statutory issues (where data must be carefully controlled and maintained by law).

What this means is that even if a version of a 3-D model and its drawing is wrong or has a mistake, it may be necessary to keep the incorrect drawing and the metadata about that drawing for a long time. The product may have been manufactured with that mistake (hopefully not to the detriment of function and safety), and copies of the product are now in the field. Thus, the incorrect drawing as well as the corrected drawing should be kept on file (according to proper record retention policies) to support that fielded application.

Therefore, the database manager software needs to control not only the records about specific items in the database files, but it needs to control versions of the various items. It would normally present to the user the latest or up-to-date version of items, but then it may also present the older versions. It may also need to present the full history of the item (not to be confused with part history of part modeling).

Beyond the support of item history for long-term product support, version management is needed if parts are re-used in a number of different design projects. For instance, if two different 3-D assembly models (such as a tricycle and a bicycle) both use the same front wheel, and the front wheel is revised for the benefit of the tricycle only, there may be a time when both the old and new versions of the front wheel need to be supported (since the bicycle is still using the older version). When a designer looks at the latest 3-D tricycle model, the newer front wheel should be shown. But when another designer looks at the latest 3-D bicycle, the older front wheel should be shown.

Of course, there could also be interchangeability issues related to whether the new front wheel could still function on the unchanged bicycle assembly. And, there may be effectivity issues in terms of exactly when changes are rolled out to the manufacturing process. So even in items that are revised in all parent assemblies (such as updating or regenerating the bicycle and the tricycle assemblies for the change in the front wheel), the older versions may still need to be tracked until the roll out of the new design to manufacturing is complete. These are all scenarios that the 3-D CAD database system needs to be able to address.

11.2.6 User Management

The 3-D CAD database management should also consider the privileges of different users or classes of users. In many organizations, there is a distinction between users that can administrate the CAD database and those that work with it. For instance, a company may want to limit the number of users that can delete certain items or records in the CAD database. This typically is done with information that is considered released or active. This information (models, drawings, BOMs, etc.) is often critical for manufacturing or product support, and it may require added security.

Controlling access to the critical information may entail security at more than one level. First, users should be prevented from deleting the operating system files (i.e. DWG, PRT, ASM, DBF files). This security is dependent on the particular operating system being used, but usually folders or directories or individual files can be protected against delete privileges for users (except for the administrator).

Beyond the operating system protection, the CAD database management software can usually be used to restrict access to information, as well. In this case, users may be working on different projects or within different departments, and it may be desirable to make sure their data is not accidentally altered by the wrong users. These different departments may be given their own libraries or drawers within the context of the CAD database, and the CAD system may allow an administrator to allow only certain users access to these various libraries. The administrator should also protect the data organization itself so that users can not inadvertently delete or alter the libraries, drawers, folders, etc. or other types of customizations.

11.3 DATA MANAGEMENT (THE USER PERSPECTIVE)

Having presented CAD data management from the administrator's point of view, this section presents CAD data management from the users point of view. Of course, administrators should become familiar with this perspective, as well.

11.3.1 Ownership and Revision Control

Probably the most fundamental issue of data management from the designer's point of view is ownership and revision control. When all design work was based on paper drawings, ownership of an aspect of product design was based on who actually had a given paper drawing in their possession. Unlike computer files that can be easily duplicated over and over again, duplicates of paper drawings would generally be discouraged (at least for use by designers). Thus, it was fairly simple to resolve ownership over an aspect of design. The designer with possession of a drawing would be able to make changes to the drawing as deemed necessary. But with the 3-D CAD system, where so many items are tracked by the CAD database software, and there is a web of interrelated parent/child relationships between the items, there are many more possible scenarios of ownership and users need to become more savvy with understanding how the 3-D CAD data management works.

The idea of a vault is helpful in understanding ownership of data. When someone needed to work on a paper drawing, it was often taken from a vault or print room. As with a library of books, this could be referred to as checking out a drawing. As the drawing was removed from the vault a record was also made of who took the drawing, when it was taken, etc. This paradigm is often carried over into the 3-D CAD system. If someone needs control over a 3-D model, then they would check it out from a virtual vault, cabinet, library, etc. Once this has been done by a designer, this particular 3-D model would become off limits for other designers to make changes. Just like a single book in a library, it can not be checked out twice at the same time. This restricted access capability can be referred to as revision control (although items may be checked out for other reasons besides an actual design revision to an item).

It is important to have this sort of data access scheme applied via database management software with the 3-D CAD system. The non-metadata approach of simply relying on the naming and locations of files in the operating system is not very reliable with respect to revision control. Once a part file is copied or renamed, there is no robust tracking of who did it, when they did it, where it was copied to, etc. This operating system file manager approach should be avoided.

11.3.2 Assembly Revision Control

Checking out a 3-D part model to control the part's design is relatively simple; there is just one item to control. The designer checks out the part model in order to change it. The CAD database software records this activity, and then no other designers would be able to then check it out.

However, in many cases, there are 3-D assembly models which are the parents of the 3-D part model (i.e. the BIKE_ASSY assembly model in the previous chapter is the parent of the SEAT part model). Now, the checkout concept becomes more complicated. For instance, one designer could be responsible for the assembly model, while another designer or designers would be responsible for the parts. The assembly designer now will need to have copies of the part models, but this designer will need to be prevented from actually changing the parts. On the other hand, the designers changing the parts will probably not be able to design these parts without seeing the overall design context (i.e. the parent assembly). Thus, the part designers need a copy of the assembly, but they must be

Managing 3-D CAD

prevented from changing the assembly model (for instance, moving parts around within the assembly).

A common solution to the issues presented by assembly revision control is the concept of reference. In paper drawing management systems, there was a stamp that could be placed on drawings that said, "For Reference Only." This was stamped on a copy of a master drawing when a designer needed to see a drawing from the vault, but he or she did not actually check out this drawing. The designer could then see the state of the design at that moment, but there was no real guarantee that it would remain current (since someone could go ahead and check it out and change it later). In order to clearly show that the copy was not the master drawing, "For Reference Only" was stamped on the drawing.

It is not difficult to transfer the concept of "For Reference Only" into the 3-D CAD system. When a designer needs to see other 3-D models for designing within the context of the overall project, copies are made of the other 3-D models for the designer but the CAD database software records the fact that these copies are "For Reference Only." Using the example of a bicycle assembly again, the designer that checks out the entire bicycle assembly could just reference the front wheel assembly. In this case, it is assumed that this designer would be able to change the position of the front wheel assembly within the context of the overall assembly. But, this designer would not be able to actually change the front wheel assembly or its constituent parts.

Although the "For Reference Only" approach is basically practical, it can be rather complicated and difficult for some users to work with. For instance, users may try to check out parts without checking out the proper parent assembly or understanding the "where used" issues. If a user checks out a part that is used by an assembly, and that assembly is checked out by another user, there may be no guarantee the changes to the part will be accepted in that assembly unless the users cooperate in the design control.

This is just one example of the complicated nature of assembly data management that can result with multiple users working on the same design. There are even more variables that can play into these scenarios such as configuration management (where different options are available for a product), constraints (where users may need to constrain assemblies together without owning all the assembly models), assembly drawings (where one user may need to just add a note to an assembly drawing, but not change the geometric characteristics of the assembly model), etc. Although this is a complex matter, it is essential that users eventually understand and work within the CAD database management system. The problems (such as who really controls the design of the bicycle) are really design control issues, not database software issues. The database software is merely trying to keep all the data properly organized and controlled.

Of course, using "For Reference Only" is only one approach to the team-based assembly design. Another approach would be to allow all the part designers and the assembly designer full access to all the models (even across the Internet). In this case, each of the designers would be changing each others view of the design context in real time. This approach may be manageable for conceptual design and other prerelease design activities. But, once a design is released to production, a more systematic approach of checkout and reference would probably be needed to demonstrate an appropriate level of design control (as required by quality system and management systems such as ISO 9001).

11.3.3 Part Numbering and Revision Levels

Another important data management issue from the perspective of the user is naming conventions. When items are created in the 3-D CAD system, they need to be given consistent names by all users. This makes it much easier for users to locate models and data. This is especially important since 3-D CAD systems contain so many kinds of files and data.

The vast majority of companies use a part numbering system as the primary tracking scheme for all their product data. The part number system may actually use letters and numbers, but it is customary to call them part numbers anyway. In this case, the part number could easily be used as the naming convention for the parts, assemblies, BOMs, etc.

Obviously, the most important characteristic of the part numbering system is that each item (part, assembly, specification, etc.) has a unique identifier. Often there is a master part numbering software that provides these numbers. Many companies will also use an intelligent part numbering system where certain kinds of items will have different starting numbers. For instance, BOMs may start with one particular number (such as "1."), drawings with another number (such as "2."), specifications with another number (such as "3."), and hardware (such as nuts and bolts) with another number (such as "4."), etc. Intelligent part numbering is very valuable with paper controlled processes, but with the use of database management software in a 3-D CAD system, it can be of less importance. In this case, items can be sorted and searched based on various keys irrespective of the part number (such as whether it is a BOM, drawing, specification, or hardware).

If users have to locate their parts and assemblies by file name alone (which again is not a recommended approach), then obviously the best solution is to use the part number as the file name. Of course, operating system file naming conventions must then be followed. This may restrict the type of characters that can be used, and it may limit the length of the file names (and thus the part numbering). The part numbering system used in conjunction with a 3-D CAD system would also preferably not use nonalphanumeric information. In other words, only numbers and letters are used. Characters such as hyphens (-), slashes(/ or \), pound signs (#), etc. should be avoided. Almost all of these characters are reserved for special meanings for the operating system.

In addition to using part numbers, virtually all manufacturers use the concept of revision level to indicate the proper version or revision of a design and its constituents. This is extremely important as most designs are produced in an iterative process. The most common schemes for indicating revision levels is to use numbers or letters. A part model at revision "0," could be considered the initial release. The first revision would be a "1," the second revision would be a "2," etc. Companies may also use letters. In this case, perhaps no letter would be the initial release, and then A would be the first revision. Unfortunately, database software may not allow a blank for the revision field, and thus some other character will need to indicate the initial release.

The CAD database software should provide for the revision level for all the 3-D items (parts, assemblies, BOMs, etc.). In other words, each released item in the vault should have a discernable revision level. As a model is checked out, changed, and then checked back in, the CAD system should make allowance for the new revision level to be indicated. Of course, revision level often should be tied to the Engineering Change Order or ECO or release process, and this process may be controlled by software other than the CAD system. Therefore, it may be unpractical to have the CAD data management automatically set the revision level (such as incrementing the level by 1 number or letter each time a change is made). Also, 3-D models may have to go through an approval process within the engineering department before it is released, and this approval cycle may generate changes and new database models (i.e. checked out, changed, and checked in again).

11.3.4 Product Data Management

Particularly since 3-D CAD systems can create and maintain Bills of Material, 3-D CAD systems can provide a level of product data management. Product data management can be thought of as a total system which allows a manufacturer to completely specify its product information. Despite the BOM capability for assemblies in a typical 3-D CAD system, the product data management activity is often managed by a separate class of software called PDM software. It is essentially a level above the CAD system (although it may appear to be merely a feature of the CAD system).

PDM would go beyond the traditional CAD system to include such items as specifications (essentially documents that cover departments like marketing, materials, production planning, purchasing, etc.), technical publications (documents such as field manuals or parts lists), configuration management (different options or special features of the product that can be purchased), etc.

The PDM system could also dictate control over design processes. For instance, the PDM system may control what designers are allowed to work on which project. It may control which users have to approve changes to the design (the types of users are often referred to as "roles"). It may control how data is controlled and/or released to manufacturing (often referred to as "states" of the data). The installation, use, and management of PDM software is beyond the scope of this work, but CAD users (and 3-D CAD users in particular) should be aware of this class of software and its capabilities.

11.4 FILE ORGANIZATION

There are many types of operating system files that would be used in the context of a 3-D CAD system. A number of them have been mentioned already (PRT, ASM, DWG). Some other possible types of files would be files for Bills of Material and database files for the metadata. Most of these files (except perhaps for the Bill of Material file) would be in binary format. Table 11.1 lists information about some of these files.

A somewhat unique file type for 3-D CAD systems is the model file. The model file is often a sort of personal database of CAD data. Depending on the user or the context, a model file may contain part data, assembly data, drawing data, analysis data or whatever a user is working on at the moment. The model file allows this dispersed design data to be brought together in a single operating system file (as opposed to working with dozens, hundreds, or even thousands of separate files that all share the same owner and privileges). As personal data, though, it is usually considered temporary, and finished work is generally filed back into the vault as the separate PRT, ASM, etc. files.

It is important that all these types of files be organized and carefully managed on the computer system or network. As mentioned previously, there may be some files that need to be protected from accidental deletion or corruption (such as released drawings, models, and BOMs). These files would probably be kept in a folder or directory or file system that has restricted access. Other files, such as the model file, are less critical. These files may not require the same level of restricted access.

Due to the widespread use of computer networks, there are some obvious approaches to organizing the 3-D CAD databased on the network architecture. For instance, model files would typically be kept on an individual's workstation. Critical or shared data such as the metadata databases obviously would be best kept on file servers. However, since the amount of 3-D CAD data for a single assembly model might be 100s of megabytes, and all this data might need to be copied from the file server to the workstation at one time, the performance of the

Managing 3-D CAD

TABLE 11.1 Typical Operating System Files for 3-D CAD Systems

Item	Typical file extension	Description
Part models	PRT	Typically each 3-D part model eventually is stored as a single operating system file. This PRT file is not standardized however. Each CAD system uses its own proprietary format. It contains all the data for 2-D sketch planes, 3-D surfaces and edges, features, part history, etc.
Assembly models	ASM	Typically each level of assembly (assemblies as well as subassemblies) is stored as a single operating system file. Again, this ASM file is not standardized for all CAD systems. It would contain information such as the part instances used in the assembly, the 3-D location of each instance, data about each instance such as whether it is translucent, pointers to parent and/or child relationships, etc.
Drawings	DWG	Drawings are generally independent operating system files as well. Unlike 2-D CAD, though, this file would contain pointers between various geometric 2-D entities (like lines and arcs) and 3-D entities (edges) found in the associative 3-D model (contained in a PRT or ASM file).
Model files	MF	Typically this is a user's temporary file for work in process. It may contain part data, assembly data, analysis data, or drawing data. Model files prevent having to make many copies of individual operating system files that are all owned by the same user.
Metadata	DBF	Typically this is standard database program files or tables. They would control or track the location and disposition of the various other operating system files. They would also contain data attributes and relational data about the various items.
Bill of material	DAT, TXT, BOM	These files are basically reports for the ASM or DBF files. They would be related to the 3-D assembly structure data in some way, but they may also contain non–CAD data (such as material specification or vendor data). The data for Bills of Material may also be contained in the metadata data bases themselves (in which case they are not separate operating system files).

network must be designed carefully. In fact, there will likely be multiple users trying to obtain that much data simultaneously.

11.5 INTEGRATION

Another issue to consider for managing the 3-D CAD system is integration. Integration in the context of this work is combining separate computer software systems to share data and work together in some fashion. Although this integration of software systems may seem superfluous (since there are such obvious benefits to their combination), one needs to realize that the separate systems were each developed independently over the years, and each system was intended to automate specific design and/or engineering activities. Furthermore, there is a great deal of complexity and expertise involved with each of these engineering activities, and no single software vendor has really managed to achieve a high level of capability for each of these activities within the same software package. Thus, the different systems have remained independent, and CAD system managers must be prepared to deal with integration.

Naturally, this software integration has been dependent on the development of system connectivity. At one time, each of the different software systems would run on separate computer hardware. However, the adoption of Local Area Networks (LANs), Wide Area Networks (WANs), and the Internet has in many cases made the integration process more practical.

11.5.1 CAD to CAE Integration

Probably the most important and direct integrations are between CAD and CAE (Computer-Aided Engineering), and between CAD and CAM (Computer-Aided Manufacturing). Indeed, these three are often combined as CAD/CAM/CAE. Although it is somewhat of a dangerous generalization, one can say that designers are concerned with the creation and management of the precise configuration of a product, while engineers are concerned with the behavior and performance of the product by analyzing it in the context of physical laws and approximating calculation techniques. This being said, the CAD system would generally be the system for designers, and CAE software would be for engineers. CAE software can include such software systems as FEA (Finite Element Analysis for analyzing solid products under forces or loads), CFD (Computational Fluid Dynamics for analyzing fluid motion and behavior), simulations (analyzing dynamic events such as mechanisms, vibrations, and impacts), and perhaps E-CAD (emulating electronic systems behavior). Also, most engineering departments would have various in-house programs for doing engineering analysis within the specific context of a company's expertise and products.

Managing 3-D CAD

Clearly there should be a close link between the CAD and CAE systems. The CAE system is going to be used to analyze and/or simulate the product being designed, and the design is encapsulated in the part and assembly models that are in the 3-D CAD system. Therefore, the engineers are going to need to get the models for analysis. In some cases, the actual PRT and ASM files would be used within the CAE software (particularly if the same vendor supplies the CAD and CAE software). In other cases, there would need to be some sort of translation between the systems based on neutral files such as IGES or STEP (refer to the Section 11.6 on translations); however, it is usually best to avoid the translation step if possible. Of course, the product being designed in the CAD system may need to be changed to reflect the results and/or optimization of performance from the CAE system. So there may also be a need for the data to be transferred from the CAE system back to the CAD system (using the native files or the neutral files). The network and/or systems architecture needs to reflect the need to transfer this data.

11.5.2 CAD to CAM Integration

The other very common integration is between the CAD and CAM software. The CAM software is responsible for the automation of the manufacturing of products. It is usually used within the context of a Manufacturing Engineering department. The CAM software will usually be closely tied to the Numerically Controlled (NC) equipment used by a company. NC machines are machines or tools that operate on parts automatically. The movement of a lathe, mill, press, robot welder, etc. is controlled by electronic systems instead of by human operators. At one time these machines worked from instructions that were read from a paper tape with holes punched into it, but most of this equipment has been replaced with a direct computer network type of connection. This direct arrangement may be referred to as Computer Numerical Control (CNC) or DNC.

The creation of the instructions for the machine tools is driven by the geometry created in the CAD system. Obviously, the tool must follow the contours and features of the part in order to create the part. This is done by creating tool paths that not only follow the direct surface of the part, but also by cutting away unneeded material in the raw material (that may not be directly on the part). The creation and management of the geometry of these tool paths is the primary function of the CAM software. Depending on the factors such as the hardness of the material being cut and the surface finish desired, the tool path information may become a very large amount of data. Although the tool path data can be done with just 2-D geometry (X- and Y-values) for flat-plate types of parts (often using machines called 2-axis mills), in many cases this is done with the 3-D part geometry (using X- , Y- , and Z-values with machines called 3-axis mills). Depending on the complexity of the geometry, the 3-D part geometry may require 4- or even

5-axis mills; in this case, there are X, Y, Z coordinates for the tool path and some angles for rotating the part or jigs that are holding the part.

The tool path data is kept in data files. The data is often in a type of language or format that is understood by CAM systems (for example, one is called APT). These files are somewhat readable; one can edit the commands for the machine and the geometric tool path data. However, these types of language files are not what is actually sent to the machine tool. Instead, the machine tools have electronic control units that receive a low-level format or data stream. This data stream is created by software called a postprocessor. The postprocessor can be thought of as being like a driver for a specific type of hardware connected to a computer system.

The integration of CAD and CAM would be specialized for each company. It needs to take into consideration the particular CAD system, the CAM system, the number and type of machine tools employed, and how these systems are to be connected. Unlike the CAD/CAE integration, there may be little need to have data transferred from the CAM system back to the CAD system. However, the CAM information is going to be more dependent on the overall engineering and manufacturing processes and management. When 3-D parts are revised in the CAD system, the revision control needs to extend to the CAM system. The changes to the parts need to be appropriately reflected in the tool paths that create the part. Thus, the CAD to CAM integration may need to be tighter or more carefully controlled from a data management perspective.

11.6 TRANSLATIONS

As with drawings, there is a fairly common need to send and receive 3-D models between companies using different CAD systems. In the context of 3-D CAD, this need is usually driven by the actual design process. For instance, as designers are creating detailed 3-D models of their assembly, they may need to see 3-D models of components that are purchased from a supplier or vendor. Of course, the designers could create 3-D models for those purchased components, but obviously time and effort would be saved if 3-D models could be received from the vendors instead (reading in or importing a file). When the designer is finished, that design can then become a product that is bought by yet another company. That company may want to see the product as a 3-D model also, so there would be a need to send out a 3-D model as well (writing out or exporting a file).

There are a number of types of 3-D translations (unlike 2-D translations). It is important to understand how 3-D modeling works to some extent (as explained in previous chapters) in order to understand the different types. Some types are very simple and produce only information that can be viewed (can not be changed); other types would be more functional.

Managing 3-D CAD

TABLE 11.2 3-D Translation File Types

Translation or neutral file	Description
IGES surfaces	Surface data. ASCII format. The most typical 3-D IGES type of file.
IGES wireframe	X,Y,Z wireframe. ASCII format.
STEP	Usually part models, part instances, and assembly structure. ASCII format.
STL	A tesselated file developed for stereo lithography or rapid prototyping.
VRML	A tesselated file developed for web pages. The file can be enhanced with tags for annotation, menus, surface appearances, etc.
VDA	A format based on German automobile manufacturers (Verband der Automobilindustrie)

The simplest sort of translation would be the tessellated file. This is a format that takes each surface of the 3-D model and creates small triangles, tessellations, or polygons to approximate the surfaces. This file can be created for individual part models, entire assembly models, or part instances. When the tessellated file is imported, the 3-D model should look like the 3-D model. However, there is no surface or analytical data (such as NURBS definitions), and there is no part history (unless the receiving CAD system infers or extrapolates this somehow). File types that are tesselated include VRML (Virtual Reality Markup Language) and STL (Stereo Lithography). The VRML file is used with web browser plug-ins to view and interact with a 3-D model in the context of websites. The STL file is used by machines that create physical 3-D models based on the 3-D part models (so called Rapid Prototyping machines or 3-D Printers). The VRML and STL file may be used as a neutral file to translate between 3-D CAD systems as well.

The next type of file would be a surface data file. This is a file that converts the 3-D part model into just surfaces. The file is basically a list of the information needed to create the surface model (i.e. using NURBS information). In this case, the surfaces are not approximated (as in the tessellated file). This imported model should look better, but the surfaces are often not stitched, so the part model will not have a bounded volume. Also, depending on the basic part tolerance of the two dissimilar CAD systems, there may be "untrimmed" surfaces. Of course, since there are real surfaces, it may be possible to manually stitch and/or trim the surfaces as needed. The surface data file would have no part history.

The most common example of a surface data file is an IGES file. The IGES file is a format that has many capabilities and has been in use for many years. It can be used for 2-D data and drawings, and it can be used for more complicated 3-D models. However, in most cases, when a 3-D IGES file is transferred between CAD systems, it contains just surface data. One can more specifically refer to this file as an IGES Surfaces file. A common use of the IGES Surfaces file is to send part data from a CAD system to a CAE system. The surface data file often contains all the data needed by the CAE type of software.

The next type of file would be a product definition file. This is a translation format that tries to preserve as much information as possible between the CAD systems; this can include nongeometric information such as attributes about what material or processing is used for a part model. Since the manufacturing of components is so often tied to assemblies (and their BOMs), the product definition type of file should support the exporting and importing of assembly models.

The basic file of this type is the STEP (Standard for the Exchange of Product data) file. The STEP file is governed by a standards body (such as NIST or ISO). The STEP file may also be related to data systems called schemas that can provide added levels of intelligence to the translation. Two common application protocols for the STEP file are AP203 (Configuration of 3D design) and AP214 (Automotive design processes). If an assembly model is translated with a STEP file, then instances should be preserved (i.e. there is a master part model and it can be copied a number of times for each use of the part in the context of the assembly).

Beyond these three types of translations, there are ways to transfer 3-D CAD data between systems that are more direct. For instance, some CAD system vendors supply software that can create part files in the exact format of another CAD system. In this case, there is no neutral file such as IGES or STEP. Also, CAD systems are often based on kernel software libraries that provide the most basic 3-D data calculation methods. Two common modeling kernels are ACIS® and Parasolid®. If two CAD systems share the same kernel, then more exact part models may be exchanged. These files are much more likely to be modifiable after the translation.

11.7 PRINTING

A significant concern for the management of 2-D CAD systems is plotting (systems and procedures for making hardcopies of drawings). This is important since 2-D CAD is really part of a paper driven system, and these copies of the drawings (whether paper or electronic image data such as TIFF files) are the master source of the design and engineering data. There is no real direct equivalent with respect to 3-D CAD (hardcopy of 3-D models is more of an attempt to make realistic pictures).

Of course, 3-D part and assembly models are used to help create drawings, but then the standard 2-D CAD plotting system can still be used for these draw-

Managing 3-D CAD

ings. The 3-D models really just created the geometry in the drawings, so the 2-D data from these models should still be acceptable to the plotting system developed for the 2-D CAD system. So, in the context of 3-D models, making a hardcopy is really just standard printing of what is seen on the computer monitor, not plotting in the special sense for drawings.

Even when a company's design data is still based on drawings, making hardcopies of 3-D models is still desirable. In this situation, hard copies of the 3-D models would be used to help show the state of the design in a general sense. Indeed, they can be most effective in demonstrating a design's appearance or functionality to those less fluent in reading drawings or blueprints. The 3-D model hard copy clearly looks much more realistic than a drawing. These hardcopies can also be valuable for presentations, brochures, and advertisements.

Probably the most common approach to getting hard copy of the 3-D models is the screen dump. In this case, the data shown on the screen is sent to the printer (instead of creating some sort of industry standard plot or picture file). Graphics adapters that run 3-D CAD systems should have a sufficiently high resolution so that simply translating the data in the graphics memory to a hardcopy format creates an acceptable result on the printer. Of course, this means that the hard copy can be no better than what is shown on the computer monitor. To improve on the screen's resolution (which is particularly important if the hardcopy is going to be blown up larger than the screen image size), then some sort of plot file is needed. Unlike 2-D plotting, the 3-D plot data is going to be bitmap-type of data where all the surfaces of the model are shaded to create specific colors at a large number of specific X,Y locations.

A typical format for encapsulating the screen dump data would be XWD (for Xwindows Dump). Once the data is in this format, then it can be translated to a format or language that is acceptable for the printer (such as Color Postscript). Another possible file (that also can be used without a screen dump) is CGM (Computer Graphics Metafile). Again, it is important that the file format supports bitmap type of data, and since the resolution is usually high (on the order of 1 megapixel), these files can become rather large (many megabytes).

Regardless of the format of the data, one needs to carefully consider how well the printer supports colors. The 3-D models shown in the CAD system generally use realistic shading on a surface-by-surface basis (although they rarely show effects such as shadows). This shading typically requires support for true color, 24 bit, or 16 million colors. So for a hardcopy to look acceptable, the printer needs to support this many colors as well. In addition to the accurate support of a large color palette at high resolution, the printer also needs to translate colors from the monitor to the hardcopy accurately. When a red shade is shown on the monitor, then it should not become an orange or pink shade in the hardcopy. Often, there are adjustments that can be made on the printer to improve this translation or correlation of colors.

11.8 MODEL DOCUMENTATION

The only real opportunity for a more advanced use of 3-D models as the master source of design data (as opposed to drawings) is to replace drawings completely. This is problematic, but it can be done in some circumstances. For a large manufacturer, there would be thousands or even millions of drawings spread across their own virtual vaults and then through all the vaults of companies that are suppliers or vendors for the manufacturer. As long as the systems are needed to support this legacy 2-D information, or if just one critical supplier or customer doesn't "do 3-D," it becomes difficult to switch to an all 3-D documentation system. On the other hand, if a company creates only new documentation and it can force all suppliers to use the same 3-D CAD system, then doing 3-D model annotation or documentation becomes more realistic.

Of course, relying on 3-D models for design data really implies that there is no need for hardcopy at all. If the 3-D models are annotated to provide all the notes/information that drawings used to provide, then it would make sense to just have everyone receiving the models just view the 3-D models with a computer system (for instance using a VRML-enabled web browser). This then means that everyone viewing the models needs to ensure that the 3-D models are up-to-date and valid. This can also be a problem for large companies and their vast supply chains. It is probably simpler to just determine the latest revision level of a drawing, and then check this against the Title Block of the drawing.

Hopefully these issues for 3-D documentation will be resolved over time and drawings will no longer be the master source of design data. 3-D models are far more intelligent and useful to all those who need the design information (they can be measured, clipped, and provide data such as volume and weight). The American Society of Mechanical Engineers (ASME) has drafted standards (Y14.41) to cover 3-D model annotation. Hopefully this can form the foundation for the wider adoption of 3-D model documentation.

11.9 CHAPTER EXERCISES

1. Study a company's process for releasing 3-D models or the data related to 3-D models. Audit the company's actual use of the system and record whether it complies with the standard process.

2. Determine the file extensions for the various items in the 3-D CAD system. Record the average file size for these various types of files.

3. Determine how data from the CAD system is "backed up." Record whether a model that is deleted during a given week can be restored the following week. If not, why not?

Managing 3-D CAD

4. Attempt to translate a 3-D part model into a neutral file format (such as IGES or STEP), and then re-import the file back into the CAD system. Record what, if any, problems appear.

11.10 CHAPTER REVIEW

1. Explain some of the benefits for using a CAD database to manage 3-D CAD data instead of just the operating system file names/directories.
2. Explain some differences between the plotting of drawings and the printing of 3-D models. Include how these hard copies are used differently and how the data from the CAD system usually differs.
3. Explain why an STL or VRML file would generally not be appropriate for import into a CAD system for changes to the part or assembly model.

Glossary

ACIS   A kernel for 3-D modeling, a set or library of routines that CAD system vendors can buy for use in their system. These routines would provide the very basic and core algorithms that create 3-D geometry. If two CAD systems both use ACIS, then they should be much more able to exchange 3-D data. It has been developed by Spatial Corp. which is owned by Dassault Systems. Another competing part modeling kernel is Parasolid.
ADAMS®   A mechanism design CAE program (kinematics and dynamics) that is often integrated into 3-D CAD systems.
Add   A 3-D part operation where different surfaces are brought together. Adding implies that the CAD system does not determine the intersection of surfaces such that a new bounded volume is created (i.e. adding avoids how surfaces would be stitched up).
AIX   IBM's version of the unix operating system.
Alphanumeric Data   Data that is made up of text such as numbers and letters. This is to distinguish it from purely numerical data. Computer programs deal with text data quite differently than with numerical data. Alphanumeric data might be considered Character Data, but Character Data really implies that non-printable codes such as LF (Line Feed) or ESC (Escape) could also be part of the data.
Anti-aliasing   A process of making sharp geometric entities (such as lines) appear more smooth on the computer monitor. Since monitors have a finite number

of pixels, a single line at an angle, or a circle, winds up with some level of jaggedness ("jaggies"). Anti-aliasing is a technique of using subtle color changes along the entities to make the entities less jagged.

Aperture Card   A mainframe punch card that includes a photographic image of a drawing. The image can be created by a computer controlled device attached to a CAD system. This Computer Output Microfilm (COM) technology is probably archaic. It has generally been replaced with on-line imaging systems.

Assembly   A collection of parts or details. An assembly is usually documented in an assembly drawing. This type of drawing shows how the parts are related to each other. The parts may be welded together (perhaps then called a weldment drawing), they may be fastened together (as with bolts), or they may just be assembled together as a working mechanism. An assembly drawing may show a Parts List or Bill of Material, as well.

Assembly Model   A model in a 3-D CAD system for an assembly. The Assembly Model contains or references Part Models as needed. Each Part Model that is included in the Assembly Model should be instanced. This means that there can be more than one copy of a Part Model, but all these copies are based on or controlled by the one Part Model (see also Assembly Structure).

Assembly Structure   In a 3-D CAD Assembly Model, the list of the parts and relationships that are contained within the Assembly Model. Each instance of each Part Model would be a separate item in the structure. An Assembly Structure can be used to automatically generate a Parts List or Bill of Material.

ASCII Code   A set of binary codes (1s and 0s) determined by the American Standard Code for Information Exchange that are equated to letters, numbers, punctuation, control codes, etc. They form the basis for the actual binary data on most CAD systems. Basically, only mainframe computers use a different code system.

ASCII File   A file that uses the ASCII Codes, but it is also generally assumed to be a file that has a structure that allows the contents to be seen with a file editor or text editor program. Basically, all the information in this type of file is plain to see. This is in contrast to the Binary File. ASCII files may have file extensions of TXT, DAT, ASC, PRN.

AutoCAD®   A CAD program from AutoDesk, Inc. It essentially was the de facto standard for PC-DOS computers. The DXF neutral file for CAD came from this system.

Axis Plane   See Coordinate System.

Backlighting ("backfaces")   A feature of a 3-D graphics system to make both sides of a surface bright or shaded. In normal 3-D part modeling or solid modeling, one side of a surface is facing outside the part and thus is shaded according to the light sources available; the other side of the surface is facing inside and would be dark. Backlighting forces both sides of the surface to be shaded. This is helpful in Surface Modeling tasks (where part models aren't always solid).

Glossary

Balloons A number in a circle with a leader line (a line with an arrowhead) pointing to an item in an assembly drawing. The item numbers generally correspond to the list of parts that are documented in the assembly's Parts List or Bill of Material.

Bandwidth Generally used in conjunction with computer networks to indicate the raw information carrying capacity of the network. Insufficient bandwidth can lead to a bottleneck where CAD workstations are trying to send data, but the network is unable to efficiently respond.

BASIC A computer programming language. As the name implies it is a basic language that is intended to be used by personnel with a minimum of training and computer systems background making it a very widely used language. In general, BASIC programs are interpretive. This means that the computer needs to digest or decode each line of the program as it is read, then convert this to Machine Code types of instructions before performing the next line of programming. This is the least efficient approach in comparison to the use of a Compiler. During the history of the PC platform, there were compiled versions of BASIC; and with the advent of the Windows operating system, GUI-based versions of BASIC were created that are event driven and making use of Dynamic Link Libraries (DLLs) (which somewhat blurs whether it is interpretive or compiled). A BASIC file would typically have a file extension of BAS.

Bills of Material (BOMs) Lists of data about how products are assembled or grouped. The Bill of Material includes Part List data (i.e. what parts are included in the assembly); but, it is also implies some nongeometric information such as the material the parts are made from, whether the items are purchased or manufactured in-house, etc. Bills of Material record the quantity of items; that is, if a specific part (identified by a specific Part Number) is used 5 times in a level of assembly, then this item is listed once in the bill, but it has a quantity of 5. Bills of Material usually have parent/child relationships where one level of assembly can include other levels of assembly. A top or final level of assembly shows the last items that are brought together to form the actual shipped product. Bills of Material can appear as notes on a CAD drawing; they can also be controlled and generated from a separate computer software system such as BOM, MRP, or ERP software.

Binary First, refers to a number system with the base of 2. The only two possibilities in this number system are 0 and 1. The 0s are 1s are an abstraction of the behavior of electronic devices that process information based on a high voltage or a low voltage. Second, can also refer to an executable file on a unix system. System-wide programs for a unix system would typically be stored in a directory such as /bin or /usr/bin where bin stood for binary.

Binary File A file that presumably uses ASCII codes, but the file does not include the codes necessary for the file to be directly edited (see ASCII File). In this case, the data is just a long stream of bytes (or 1s and 0s) that must be interpreted by a program that knows the structure of the data. Binary Files are not universally transportable because different computer systems order the bits in a byte in different directions (big-endian versus little-endian).

BIOS or Basic Input/Output System, (a PC-type of system) A special integrated circuit device (or chip) and its associated programming that stores pertinent setup information for the configuration of the system. This is information that allows a specific computer system to be unique. For instance, the BIOS setup can indicate what types of peripherals are attached to the system. For example, some computers may have 8 gigabyte disk drives, while others may have 16 gigabyte disk drives. Through a special program that is setup by the computer manufacturer, this information can be modified and stored on the BIOS chip.

Bit A binary digit. It is a single value in a number system with the base of 2. The only two possibilities in this number system are 0 and 1; therefore bits are either 0 or 1. Dealing with numbers and data via individual bits is usually too cumbersome; this type of data is usually converted to Hexadecimal (base 16).

Bitmap A set of data that indicates the color of individual Pixels on the computer monitor. A graphical image can be captured in a Bitmap, but it is not able to be modified in the sense that geometric entities (such as lines and arcs) can be individually discerned by computer software (although object recognition or OCR software can be useful in attempting to do this somewhat). This is in contrast to the Vector Graphics or Vector Data that is made up of information on a geometric entity basis. Types of files that are based on Bit Maps include JPEG, PCX, and TIFF files.

Block See Groupings.

BOM See Bills of Material.

Boolean Refers to logic that compares sets or items (named after English mathematician George Boole). In terms of databases and searching, Boolean operators are used to find items that meet one criteria AND another, one criteria OR another, one criteria but NOT another, etc. In 3-D CAD systems, Booleans or Boolean operators alter the 3-D model by joining separate solids together (as in an OR of volumes), cutting solids (as in a NOT of volume), or intersecting solids (as in an AND of volumes).

B-Rep Boundary Representation. This is a basic theory for construction of solid models based on edges, faces, surfaces. This technology is basically hidden from the user. The 3-D CAD system may use this method, another method (such as CSG), or a combination of them.

Byte A sequence of 8 bits. It is used to make working with binary information easier. There are 256 combinations of 1s and 0s in a Byte (starting from 00000000 to 11111111). Hexadecimal is usually used to work with binary data. In hexadecimal, a Byte is just 2 characters (each hexadecimal digit holding 4 bits; 4 bits sometimes being referred to as a "nibble.")

Glossary

C/C++   C is a computer programming language. It was extremely popular with the unix and PC platform until the development of GUIs. It is a compiled language that is very efficient for managing the computer's resources and user interface. C++ (pronounced C-plus-plus) is a sort of "super-set" language for C. It uses object oriented techniques which allows libraries of cross-functional routines to be developed which offer a higher level of abstraction versus the compiled C. C++ became particularly popular as GUIs were adopted.

CAD (Computer-Aided Design)   A computer program or a system of programs that assist and enhance product development. At times, it has supposedly meant just Computer-Aided Drafting (limiting CAD to just drawings), CADD meaning Computer Aided Design and Drafting. However, CAD is a term that has continued to be applied to the large software systems that automate and manage design activities of all kinds. A CAD system does 2-D design, drawings, 3-D design, and data management. As the CAD system is integrated with other engineering-related systems, they are usually referred to as CAD/CAM/CAE.

CADAM®   A CAD program sold by IBM for their mainframe computers. It was the de facto CAD program for mainframe computers. It was developed by Lockheed, an aerospace company.

CAE (Computer-Aided Engineering)   A computer program or a system of programs that assist with engineering analysis. These programs use engineering methods/formulae/algorithms to predict or estimate how products will perform. Some of the most important types of CAE software are FEA (predicting the behavior of solid materials) CFD (predicting the behavior of fluids), and Dynamics (simulating physical events such as mechanisms and impacts).

CAGE Code   Part of the U. S. Standard Title Block. It is a number assigned by the Federal Government to identify the type of item or business associated with the manufacturer. It is a five character code.

CAM (Computer-Aided Manufacturing)   A computer program or a system of programs that assist and enhance manufacturing activities. Typical tasks for a CAM system are determining how machine tools are positioned to create parts (NC tool paths), factory floor visualization, nesting (figuring out how different parts can be combined onto single pieces of stock for simultaneous manufacture), and managing the various types of data generated by the software.

CATIA®   A high-end CAD system that is sold in the U. S. by IBM. It is developed by Dassault Systems, an aerospace company.

CDE (Common Desktop Environment)   A set of standard appearance and functionality for the Xwindows GUI that is popular for the unix platform. Computers from two different computer manufacturers that both use CDE would look totally similar to their users.

Centerline   A type of Line Font that indicates the presence of circular features in a drawing. They are lines with long solid segments and then some shorter segments. When put on a hole that is seen as a circle (not seeing the hole from the side or at an angle), then a bull's-eye is shown where two shorter segments of the Centerline form a cross.

CFD (Computational Fluid Dynamics) CAE software that attempts to predict the behavior of fluids (liquids and/or gases) for products. Because the flow of fluids are more chaotic and nonlinear in comparison to solids, these programs tend to be extremely specialized for particular types of problems. Thus, separate programs would be typically be used for pump design versus compressor design versus the flow of air around a truck body. CFD programs are often the most computationally intensive programs a company will use. In some cases, they can run calculations for days or weeks.

CG (Center of Gravity) Meaningful for knowing the point in a physical object that is at the center of the object. The object would rotate about this point. However, CG is also used to indicate the center of an area, and the center of an assembly. However, more accurate terms such as centroid or center of mass should actually be used. A 3-D CAD system can actually calculate the CG of an assembly model if all the parts are modeled accurately (as solids) and assigned material properties (or at least a density). The CAD system would find the weight of each part by multiplying the volume of the solid times the density. Then the positioning of each part instance with respect to some origin would allow each part's distribution of weight to be accounted for.

CGM File A type of plot file or neutral graphics file. CGM stands for Computer Graphics Metafile. It is generally used as a middle step between the CAD system and specific instructions for a hard copy device. It can be used in both 2-D plotting and 3-D model hard copy production.

Character Data A set of data that uses text and codes. Beyond alphanumeric data, it could contain any ASCII codes (including control codes or nonprintable characters such as LF (Line Feed), CR (Carriage Return), ESC (Escape), etc.). Any data in a file can be loaded as character data; however, it may not be formatted unless it is in an ASCII file. A single character would be a letter A, and it would be read as a single ASCII code. A set or sequence of characters such as the word AXIS would be called a character string or just a string. Strings may also assume that an appropriate ASCII code be placed at the end of the string (to indicate the end of the word or line of data).

Character String See Character Data.

Client Usually refers to a specific set of software on a workstation that is operating as a node on a network of computers. The Client is assumed to work with the assistance of the Server (which has other software that the Client needs or uses). This Client/Server arrangement has meant other arrangements over the years (particularly with respect to Xwindows). Sometimes, Client might also refer to the entire computer on the network (although Node would be more appropriate).

Clump See Groupings.

Glossary

CNC (Computer Numerical Control)   The networking of computer controlled machine tools in a manufacturing environment. This was an extension of basic NC (Numerical Control) machine tools (which were not really networked together). The CNC approach allows tool-path and other specific machine operation instructions to be cataloged on server-type computers that can then download to the machines as appropriate.

COM Card   See Aperture Card.

Compiler   A program that converts high-level programming code into low-level instructions. A compiler is used with languages such as FORTRAN and C to create a standalone executable file that is efficiently loaded and run by the computer. This process becomes less clear with languages that use dynamic linking; in this case, the high-level programming is compiled, but it becomes object code that is later loaded on demand into Memory for use by the CPU (and converted to machine instructions as needed).

Constraint, Assembly   A constraint for part instances in an assembly model. The constraint fixes or controls the location of the part instance with respect to the overall assembly model or with respect to other part instances. Assembly Constraints could be dimensions (such as a distance from one face to another, or from one edge to a face, etc.), or geometric constraints (such as one face is perpendicular to another face), or logical constraints (such as one part instance is welded to another, or a group of part instances are to be a rigid body).

Constraint, Sketching   A constraint for a 2-D sketch that is created to control the geometry of a feature or part of a 3-D model. It could be used with a 2-D CAD system, but they are typically only used with 3-D CAD systems. This constraint fixes or controls the location of 2-dimensional geometric entities (such as line segments, arcs, splines, etc.). These constraints could be dimensions (such as the location or size of a hole), or geometric relationship constraints (such as one line segment is perpendicular to another), or logical constraints (such as the end of a line is grounded or immovable).

Coordinate System   A mathematical entity that specifies the orientation of the 2-D or 3-D space. In 2-D space, a coordinate system has an X- and Y-direction (or axis). In 3-D space a coordinate system has X, Y, and Z. The origin of the coordinate system is where these directions have a value of 0. For a 3-D coordinate system, an axis plane is formed by 2 of the 3 directions. That is, the XY-axis plane is a plane that contains the X- and Y-direction (while Z is always 0).

CPU (Central Processing Unit)   An Integrated Circuit or chip that runs the computer. It controls the data that is being processed and all calculations (in support of the application program).

Cross-hatching   Cross-hatching is the use of a pattern of spaced apart lines (typically at an angle) that fills a 2-D area. This is used in a drawing to indicate that a section of the object in the drawing has been cut open to reveal the inner material. Thus, it is usually used with Section Views in a drawing.

CSG (Constructive Solid Geometry) A basic theory of the construction of 3-D solid part models. This technology is really hidden from the user. The 3-D CAD system's operations may be based on this technique or other techniques (such as B-Rep), and it may also allow a mixture of techniques.

Cut Plane (Cutting Plane or Cut Line) Indicates on a drawing where objects are cut for a Section View. It is a heavy line that has a letter shown at each end of the line. The letter identifies the particular cut for a particular Section View.

Degrees of Freedom DOF is associated with the constraining of sketches and assembly models. Each element in the sketch or part instance in an assembly model initially is able to move freely. As constraints are added, DOFs are removed, and the element becomes more restricted until it cannot move. At this point, it is considered fully constrained. DOFs are also important with mechanism simulations for assembly models. In this case, DOFs that are not removed are driven by forcing functions.

Detail A single part that is modeled or shown in a detail drawing. It is best to refer to them as Parts in the 3-D CAD system, however.

Dialog Box In the GUI environment, a form or small window that appears from the computer program. It is usually a presentation of information to the user, or it is a means for entering new data into the computer program.

Dimension A type of constraint that fixes the distance between two geometric elements. On a drawing, dimensions can be indicated between points and/or lines. However, in a drawing, dimensions do not drive or constrain the 2-D geometry. Instead it is just a measurement of the distance between the elements. If the number shown in the dimension is changed, the shape of the geometry in the drawing would not change. In a 3-D model, dimensions are usually constraints and they can drive changes to the drawn geometry. Dimensions in an assembly model would drive changes between entire parts.

Directory A logical partitioning of a Storage system (such as a Disk Drive). Different directories usually contain different types of files. A program file directory would contain files for programs loaded on the computer system, for instance.

Disk Drive A Storage system that can store and more or less permanently save the data. They use a spinning disk and tiny regions of magnetism to record the information. A Disk Drive may also be referred to as a hard drive, C drive, hard disk. This is to distinguish it from a Diskette Drive.

Diskette A portable medium for exchanging data files between computers. It is inserted into a Diskette Drive (or floppy drive), data is written on it, and then it is removed. They can also be referred to as floppies.

DLL (Dynamic Link Library) A type of file that works in conjunction with a computer program that does not load all its instructions into the Memory system when the program starts. Instead, pieces of instructions are loaded and executed

Glossary

on a more as needed basis. DLLs are also often shared between totally different programs (which may be part of the operating system or written by users).

DOS   The de facto operating system for the PC platform prior to the use of GUIs on PCs. It was the primary product from Microsoft Corporation that was instrumental in the standardization of PCs. It operated in a character or command processor mode where commands were typed into the system using the keyboard. It has been replaced by the Windows operating system and its derivatives. There was also a DOS (Disk Operating System) that was used by mainframe computers, but this is rarely what is meant by DOS.

DOS Prompt   The means for entering commands to the DOS operating system. It has come to mean entering commands into any computer system directly (as opposed to interacting with the GUI).

Download   The retrieving of data files from a remote source. This could mean obtaining a copy of a file from a file server, mainframe, or from the Internet.

Drawing Number   A field in the Title Block of a drawing. It indicates the unique number assigned to that drawing, and usually by inference to the object documented in the drawing. In many cases, the Drawing Number would be the Part Number, as well.

Drawing Package   A set of drawings that are managed as a group. Typically they are used during drawing release, and the list of drawings in the package usually share a release number of some kind (for future reference). Drawing Packages may also be used in exchanging a set of information between groups or companies.

Drawing Size   An indicator of the overall size of a hard copy of a drawing. In the inch system, they are letters such as A, B, C. In the mm system, they are the letter A followed by a number (e.g., A1, A2, A3, etc.). ASME (ANSI) and ISO standards specify the standard sizes. It is included in the Title Block of drawings.

Driver   A small amount of programming in a file that enables a device or system to communicate with a larger system. For example, a plotter expects a certain set of codes to create drawings. The CAD software may not produce these codes directly, but rather rely on a Driver to do the transformation. Drivers are generally employed to make a program less dependent on specific computer hardware.

DXF   A CAD drawing neutral file format. It allows a CAD drawing (generally presumed to be 2-D) to be exchanged or translated between two different CAD systems. It was developed by the company that produces AutoCAD. It has been adopted by many other CAD system vendors.

Dynamics   Refers to the analysis of mechanisms or assembly models where the forces and torques on the system are applied, evaluated, and/or predicted. A Dynamics simulation can predict how an assembly model will behave based on an actual physical environment. Although this capability may appear in a CAD system, it is really a CAE function. Dynamics is different than Kinematics; Kine-

Glossary

where parts of an assembly model move, based on some driving motion of forces).

E-CAD A generic term for CAD-type systems for electronics design. Although E-CAD may be used to lay out the physical arrangements of components on a PCB (Printed Circuit Board), most of the effort in E-CAD is concentrated on the interconnection and simulation of the components (i.e. the schematic).

ECN Engineering Change Notice or Engineering Change Number. See Release.

ECR Engineering Change Record or Engineering Change Release. See Release.

ECO Engineering Change Order. See Release.

ERP (Enterprise Resource Planning System) A system of software that is supposed to manage many aspects of a manufacturer's operations (i.e. enterprise-wide). These systems are usually considered an expansion on MRP (Material Resource Planning) software. Of particular interest is that the ERP system may be responsible for the Bills of Material that are relevant to the Parts Lists or BOMs shown on drawings or in 3-D assembly models.

Ethernet Ethernet is a basic computer networking technology based on the "collision–retry" technique. It can really be thought of as broadcasting through a wire. Computers on the network listen for their data and react accordingly. If the transmission is interrupted; it simply tries again.

Ethernet Address Uniquely identifies a computer on a network at the hardware level. These addresses are assigned to each individual Network Interface Card (NIC) or ethernet card. The NIC is the circuit board and componentry that implements the networking capability. An Ethernet Address is a sequence of 8 bytes (shown in a form such as xx:xx:xx:xx:xx:xx).

Export (of data) Implies that data is being prepared for use by another software system or another user. With respect to CAD, neutral files are often Exported from the CAD system. This would include files such as DXF, IGES, STEP, VRML, etc. Obviously, the receiving system Imports the data.

FCS (Feature Control Symbol) It is a set of graphical symbols, letters, and numerical values that indicate special geometric requirements associated with dimensions. For example, an FCS on a drawing next to a dimension for a hole in a plate may indicate that the axis of the hole must be perpendicular to the face of the plate within a number of degrees.

FEA (Finite Element Analysis) CAE software that is used to predict the behavior of designs. It is most often used for stress analysis. Stress analysis calculates the stresses in models; stress is a mathematical representation of how the loads (forces) may deform an object. FEA computer programs can also solve other kinds of problems. FEA programs work with a Mesh. A Mesh is a model that is broken down into a large number of small regions where the actual mathematics is applied. CAD data (2-D or 3-D) is often used as the starting point for these Finite Element Models. Some CAD systems will have a direct connection between CAD and FEA; others will use neutral files (such as IGES Surfaces).

Glossary

Feature A segment or region of a part. For example, a hole, rib, or boss on a part would be referred to as a feature. With respect to 3-D part modeling, an operation to create one of these regions is obviously referred to as a feature. However, a 3-D part modeling technique can create many holes at one time. Now, the 3-D CAD system may refer to a feature of the part meaning a single operation (or a step in a part history), when in fact many geometric features are created at the same time.

Features-based Modeling The overall method of 3-D part modeling where the part model is built in a series of steps (captured in a part history), and each step begins by sketching 2-D geometry on an existing plane on the part model.

File Extension A means of quickly identifying a type of computer file by the end of the file name. For some operating systems, they are of a fixed format and/or length. For others, they are at the discretion of the programmer or user. Most operating systems recognize the File Extension based on the presence of a period or dot; however, GUIs often just portray the File Extension as a column in a table of file names. Some File Extensions are de facto standards; others are not standardized at all. For example, a file ending with the letter C (e.g. FIXIT.C) would be universally recognized as a file with C programming source coe. However, a file ending in DAT (e.g. FIXIT.DAT) could be any number of data files for many purposes.

File Format Indicates the basic structure of the data in a file. The most important distinction for File Formats are ASCII versus Binary. In a FORTAN program, a formatted file that is created by the program is basically an ASCII file, and an unformatted file is basically a Binary file. Therefore, sometimes a file may be referred to as formatted versus unformatted (but this really means ASCII versus Binary).

File Server A computer on a network that is primarily acting as a source of files for other computers. It allows files to be stored and managed centrally.

File System A distinct set of files that are part of an overall directory structure for a computer. File Systems generally allow multiple Disk Drives to appear to a user as a single system. For example, in a unix system, all files under a directory /usr may use one Disk Drive (or a partition of the drive), while all files under a directory /home may use another Disk Drive (or a partition of the drive). In this example, the /usr and /home File Systems are then mounted to the root directory (or /). If this mount point is actually implemented across a network to another computer, this is referred to as a Network File System.

Fillet (pronounced "fill-it") A type of feature on a part that rounds a corner or blends surfaces meeting at an angle. Virtually all physical parts have fillets when surfaces meet; it is basically impossible to create a perfectly sharp corner in a physical part. Sometimes the fillets are specifically modeled or dimensioned in a drawing. In other cases, notes are used to indicate that all corners are filleted to some standard amount.

Filter In a CAD system refers to limiting the appearance of certains types of entities, attributes, or database items. For example, to filter for dimensions in a drawing means to have the CAD system select all the dimensions (so that they can all be changed in some way). Another example, would be to filter by color. In this case, the CAD system would select or operate on all the surfaces in a 3-D model that are a certain color. If one is searching for a specific Part Number in the CAD system data base, then to more easily identify a match, the filter may indicate that specific Part Number.

Firewall A computer program that monitors the data and/or network traffic that is being received by a computer (particularly servers) and tries to identify malicious code or unauthorized access. It is a basic security procedure that may also be bundled with scanning. Scanning may be looking for specific file names (such as MELISSA), file types (such as VBS), strings of binary data (often called footprints). These files are expected to be quarantined and/or destroyed and not passed on to other systems or users.

First Angle Projection A standard for indicating where views are placed on a drawing. In First Angle, the Right View of the object is placed on the left of the primary or Front View. This is the standard in European countries; Third Angle is used in the United States.

Flat Pattern 2-D geometry that is cut from a flat plate, where the flat plate will eventually be bent into another shape. The Flat Pattern has to account for the growth of material in the flat form versus the bent. This is actually related to the location of the neutral axis surface within the part that is between the inner and outer face of bends. CAD systems which have the functionality to create sheet metal parts should have the ability to create Flat Patterns. This can save a significant amount of time and calculation in getting parts manufactured.

Floating Point Calculations Calculations based on numbers that are not integers. For instance, calculating the area of a circle involves the number pi (3.14159...). pi is a floating point number. Floating Point Calculations are performed quite differently from integer calculations (such as adding 314 and 159). Floating Point Calculations take the computer system much longer, particularly if the numbers use high precision (such as 14 significant digits or a number such as 6.2940129993456). A CAD system (and even more so a CAE system) uses a great deal of Floating Point Calculations; therefore the computer system must perform well for this type of calculation.

FLOPS A measure of performance for Floating Point Calculations; it represents the number of Floating Point Calculations that can be performed per second. A VAX minicomputer from the early 1980s generally performed 1 million FLOPS, or a MegaFLOP. Sometimes, the performance of that VAX system is used as the basis for a standard MegaFLOP. A PC using the 80486 CPU and 66 MHz clock speed could also be similar to a MegaFLOP. Obviously, systems are far beyond this performance now. Indeed, almost any computer generally has the

Glossary

Floating Point performance needed for a midrange CAD system. However, CAE systems continue to demand as much performance as is available.

Folder A logical partitioning of a Storage system (such as a Disk Drive). Different folders usually contain different types of files. A program file folder would contain files for programs loaded on the computer system, for instance. This is basically the same as a directory.

Fonts, Line The type of line shown in a drawing. A solid line is a Line Font; a dashed line is another Line Font. However, in a drawing, the Line Font also has meaning for the representation of the object. A solid Line Font indicates a visible edge; a dashed Line Font indicates a hidden edge (thus it may be called the hidden font).

Font, Text A lettering style (similar to styles found in word processing software). However, in a CAD system, the Text Font is usually constructed from simple lines and arcs; so many fancy fonts that would be found in a word processor are not used.

Format, Drawing The standardized border or layout of drawings. It is the information and structure that all drawings for a company are expected to have. The specific geometry and documentation for an object is then placed within the confines or border of the format. The Drawing Format would include components such as the Title Block, Revision Block, Zone indicators, etc. ASME (ANSI) and ISO standards are available for standardizing of Drawing Formats.

FORTRAN (FORmula TRANslation) A computer programming language. Its primary function is for programming engineering and scientific calculations. Obviously, this can apply to many functions of a CAD system. However, a CAD system is also even more dependent on computer graphics and user interface programming, so it is typically not used for most CAD system programming (although specific calculation functions within the CAD system may still use FORTRAN). User-written programs that perform only calculations (and little or no user interface) still may use FORTRAN. The input data for the program may be created interactively, and the results of the analysis may be shown graphically, but often these functions are simply performed by distinct computer programs that are in a more suitable language such as C++ (these other programs being called Preprocessors and Postprocessors respectively).

Frame Buffer A final stage in a long set of calculations needed to create complex computer graphics on a monitor. The Frame Buffer is Memory (RAM) where individual Pixels on the monitor have an equivalent bits or Bytes of memory assigned to it. If multiple Frame Buffers are used, and the monitor quickly switches between them, then computer animations can be accelerated. Although it is a simplification, other terms for this component of a Graphics Adapter would be video buffer or page buffer.

Free Edges An edge in a 3-D model that does not meet and stitch up to any other edges of other surfaces. This indicates a model which is not solid. In a solid model, all the surfaces meet so that a specific volume is bounded or trapped.

Free-Form   Pertains to geometry that is quite arbitrary. A Free-Form surface is one that is sculpted (such as a car body panel). A Free-Form CAD system or other modeler software would work with Free-Form surfaces without being limited to geometry or constraints that are sketched on a plane.

GD&T (Geometric Dimensioning and Tolerancing)   A system of symbols and values that indicate specific accuracy of the geometry shown for an object. Although a standard dimension can indicate a distance between two features of a part (and a number of decimal places for precision), GD&T can indicate such behaviors as perpendicularity, alignment, parallelism, etc. GD&T should be able to create a more accurate part that more carefully meets the need of the design. ASME Y14.5M is the GD&T standard for the United States.

General Arrangement (GA)   Shows the highest level of design for a product. It shows the product as it would appear in the field. It is often used as a basic reference for the product (as opposed to a drawing that manufacturing would build to). A 2-D representation of this would be a General Arrangement Drawing. A similar concept would be a final, final assembly, or top level drawing.

Gouraud Shading   A common process or algorithm for displaying 3-D models in shaded mode (where the model looks real). This algorithm is assumed to make each surface of the part shaded as if it was the only surface shown. For instance, there would be no shadows from one surface to another. This may seem less accurate, but it is beneficial to the CAD system. It allows the total model to be seen and worked on more easily.

Graphics Adapter   The device in a computer system that provides the computer graphics data for display on the Monitor. The CAD system contains the true size and characteristics of the objects. In order to see them, the CAD system utilizes the Graphics Adapter to approximate, rasterize, manipulate, etc. the model as shown to the user. There are many calculations that need to be performed by the Graphics Adapter; therefore, the more that the Graphics Adapter does (using coprocessors), the less the CAD system and the main CPU need to do, and the faster the system's performance. The Graphics Adapter can also be known as the video adapter, graphics card, video card, etc.

Group Trapping   See Selecting

Groupings   A set of 2-D entities that are combined as a unit. This allows the set of entities to be instanced (re-used), moved, rotated, scaled, exported, etc. In different CAD systems this capability is referred to as blocks, clumps, symbols, etc.

GUI (Graphical User Interface)   Interaction with a computer system based on graphics instead of based on typed commands. Windows and Xwindows are examples of a GUI. However, Windows is not only the GUI, but the overall operating system. Xwindows is just a GUI, with unix being the usual operating system that accompanies it.

Hexadecimal   Refers to a number system with the base of 16. This number system goes from 1 to 9, and then uses letters for the 10th, 11th, 12th, 13th, 14th,

Glossary

and 15th numbers (A, B, C, D, E, and F respectively). Thus a 10 in Hexadecimal is the number 16 in "standard" decimal numbering. Hexadecimal is very valuable for dealing with binary data in a compact form. In order to prevent confusion with standard numbers, Hexadecimal number may have prefixes such as &H or $. Thus, $10 would be distinguished from 10.

Hollerith Data   An archaic reference to Character Data. The FORTRAN programming language known as Fortran-66 used an H as a prefix for Character Data with a number before the H indicating the length of the Character Data (or string). Thus, 7HSCHOONY would be recognized by the system as the string "SCHOONY." Some neutral CAD files still use this method.

HPGL (Hewlett-Packard Graphics Language)   A plot data format that was used by H-P's pen plotters (which were very popular for CAD systems before laser printers). It was a de facto standard for pen-based, vector CAD data for plotting. Many laser-type printers can still accept this format through emulation. HPGL, however, would only be appropriate for CAD drawings (not 3-D models).

HP-UX   Hewlett-Packards version of the unix operating system.

HTML (Hyper Text Markup Language)   The most basic format for web pages. It indicates what text and graphics is shown on the page, where to locate them, the font to use, etc. It is one of the languages that a web browser uses. Probably its most valuable capability is the hyperlink, where a user can click on the web page and go to other pages that are relevant to the user. This is an excellent capability for ancillary materials for a CAD system such as on-line help, documentation, and tutorials.

IC (Integrated Circuit)   The basic, digital technology that functions in a computer system. The electrical circuit components (millions of them) are etched onto wafers of silicon. They are often referred to as chips. Examples of ICs would be the CPU, RAM, ROM, graphics coprocessors.

Icon   A small graphical image that symbolizes a program or option in a computer program. They are clicked on, selected, or opened to have the program perform a function. They appear in the GUI of operating systems (click on an icon to start a program such as a CAD program) and in the user interface of CAD programs (in an icon panel).

I-DEAS™   A high-end CAD system that was sold by a company called SDRC. It was also available on IBM mainframes as a system called CAEDS. In the 1990s, it was totally reorganized as a system called I-DEAS Master Series. This CAD system had a somewhat unique level of integration between a FEA program and the 3-D CAD modeling. It also used a rather sophisticated set of database tools to manage the 2-D, 3-D, FEA data.

IGES   A CAD neutral file. It can be used for 2-D (drawings) as well as 3-D (models). It can contain data for a wide range of geometric entities, therefore. However, a 3-D IGES file can be considered a file with just 3-D trimmed surfaces (but not solid models). IGES stood for Initial Graphics Exchange Standard.

Import Data that implies that data is being received for use from another software system or another user. With respect to CAD, neutral files are often Imported to the CAD system. This would include files such as DXF, IGES, STEP, VRML, etc. Obviously, the sending system Exports the data.

Instance Refers to a somewhat intelligent copy of an aspect or feature of the computer software or system. In a GUI different windows can be different instances of the program (i.e. running the program two times at the same time). In a 2-D CAD system, the Groupings of entities can become multiple instances in the drawing. In a 3-D CAD system, parts shown in an assembly model should be considered instances of the parts. Instancing can imply that all the copies of the item can be automatically changed by modifying a master copy of the item.

Interchangeability The ability to use a new Revision of a part without concern for the older Revisions. If a part is interchangeable, then either Revision of it can be used without any problems. If a new Revision of a part is not interchangeable, then it is often assigned a new Part Number. The handling of these situations are referred to as Rules of Interchangeability.

Internet A global system of computer networks. It was related to the Arpanet (connected with the U.S. Defense Advanced Research Projects Agency [DARPA]) which connected government, industry, and educational institutions. Aspects of the Internet would include the World Wide Web, global e-mail, ftp (File Transfer Protocol) sites (where files could be uploaded and downloaded), and use-groups (an electronic bulletin board system based on topics). Many aspects of computer networking were developed in conjuction with the Internet (for instance, IP).

Internet Address See IP Address.

Intranet An Internet-like or web-like approach to linking together computers at an individual company or institution for use within the company or institution only.

IP Address (Internet address) A numerical address assigned to a computer on a computer network. Unlike the Ethernet address (which is assigned to hardware), the IP Address can be configured or assigned. It is a set of 4 numbers separated by a period. The range of the numbers in the different positions can indicate a scheme of subnetworking; these schemes are referred to as Classes.

Item Master Refers to a master list of all Part Numbers used by a company. It forms a master database of much of a manufacturer's current and past products. These Part Numbers may be assigned to parts, assemblies, BOMs, specifications, technical publications, material specifications, etc.

JAVA A computer programming language that has many of the advantages of C++, but it also is designed to assist with programming for the Internet. A basic feature of JAVA is that GUI types of programs are made to work regardless of the computer hardware and operating system (so called "interoperability"). It was developed by Sun Microsystems.

Glossary

JPEG A type of image data file. It contains Bitmap data. It does not contain Vector data; therefore, it is not appropriate for translation to a CAD system. JPEG files are often used in conjunction with a web browser.

Join A 3-D part creation operation where different sets of 3-D data are brought together. Joining also can imply the determination of the intersection of surfaces such that a new bounded volume is created (i.e. Joining determines how surfaces must be stitched up). Join can be considered a Boolean operation for the 3-D model.

Kernel A feature of an operating system. In this case, the Kernel is like a master program that always runs or is resident in Memory. The Kernel loads and controls all the other program's use of Memory. This approach tends to be stable since the Kernel program can keep programs from corrupting each other's use of Memory. Second, a Kernel is a central or core set of programming in an application program. For 3-D CAD systems, there are Kernels that CAD software vendors can buy or license to provide the basic algorithms for making 3-D surfaces, features, etc. If CAD systems standardize on the same Kernel, then these systems will interoperate to some degree.

Key Used in database programs to identify the most important fields of the data. For instance, a database of all drawings in a CAD system could use the Drawing or Part Number field as the Key. This means that the database can be most easily sorted, searched, etc. using this field. If there is more than one Key, then the Key with the highest precedence is called the Primary Key, the next one is the Secondary Key, etc. Just because a Key is used, does not mean that the data base can't be searched or sorted based on another field, however.

Kinematics Refers to the analysis of just the motion of parts in an assembly model. Unlike Dynamics (which figures out motion based on forces and torques), Kinematics assumes that some forcing function is applied somewhere to make some part or parts move. Then the Kinematics are solved to figure out where all the other parts move in relation to that applied forcing function.

LAN (Local Area Network) Typical for creating a somewhat localized computer network. At one time they would be limited to a single office or building. Over time, the extent of LANs has increased. Different LANs are often connected together via a backbone. Also, see WAN.

Layout Refers to a high level of design work. At this stage of design, basic or major components are placed in the context of the overall product design. Subsequently, more specific design is performed for all the components (known as *detailing*). A drawing that shows the design at the layout level is referred to as a Layout Drawing.

Leader A line on a drawing that points from a note or other text-based entity to an appropriate segment of geometry. It may have an arrowhead to assist with indicating the proper geometry.

Levels of Detail (LOD) Indicates how much detail is generated for viewing a 3-D model from a certain distance (or zoom). This could apply to the appearance of surface textures or the tessellations on the model. Usually, the Graphics Adapter allows a 3-D model to be manipulated interactively; however, when the model appears too crude (for instance, circular-type surfaces show up as coarse polygons), then the CAD software must regenerate the data stored in the Graphics Adapter. This creates a new Level of Detail. These Levels of Detail can be remembered in the system to accelerate performance. They may also be stored in the actual CAD files for later use.

Library In computer programming, a set of routines or functions that can be re-used (as in a DLL). In a CAD system, a Library would typically mean a logical organization of CAD data files (as in "check drawings into a Library").

Lighting Refers to the light sources that a Graphics Adapter assumes to figure out the Shading of a 3-D object. The position, intensity, "diffuseness," color, etc. of Lighting can change the appearance of the 3-D object.

Line A geometric entity that follows the formula of a straight line; this formula extends the line infinitely. However, in a CAD system or drawing it really refers to a line segment (that is, some finite amount of the line). *Line Work* refers to the entities in the drawing that are supposed to be on the object being documented (as opposed to notes, dimensions, etc.).

Line Weight Refers to the thickness or heaviness of lines shown on a drawing. A heavy line is thick and dark and is more obvious; a light line is thin and more subtle. Lines that are supposed to be on the object being documented (such as edges of the part) are to be heavy. Lines that are not really on the object (like centerlines) are not.

Linux A version of unix that is considered open source. This means that no one company really owns or controls the program.

LOD See Levels of Detail.

Loft A 3-D part modeling operation that creates surfaces that try to follow and blend among some cross sections. Usually these cross sections are somewhat dissimilar (i.e. blending from a rectangular cross section to a circular cross section).

Machine Code See Machine Instructions.

Machine Instructions The most low-level commands that a computer system uses. These instructions are specific codes that a particular CPU understands. These instructions would include moving data from one address to another; or adding one number to another number and storing the result in a register, etc. All computer software on a system must eventually be turned into this basic set of instructions. Only very rarely are programs written at this level.

Mainframe A computer that is expected to control a large number of users. Furthermore, the users are assumed to have little or no processing power. In this scheme, there is really just one CPU and users compete for use of that CPU. In the computer networking approach, very large computers that may also be file

Glossary

servers might be referred to as mainframes. However, the adoption of computer networking (or client/server technology) pretty much negates the need for mainframes (particularly for CAD).

Manifold Refers to a part model that has all edges of all surfaces meeting just one other surface's edge. If more than two surface's edges meet at the same location, then a nonreal part or abstraction results. In a real part, the insides of surfaces must bound material, not other surfaces.

MDA An acronym for Mechanical Design Automation. It has been used by CAD system vendors to distinguish their more sophisticated systems, since the acronym CAD can still be associated with just creating CAD drawings. However, most users still refer to all these systems as CAD systems.

Memory The volatile data medium used directly by a computer system. It can also be thought of as just memory chips (ICs). It should not be confused with the nonvolatile Storage system that can permanently store data (such as a Disk Drive). There are a variety of types of Memory, such as RAM, ROM, DRAM, etc., but in general referring to Memory of a computer system implies the main, system, or core RAM or memory that the CPU uses with all its operations.

Mesh A grid of geometric data that is created to assist with the calculations of a CAE program. The Mesh may be 2-D or 3-D. It contains individual points (called nodes) and regions that surround the points (called elements). The Mesh breaks down designed parts or assemblies into many small regions.

Metadata Data that gives information about other sources of data (i.e. "data about data"). Probably the most common use of Metadata in a CAD system is with databases that control and/or track the location of the operating system files that represent CAD data (part, assembly, drawing, and BOM files).

Minicomputer A computer that is larger than a microcomputer (i.e. a PC), but it is smaller than mainframes. They were instrumental in the development of a number of CAD systems in the 1980s; however, most CAD systems now use a client/server architecture or just standalone PCs.

MIPS (Millions of Instructions Per Second) A performance parameter for computers that are not doing Floating Point Calculations (they are doing "integer" calculations). It gives an indication of the most basic performance of the computer system. However, Floating Point Calculations are usually a much more important performance characteristic for computers running CAD systems (so MIPS can be ignored).

Model File Used by a number of 3-D CAD systems to help manage the large amount of data that is generated by the 3-D models. An assembly model may contain thousands of part models, and instead of having a user work with the thousands of operating system files, one large Model File is created and managed.

Monitor The screen device that displays the software's graphical information. It may also be referred to as a display or screen. It should not be referred to as a terminal.

Motherboard The main circuit board that the computer system is built around. It has other circuit boards attached to it that may be referred to as daughter cards, expansion cards, etc.

MRP Materials Resource Planning. This activity attempts to plan overall flow of raw material, parts, assemblies, etc. through a manufacturing operation. This activity was automated by MRP software. This concept was expanded to include other functions of a company under ERP software.

NC (Numerical Control) An automated function for manufacturing tools (such as 2-axis mills, 3-axis mills, lathes, machining centers, drill presses, etc.). NC instructions are a set of data that gives geometric instructions on how the part or tool should move during the manufacturing operation (there may also be commands for machine parameters such as "feeds and speeds"). These instructions are also referred to as tool paths. NC data is usually kept in computer files; however, they can also be stored as punch data in rolls of paper called paper tapes. The NC instructions do not directly control the machine tool device; usually the NC data is converted to control commands by a postprocessor.

Network A system of somewhat independent entities interconnected so that they can operate as a unit in some fashion. Unless otherwise indicated, it can be assumed that Network means a computer network. In this case, the computers (which can usually function on their own) are connected together so that they can share data and resources. Network may also be used in reference to constraints in CAD models (i.e. a "constraint" network).

NFS (Network File System) A File System is a distinct set of files, Disk Drives/partitions, and their related directories that can appear as part of a workstation's overall directory structure. A Network File System also appears as part of a workstation's directory structure, but the actual files and storage devices are on a File Server. It is developed by Sun Microsystems, but many computer system manufacturers use it or something similar in their networking technology.

NIC (Network Interface Card) The circuit board and/or devices that actually communicates with the computer network. The vast majority of them use a networking technology called Ethernet. Thus, they are often referred to as Ethernet cards or network cards.

Node, Network One way to refer to a computer on a computer Network. Computers are assigned Node Names so that they can be identified by users and administrators. However, each computer also has a numerical-type of address that is actually used by the network control software.

Node, Part A term that can be used for a single step in a 3-D model part history. Often these steps in the history are referred to as Features, but in some cases a single step in the history would not create an entire feature, or it may create multiple geometric features on the part model. Therefore, Part Node can be a better term to apply to a single step in a part history.

Glossary

Nonmanifold A part model where more than two surfaces meet at the same edge. A normal solid model has just one other surface meet a given surface's edges. To have more creates a nonreal or abstraction of a part.

Normal See Surface Normal.

Null Part A term that can be applied to a 3-D part model that has no geometry. Although most steps in a part history actually create or work on geometry, there are some that do not. They may set attributes for a part (such as material), or they may set construction or reference geometry. If a part model just has these sorts of steps (or it has lost or corrupted all its geometric features), then it might be referred to as a Null Part. They are often used as place holders in a 3-D modeling scheme.

NURBS Non-Uniform Rational Basis-Spline. This is a common mathematical formulation for the geometric entities that are found in a 3-D CAD system (such as surfaces). Therefore, some users may use NURBS interchangeably with Surface.

Object Lines The line segments that are actually derived from the physical object being documented by the drawing. This would distinguish them from line segments that are drawn as annotation of some types (such as centerlines, leader lines, etc.).

OCR (Object Character Recognition) A computer program or algorithm that attempts to find entities within Bitmap types of data. If a drawing has been scanned, it results in a Bitmap type of file where all the geometric entities (lines, arc, etc.) are reduced to dots. In order to recover the fact that some entities were lines, arcs, letters, numbers, an OCR capability may be applied. It can be quite inaccurate, however.

OGL See OpenGL.

OpenGL A standard programming interface for graphics. It is the de facto standard for high performance 3-D graphics. Many 3-D CAD systems will require the use of OpenGL or OpenGL-compatible Graphics Adapters. OpenGL allows the CAD system vendors to write their programs in such a way that they do not have to worry about the specific graphics hardware users may have. It was developed by Silicon Graphics, Inc.

Open Part A 3-D part model that does not have all surfaces stitched up so that a volume is bounded.

Operating System The main controlling software for the computer. It manages the devices, programs, users, etc. of the computer.

Origin The point in a geometric system at which coordinate values are 0. Often this point is coincident with a Coordinate System (which indicates the direction of coordinates). The Origin of a modeling space, part model, or assembly model, or drawing view is an important basis with which to assess the location of geometry of various types.

Orphan Part A term that can be used for a 3-D part model that has no part history. This part may still have geometry (the surfaces are shown), and it may even be solid (all surfaces are stitched up), but there would be no way to modify or assess individual features (such as by modifying a dimension or constraint). The part is just a list or set of unintelligent surfaces. Orphan Parts are typically the result of translating 3-D parts between CAD systems.

Package See Drawing Package.

Packet An individual chunk of data that transmitted across a Network. For example when a file is transferred from one computer to another, it is actually broken into a large number of pieces that are then reassembled. These pieces can be considered Packets.

Paging A method of handling a limited amount of Memory in a computer system. The operating system may move programming instructions between parts of Memory or other devices in order to make room for new programming. The operating system keeps track of this movement and can then restore the shifted instructions when needed. It should generally be considered a different operation than Swapping, but these terms are often used interchangeably.

Parasolid A kernel for 3-D modeling. This can be thought of as a set or library of routines that CAD system vendors can buy for use in their system. These routines would provide the very basic and core algorithms that create 3-D geometry. If two CAD systems both use Parasolid, then they should be much more able to exchange 3-D data. It has been developed by Unigraphics®, one of the larger CAD system vendors.

Part An individually designed item. It refers to a single physical object that the manufacturer uses. At one time, these individual items were referred to as Details; they are documented in a detail drawing.

Part History In standard 3-D part modeling, the model is made from the execution of a set of sequential steps. This list of steps is the Part History. It may also be referred to as the part structure, hierarchy, or feature list.

Parts List A table of notes in an assembly drawing that lists the individual items needed to be put together for the assembly shown in the drawing. This is similar to a Bill of Material, but the Bill of Material usually has more information about the assembly and its control in the manufacturing process.

Part Model The 3-D model for a Part.

Part Number A master tracking number for information that is used by a manufacturer. Obviously, individual part and their models are assigned unique Part Numbers. However, the same numbering system may be applied to assemblies, specifications, manuals, procedures, etc. Part Numbers are usually accompanied by Revision Levels to fully specify and control the information.

Part Tolerance The small amount of distance that indicates whether the edges of surfaces are close enough to consider the two surfaces stitched up. This may be a value such as 0.01 mm. In some cases, it is set by the 3-D CAD system; in

Glossary

other cases, it can be changed by the user. It becomes an issue for the user when very small parts are modeled, or when 3-D models are exchanged between CAD systems. If the Part Tolerance is different between the systems, part models do not become solids properly.

PCB (Printed Circuit Board) Sandwiched material that electronic components are connected to. These are also often referred to as circuit boards or just boards. Some CAD systems allow for special techniques for laying out or designing PCBs. Particularly if these systems also simulate the electronic behavior, the software may be referred to as E-CAD.

PC-DOS See DOS.

PDM (Product Data Management) Entails the activities related to managing all the data necessary to specify and/or document a product that is manufactured. It is concerned with CAD data, Revision Control, the state of designs, and perhaps the roles of the people working with the data.

Perspective How a 3-D model is viewed by an observer. In a Perspective view, the 3-D image is more realistic since the image seems to shrink away (as a real object would do). However, this is not always desirable when designing since geometry does not line up as it should. Also, whenever a CAD system displays a model, there is a Perspective transformation that occurs to figure out how the model should appear based on how far the virtual observer is from the virtual object and how the object is rotated in modeling space.

Picking Using a pointing device to make a selection of a displayed entity. It is similar to Selecting, except that Picking presumes that the pointing device is used.

Ping A program or utility to see if a computer (or Node) is still actively connected to the Network. Presumably it is based on the idea of sonar, where a signal is sent out and then it "listens" for a response.

Pixel Picture Element. It is one of the individual "dots" or regions of the Monitor that can be turned on or off or changed color. It is very common in expressing the Resolution of a Monitor.

Plane A geometric entity that extends a flat surface to infinitely long distance. A Planar Surface is a surface that is perfectly flat.

Platform Encompasses a computer system's hardware, software, and peripherals. Some operating systems, in particular, can be used on different kinds of computer hardware; therefore, platform can be a better distinguishing term than the CPU or operating system for a CAD workstation.

Preprocessor A computer program that works on data files prior to some more significant operation. Many Compilers use a Preprocessor to condition source code before doing the real compiling. A Preprocessor for a CAE program (such as FEA or CFD) is used to prepare data to be analyzed. The Preprocessor activity could, in fact, be more intensive than the calculations for the analysis; getting good results is often wildly dependent on the proper preparation of the input data (i.e. preprocessing).

Postprocessor   A computer program that works on data files after a more significant operation. In a Computer Aided Manufacturing (CAM) environment, a Postprocessor translates the NC data from a CAM program to actual control codes for a specific NC machine. In a Computer-Aided Engineering (CAE) environment, a Postprocessor is used to view and analyze the large volume of results from the run of an analysis.

Product Structure   A logical arrangement of information that is needed to fully manufacture and/or assemble a product. It includes the Assembly Structure to indicate what part models are needed in the assembly model, as well as non-geometric information related to the assembly. A Bill of Material that includes drawings, specifications, manuals, material specifications, etc. can be assumed to fully define the Product Structure.

Pro/ENGINEER® (often called just "Pro/E")   A high-end CAD system that is sold by PTC. It was probably the most instrumental system for the rise of 3-D CAD systems in general.

Prototype   An evaluative model of a product. It is assumed to demonstrate whether the product is practical or meets requirements. The most common type of Prototype is a physical model of a product. A car manufacturer, for example, may manufacture a new model of car just for test purposes. However, Prototype may refer to other kinds of products such as logic systems or software.

Prototype Drawing   A drawing that is used as a template for other drawings that are subsequently created in the CAD system.

RAM (Random Access Memory)   Made up of banks of memory chips (ICs). Although RAM chips are used in a number of subsystems of a computer, RAM often is assumed to be just the main Memory of the system that the CPU communicates with.

Rapid Prototyping (RP)   The technology related to the creation of physical models based directly on 3-D CAD models. Rapid Prototype machines use the 3-D CAD model data to guide computer-controlled material sintering or deposition devices.

Raster Graphics   The type of graphics that has been broken down into discrete elements. When a Monitor displays an entity (such as a circle), it has been rasterized. This was the process of taking the pure circle model (the center of the circle plus a radius) and translating it into the appropriate series of Pixels to be turned on on the Monitor. This type of graphics would be different than Vector Graphics (which does work with the pure data). There were computer graphics displays at one time that relied on the use of oscilloscope-type, analog signals that would not necessarily involve the rasterization process.

Raster-to-Vector   Techniques that attempt to convert Raster Graphics into Vector Graphics. This is similar to OCR technology.

Glossary 309

Ray Tracing An advanced computer graphics technique that allows a 3-D model to have effects like reflections. It involves an extremely large amount of computation to trace virtual lines of light (or rays) as they go from the observer to each surface on each object in a scene, and then it sees if the end of this process ends at a light source.

Release Indicates when a drawing or other engineering product is finished, approved, and shared throughout the company. A released drawing is one that is available for vendors, manufacturing, etc.; while a drawing that is considered pre-release is not available for use to these areas. Since the changes to the released drawings may have to be made simultaneously, they are then often grouped into a single event of release. The group of drawings is often referred to as a Drawing Package, and the event may use such terms as Engineering Change Order (ECO) or Engineering Change Notice (ECN) or Engineering Release Notice (ERN), etc.

Resolution Indicates how refined or precise a device may be using a numerical limit or capability. This is particularly true for Monitors. In this case, Resolution is measured as the maximum number of Pixels that can be accommodated. For intance, a Resolution of 1024 Pixels (in the X) by 768 Pixels (in the Y) is a higher resolution than 800 by 600.

Revision Block A set of notes on a drawing that shows the Revision Level of the drawing and the past Revision history.

Revision Level The current state of a drawing, part, Bill of Material, etc. Manufacturers tend to avoid recreating designs and documentation (instead favoring re-use). In order to handle the evolving nature of the items, Revision Levels are applied and managed. It is usually a number or letter that is incremented as the item is changed over time.

Revolve A 3-D modeling technique whereby a set of existing geometry is revolved about an axis. A doughnut shape can be the result from revolving a circle about a central axis.

RGB The primary colors for emitted light—red, green, blue. Colors shown on a Monitor are generated by varying the level or intensity of each of the constituents. RGB also referred to a type of connection between a Monitor and its Graphics Adapter. These were special coaxial cables for the colors.

ROM (Read Only Memory) ICs or chips that have information permanently written on them (they are not read/write or Random Access). They are used for functions that are static for the life of the computer system or component. For example, if a Disk Drive needs to use some programming to help calculate the location of data on the drive, this programming may be stored on a ROM chip that is attached to the drive (instead of using up main memory).

SAT® File A geometry exchange file format for the ACIS kernel. SAT files should be able to transfer 3-D geometry and data accurately between different CAD systems that use the ACIS kernel.

Scale The relationship between a physical object and its appearance in a model or drawing. In drawings, a view scale is a ratio between the physical object size and the dimensions shown in the drawing. A Scale of 0.5 indicates that the 2-D geometry in the view is half the size of the real object. Scale also refers to changing the size of a 2-D or 3-D model by some scale factor.

Screen Dump The process of making a hard copy of exactly what is shown on the Monitor. It is generally not used for drawings since they have precise paper size and release issues. However, it is not uncommon for hardcopies of 3-D models to be done via a Screen Dump.

SDRC (Structural Dynamics Research Corporation) The company that developed the I-DEAS CAD system. SDRC and I-DEAS are somewhat interchangeably used to refer to this CAD system.

Section View A view in a drawing that is based on cutting through an object. The cut object is shown with cross-hatching. They are identified by a letter, and the path for the cutting is shown by a Cut Plane or Cut Line. A Section View that uses a Cut Plane that starts and ends with a letter A would be identified as Section View A-A.

Selecting The identification of items that are available in the CAD system. This may refer to Picking items on the Monitor; it may refer to Selecting from a list of choices on a form or in a menu.

Server A computer on a network that provides a capability or service for the other computers on the network. The most common application would be the File Server that provides files from a central location. Other types of servers would be a print server or a license server.

Sewing Up See Stitching.

Shading The process of displaying 3-D models in realistic form. This involves analyzing the surfaces in the 3-D model and the virtual light source and then changing the colors (or shades) all over the surface to give the appearance of the 3-D nature of the model.

Shell First, a technique that is used in 3-D modeling. Shelling adds a thickness of material by projecting outwards (or in the normal direction) from the surfaces in the 3-D model. Second, refers to the command processor used in unix computer systems. The Shell is the program that is constantly monitoring the keyboard for commands that are typed by the user.

Shell Script A file with commands to be run by the operating system (typically a unix system, but it can generically be applied to other operating systems as well).

Sizes See Drawing Size.

Sketch A simple 2-D drawing that is not expected to be released. It is typically used by engineers or designers to document the basic geometry of the design. This idea is adopted by 3-D CAD systems to refer to the creation of basic or foundation 2-D geometry that forms full 3-D features. This Sketch is usually

Glossary

given the ability to have advanced geometric behavior (such as being constrained or even animated).

Sketch Plane The plane that is used to create the Sketch used for the basic or foundation 2-D geometry that is needed for some full 3-D features. Careful selection and tracking of these planes is an essential ingredient of making full use of features-based modeling.

Solaris® Sun Microsystem's version of the unix operating system.

Solid Edge® A midrange 3-D CAD system that was bought out by Unigraphics.

Solid Model A 3-D part model that bounds a volume. It is a normal part model, in some respects. There are technologies that only use solid models, but most 3-D CAD systems supply surface-based modeling as well. Indeed, these methods are usually combined and invisible to the user.

SolidWorks® A midrange 3-D CAD system that was bought out by Dassault Systems.

Source Code The real computer programming that software is built from. Files that contain, BASIC, C, FORTRAN, Java, etc. that can be seen, studied, and modified are examples of Source Code. Software vendors do not supply Source Code; they supply object libraries or executables which customers can only run (customers are unable to see how the software was actually created).

Spline A geometric entity that can follow rather arbitrary paths. A wire or hose that bends and turns is an example of something that can be modeled using a Spline. A Spline was originally a long thin piece of plastic that could be bent to create a smooth 2-D path on a drawing. The path was constrained by lead weights (ducks or whales) with a thin wire that protruded from them to hold down the plastic at specific points. The mathematics of Splines are related to Bezier Curves (which use control points) and B-Splines. Sometimes, the tension of a spline refers to how tightly the path follows the points. Splines are now available in 3-D as well as 2-D.

STEP (Standard for the Exchange of Product Data) A 3-D file format for exchanging part and assembly models between CAD systems and/or companies. It is based on an ISO standard process (TC184/SC4). Refer to ISO 10303—"an International Standard for the computer-interpretable representation and exchange of product data." It has many parts that are meant to assist with data exchange for various types of software and industries. For instance, Part 214 is for "core data for automotive mechanical design processes."

Stitching Also called Stitching Up or Sewing Up, refers to the connection of the edges of surfaces that meet or touch. It uses the Part Tolerance to determine if the edges are close enough. A solid model can be created from separate surfaces if enough surfaces are Stitched to each other.

STL File A file that contains 3-D part data for a Rapid Prototyping device that uses a technology called stereolithography. This file contains just Tessellations of the 3-D model; not the real surfaces. The STL File can be considered a de facto

standard for this type of data. It is used in various schemes to view 3-D models (as well as with Rapid Prototyping applications).
Storage Devices in a computer system that permanently save data (i.e. it is nonvolatile). The most common device of this type is the Disk Drive.
String See Character Data.
Subdirectory A directory that is found within another directory. Subdirectories are used to carefully organize and manage the vast amounts of data that might be found in a Storage system (i.e. Disk Drives, Network File Servers, etc.).
Surface A distinct region of the outer skin of the part. Surface finish indicates how rough or smooth it is. It is also assumed to be a region of the part that can be well described by a mathematical relationship. A flat surface is a "planar" surface and can be described by a formula for planes; a cylindrical surface can be described by formulae for cylinders, etc. Surfaces can also be "free form" in that they use very arbitrary formulae that allow for virtually any shape (this typically uses the mathematics of NURBS).
Surface Finish Describes how rough or smooth a surface is after manufacturing. It is indicated on drawings by Surface Finish symbols.
Surface Modeling A set of 3-D CAD techniques that works with parts on a surface-by-surface basis (as opposed to a feature-by-feature basis). This may include open part modeling, lofting, sweeping, etc.
Surface Normal A direction or vector that is totally perpendicular to a surface or curve. At any point on a surface, a tangent plane can be calculated that just touches the surface. The Surface Normal is then the direction that is perpendicular to that plane.
Surface Texture The application, projection, or mapping of Bitmap data to a 3-D model's surface. If Surface Texture is used, the normal shading process for surfaces is augmented with texture. This texture could be a bit pattern for things such as wood, concrete, etc. But it can also be used with pictures such as TIFF or JPEG files. Surface Texture is rarely used in production CAD work (particularly since it degrades graphics performance by using a lot of memory).
Swapping The process of making room in the Memory of a computer system by moving programming instructions and/or data from Memory (i.e. system RAM or "chips") to Storage (such as a Disk Drive). This process allows programs more than the physical amount of Memory to be run, but the computer runs much more slowly. The higher possible amount of Memory permitted is referred to as Virtual Memory.
Sweep A 3-D part modeling operation that creates surfaces that apply a single cross-sectional shape all along a path.
Symbol See Groupings.
Terminal The display device attached to a mainframe computer. It did not have any substantial processing power (i.e. a CPU) or graphics capability (i.e. a Graphics Adapter). The most common Terminal type related to current systems

Glossary   **313**

was the VT100 from the former Digital Equipment Corporation. Occasionally systems will offer the option to emulate a VT100, or VT101, or VT102. There were also graphics terminals. Most of these were developed by a company called Tektronix. They developed graphics standards such as PLOT10 and TCS.

Tessellation   The process of taking an analytical or pure surface (such as a cylindrical surface) and breaking it down into sufficiently small surfaces that the surface still appears curved. Each of the small surfaces are usually called triangles or perhaps polygons. A Tessellated File would be a file that only contains data for these triangles (instead of the analytical surfaces). Thus this file would not be appropriate for exchange of CAD models. Examples of Tessellated Files are VRML and STL.

Texture   See Surface Texture.

Third Angle Projection   A standard for indicating where views are placed on a drawing. In Third Angle, the Right View of the object is placed on the right of the primary or Front View. This is the standard in the United States.

Title Block   A set of notes on a drawing that indicates the basic identification information for the drawing (such as the drawing Title, Number, Size, Current Revision, etc.).

TIFF™   Image file format. It was originally published by the Aldus Corporation in cooperation with scanner manufactures. The latest standard is from Adobe Systems Incorporated. It contains Bitmap type data, and is not appropriate for the exchange of CAD data. It is often used with on-line imaging systems for storing and retrieving electronic copies of drawings.

Tolerance, Drawing   Indicates the precision desired for the manufacture of parts and assemblies. There are a few information systems on a drawing to indicate Drawing Tolerance. A simple system is the number of decimal places for dimension values. The more decimals shown, the higher the precision (e.g., 3.9 versus 3.875). Another system shows dimensions with upper and lower allowable limits. For a more sophisticated system see GD&T.

Tolerance, Part   See Part Tolerance.

Tool Path   A set of geometric locations that indicate the motion of an NC (or CNC) machine tool. These paths can be 2-D or 3-D (depending on the type of part and machine tool used). In order to also dictate the orientation of tables, jigs, or the sides of tools (bits), there may also be more than just X, Y, Z data for the Tool Paths.

Translucency   A characteristic of surfaces that can be seen through. Often translucency is a feature of the graphics system for 3-D CAD.

Trapping   See Selecting.

Trimmed Surface   A surface that was originally a larger size, but has been given a boundary that reflects the design of a part. Many types of surfaces (such as NURBS) always have a rectangular form to their edges. Trimming changes them to the needed shape.

UI See User Interface.

Unigraphics A high-end CAD system. It has been developed by a company by the same name. This company is owned by Electronic Data Systems Corp. (EDS®).

Unit Vector A Vector with a magnitude or value of just 1. It is a means of indicating direction alone.

UNIX or unix An operating system that is most popular with workstations. It was originally developed at Bell Labs® in the 1970s, and it was made available in source code form to computer manufacturers. There were two main flavors to the operating system—BSD (Berkeley Standard Distribution) and System V (System Five). It was most popular with scientific and engineering software development, so it became a de facto standard for the original high-end CAD systems. Currently, unix and Windows™ are typically used for CAD systems.

Untrimmed Surface A surface that has not had its boundary adjusted to fit the 3-D part model. Many types of surfaces (such as NURBS) always have a rectangular form to their edges. Trimming changes them to the needed shape. Untrimmed Surfaces are most typically seen after translating a 3-D model between different CAD systems.

User Interface The computer programming that the user interacts with. A text-based User Interface would just use commands that are typed. A Graphical User Interface allows the user to select commands and perform actions by using a pointing device (such as a mouse).

VAX An instrumental minicomputer in the early development of CAD systems; it was manufactured by the former Digital Equipment Corporation (now part of Compaq®). It pioneered virtual memory and controlling multiple programs and users without the mainframe architecture. VAX/VMS is an operating system based on the function of the VAX.

Vector A mathematical device for handling data that has both value (i.e. a scalar) and direction. Vectors are used in CAD systems to indicate the direction and magnitude of a function. When a user measures a distance from one point in a 3-D space to another point in 3-D space, one can imagine a Vector that points from one to the other. The length of the Vector is the distance between them. The direction of the vector points along the line that goes from one point to the other. Vectors have components that equal the change in distance along the X, Y, and/or Z axes of the 3-D space.

Vector Graphics A type of computer graphics based on vector data. A file which contains vector data for a line would have the actual endpoints of the lines. A program or plotter which uses this data can draw a perfect line based on those values. This is in contrast to a Bitmap which contains a set of dots that would approximate the line. Vector Graphics also could be used with the original oscilliscope-type graphics devices.

Glossary

Vendor A generic term applied to companies that provide parts, services, or other manufactured goods to a manufacturer. It can be used somewhat interchangably with supplier.

View Regions of a drawing that show the object of the drawing from various viewing angles. A Front View shows the object from a front view; the Right View shows it from the right, etc.

Viewport A computer graphics technique that creates a container for 2-D graphics. A Viewport can have its own mapping so that X- and Y-values are uniquely applied to it. Viewports also clip so that data is not shown beyond their boundaries. All these characteristics make Viewports rather ideal for use as a means of implementing drawing Views in a CAD system.

Virtual Memory A method of extending the amount of Memory that the CPU can utilize. The added Memory is attained by using the Storage system (i.e. Disk Drive) to hold programming that is not currently active in Memory. The process of moving the programming to the drive is called Swapping. Although, Virtual Memory permits larger programs to run on a computer system, the Swapping drastically slows down the performance.

VRML Virtual Reality Modeling (or Mark.up) Language. It is a file format that can allow 3-D models to be shown in a web browser. Furthermore, if the web browser has an appropriate plug-in, the 3-D model can be manipulated.

WAN (Wide Area Network) Allows computers to communicate across a network that is more remote or dispersed than a LAN (Local Area Network). A WAN may connect computers that are at facilities a company may own country-wide.

Web Browser A program that is used in conjunction with a connection to the World Wide Web. The Web Browser reads the HTML data format (as well as others) to determine how the web page is to appear.

Weight, Line See Line Weight.

Weld Symbol A notation on a drawing that indicates how items are welded together. The symbol indicates what type of weld, the size or thickness of the weld, etc. In the United States, the AWS (American Welding Society) standardizes the symbols.

Windows The primary operating systems for PCs that use Intel and Intel-clone CPUs. At one time, Windows was limited in its ability to efficiently process 32-bit programs, and some versions did not use a kernel structure; however, later versions have these features and would be similar to the unix operating system.

Wireframe 3-D entities in the CAD system that do not use surfaces. Instead the 3-D part model only uses lines, arcs, splines, etc. that do not form the basis for surfaces. At one time, Wireframe was what 3-D CAD systems provided; however, over time the use of surfaces and solids replaced Wireframe. If a solid model is displayed on the Monitor (using normal shaded surfaces), then Wireframe can mean changing the display to just show the edges of the surfaces. This looks similar to the older 3-D wireframe models.

XWindows A GUI for unix-based systems. It is a distinct program that runs on the system and then provides the means for the user to interact with the system graphically. It was developed as part of a Project Athena funded by the U. S. government, and was generally available to computer manufacturers. Xwindows was somewhat unique in that the interface's development was closely tied to the use of a computer network (and the concept of client/server).

Z Buffer Allocated memory in the Graphics Adapter for keeping track of the depth of individual pixels in a 3-D scene. This depth is a distance from the observer in the scene to the pixel (a Z distance in comparison to the X and Y shown on the monitor). A Z buffer allows the Graphics Adapter to know which pixels or part of objects are closer than others. If a pixel 1 is behind pixel 2 (and the pixel 2 is not translucent), then pixel 1 can be ignored in terms of display. This improves graphics performance and capability.

Bibliography

American Society of Mechanical Engineers. ASME Y14.5M-1994. Dimensioning and Tolerancing. ASME, New York, 1995.

American Society of Mechanical Engineers. ASME Y14.3-1994. Multiview and Sectional View Drawings. ASME, New York, 1994.

Apple Computer Inc. Apple II: The DOS Manual. Cupertino, CA, 1981.

Aziz S, S Schoonmaker, K Watson. Leveraging 3-D CAD for Integrated Product Development, ASME Management Division, 1999 Mechanical Engineering Congress, ASME, New York, 1999.

Giesecke F, A Mitchell, H Spencer, I Hill, R Loving. Engineering Graphics, 2nd ed. Macmillan Publishing Co., Inc., New York, 1975.

Intergraph Computer Systems. Graphics Supercomputing on Windows NT. Intergraph Computer Systems, Huntsville, AL, 1998.

International Organization for Standardization. ISO 128, Technical drawings—General principles of presentation. ISO/TC10, Geneva, Switzerland, 1982.

Lawry, MH. I-DEAS Master Series Student Guide. SDRC, Milford, OH, 1999.

Schoonmaker SJ. Computing the future of lifting. CraneWorks, March-April 2000, pp. 8–10.

Schoonmaker SJ. ISO 9001 For Engineers and Designers. McGraw-Hill, New York, 1997.

Schoonmaker SJ. Reports of the death of drawings are greatly exaggerated. Machine Design, April 19, 2001, p.152.

Schoonmaker SJ. Techniques in Engineering Software Quality Management, Control and Dynamic Systems. Vol. 2, Academic Press, San Diego, CA, 1993.

Schoonmaker SJ. What good is 3D CAD anyway? Machine Design, April 6, 2000, p. 202.

Index

Abstraction (*see* Model, abstraction)
ACIS, 280
Add, 202–203, 230
API, 163
ASCII (*see* File, ASCII)
Assembly:
  animation 257–258
  constraints (*see* Constraints, assembly)
  interference, 255–256
  model (*see* Model, assembly)
  structure, 244–245, 247, 259–261, 265, 266
  sub-, 247–249, 260

Backups, 19–20, 42, 52, 165–166, 267
Balloons, 76, 99–100, 103
BASIC, 66

Bill of Material, 65, 75–76, 78, 85, 100, 244, 257–258, 262, 265–266, 267, 274
Bitmap, 38–40, 44, 120–121, 149, 151, 159–160, 281
Block (*see* Groupings)
BOM (*see* Bill of Material)
Boolean, 192, 216, 218–219, 222, 223, 230
Bubbles (*see* Balloons)
Byte, 57, 60, 141

Cache, 22
CAE, 3, 138, 276–277, 280
CAGE Code, 81
Callouts, 76
CAM, 276–278
CFD, 138, 276
CG, 139, 174, 255

319

CGM 148, 149, 281
Clump (*see* Groupings)
CNC, 277
Constraints:
   assembly, 245, 251–252, 254–255, 257–258, 262
   overconstraining, 209
   sketch, 203–211, 218, 232
   underconstraining, 210
Coordinate System, 117, 179, 194–195, 198, 222, 245–247, 252
CPU, 7–16, 22–25, 27, 29–30, 33–34, 43, 45–48, 51, 69–70
Crosshatching, 90, 101

Data acquisition, 68–69
Data management, 77–79, 85, 153–156, 165–167, 258, 269–273
Datum:
   drawing, 98, 115, 145
   model, 180, 189, 193–194, 252
Degrees of freedom, 206–207, 210, 251, 258, 262
Dimension, 89, 93–98, 105, 107, 112–116, 130–131, 135, 143, 158, 181, 195, 203–206, 209–213, 241, 252–254, 257
Directory, 22, 29, 50, 52, 55–57, 62, 63
Disabilities, 43
Disk drive, 9, 11–12, 16–22, 25, 27, 29–30, 43, 53–56, 62, 70
Drawing:
   checking, 81
   detail, 85
   general arrangement, 86
   layouts, 85
   number, 76–77, 79, 154–157
   package, 104
   schematics, 86

[Drawing:]
   size, 76–77, 80, 93, 105, 109, 111, 153, 157
   standards, 86–87
   word, 76, 83, 86
   zone, 74, 83
Driver, 46–47, 147–148
DXF, 59, 61, 134, 142, 152, 158, 159, 167

E-CAD, 72, 276
ECO, 82, 150, 152, 164, 265, 273
ECN (*see* ECR)
ECR, 82, 83, 150, 151, 154, 156, 167
Edge, 179
Ergonomics, 43
Ethernet, 25–29, 56, 166

Face, 179, 193
FCS, 98, 115
FEA, 47, 138, 276
Feature, 97, 100, 172, 185, 187, 190, 192–193, 195–199, 203, 214–216, 219, 221–222, 240, 246
File:
   ASCII, 57–59, 61, 70, 141, 142
   binary, 60–61, 70, 141, 142
   compression, 60–61
   format, 57, 141
   management, 54–57, 274–276
   model, 274
   size, 142–143
   system, 29, 56, 164
Fillet, 97, 125, 126, 172
Flat pattern, 225
Font:
   centerline, 100
   line, 94, 133, 143
   text, 160–161
FORTRAN, 9, 65

Index

GA (*see* Drawing, general arrangement)
GD&T, 97–99, 104, 115, 161, 193
Graphics adapter, 31–37, 43–44, 257
Groupings, 129–131, 158
GUI, 47, 56, 63–66

Hardcopy, 37–41, 53, 133, 146–151, 167, 280–281
Hexadecimal, 133
Hiding, 134
HPGL, 38, 40, 147–149

IGES, 59, 134, 138, 142, 143, 152, 158, 159, 167, 277, 279, 280, 283
Imaging systems, 151–152
Instance:
    application, 65
    assembly, 248–249, 251, 280
    groupings, 130–131
Interchangeability, 80, 268
Internet, 24–25, 27–28, 67, 276

Java, 9, 66, 163
JPEG, 5, 60, 121, 152, 160

Kernel:
    graphics, 280
    operating system, 47, 49
Kinematics, 257

LAN, 24–25, 27, 276
Layers, 118–119
Layout, 85, 102–103, 137–138, 141
Leader, 98, 100
Line weight, 94, 133, 134, 148–149
Lofting, 231–232

Memory (*see also* RAM; ROM):
    address, 12
    configuration, 14–15

[Memory (*see also* RAM; ROM):]
    graphics, 33–39
    main, 7, 9, 11, 27, 43, 47–49
    map, 13–14, 44
    paging, 12
    virtual, 12–13, 15, 21–22, 30
Mesh, 138–139
Metadata, 153–154, 164–166, 264–265, 274
Model:
    abstraction, 238
    assembly, 173, 187
    documentation, 282
    sheet metal, 225
    solid, 173, 182, 225–227, 237, 239
    space, 111–112, 143, 245
    surface, 173, 182–183, 187, 224, 229, 241
Modeling:
    features-based, 185, 189, 218, 221–222
    open part, 229
    solid, 231
    surface, 224
Monitor, 7, 31–41, 120, 179, 237, 281
Motherboard, 14, 15, 25, 32

NC, 225, 264, 277
Network (*see also* LAN *and* WAN):
    architecture, 27–29, 165, 274
    computer hardware (*see* NIC)
    management, 51–53
NFS, 29, 30, 56
NIC, 23
Node:
    network, 26, 27, 28, 38, 43, 52
    part, 192, 246
NURBS, 177, 227–228, 279

OCR, 120
Odometer, 123
Operating system, 46–48, 51

Packet, 26, 28
Paper space, 109–111, 114, 143, 158
Parametrics, 212–213
Part:
   history, 183–186, 190, 202, 216, 218, 240, 241, 246, 279
   manifold, 239
   model, 103, 172, 187, 188, 220, 221
   nonmanifold, 239, 241
   number, 85–86, 141, 153, 154, 167, 254, 272–273
   open, 183, 229
   orphan, 240
   properties, 220–221
   tolerance, 199, 237–238, 241, 279
Parts List, 75
PDM, 155, 267, 273, 274
Picking (see Selecting)
Pixel, 14, 31–36, 120
Plane:
   axis, 189,195
   sketch, 189, 193–194
Plotting (see Hardcopy)
Printing (see Hardcopy)
Projection:
   angle, 88–89, 90, 105, 116
   geometry, 103, 131–132
   views, 75, 87, 138

Queue, 53–54

RAM, 11, 12, 14, 27, 35, 39, 41, 43, 44, 48, 164
Rapid prototyping, 279
Raster, 120, 159–160
Resolution, 31–32, 35, 36, 39, 44, 146, 281
Revision:
   Block, 75, 80, 82–83, 86, 162
   control, 156, 264, 269–271, 278
   level, 79–80, 82, 105, 156, 167, 265, 272–273

Revolve, 197
RGB, 133
ROM, 11, 26, 62

Scale (see View, scale)
Screen dump, 39, 281
Section, 174, 214–216
Selecting, 134–135
Server:
   file, 24, 27, 30, 52, 56, 62, 142, 165, 167, 274
   license, 52–53, 67, 166
   print, 40, 54
Sheet metal (see Model, sheet metal)
Shell script, 51, 70
Sketch (see Plane, sketch)
Spline, 128–129, 142, 227–228, 234, 258
STEP, 59, 277, 280, 283
Stitching, 199, 201, 223, 237, 239
STL, 279, 283
Storage, 7, 9, 11, 12, 16, 17, 21–23, 27, 29, 38, 42, 43, 48, 50, 51, 54, 70
Surface:
   finish, 104, 277
   free form, 224, 226
   information, 226–227
   model (see Model, surface)
   qualities, 229
   trimmed, 228, 240
   untrimmed, 229, 236, 241, 279
Sweep 225, 229, 231–234
Symbol (see Groupings; FCS; Surface, finish; Weld symbol)

Tesselation, 279
TIFF, 39, 60, 121, 152, 160, 280
Title Block, 74, 77–82, 97, 104, 141, 167, 282
Tool path, 277–278

Index

Translations:
  drawing, 112, 114, 157–160, 167
  model, 277–281

unix, 7, 9, 10, 45, 49–53, 56, 59–64, 139, 141, 153
User Interface, 47, 63, 162

Vector, 119, 159, 174, 176, 187
  graphics, 119
  normal, 176, 187, 194
  unit, 176, 187
Vertex, 179
Video Card (*see* Graphics Adapter)
View:
  auxiliary, 90
  detail, 91
  drawing, 75, 87–90, 94, 103, 105, 109, 116, 131, 132, 138, 158

[View:]
  origin, 117–118
  scale, 92–93, 96, 105, 111, 114, 137–138, 158
  section, 90, 101
  true, 89
Viewport, 116, 118
VRML, 279, 282, 283

WAN, 24, 25, 276
Web Browser, 67, 152, 279, 282
Weld symbol, 104
Where used, 248, 267, 271
Wireframe, 173, 177, 234, 236
World Wide Web, 25, 67

Xwindows, 47, 64, 281

Z buffer, 36